Food Chains, Yields, Models, and Management of Large Marine Ecosystems

The papers in this volume were presented at a symposium on "Food Chains, Yields, Models, and Management of Large Marine Ecosystems" convened during the 1989 annual meeting of the American Association for the Advancement of Science.

Food Chains, Yields, Models, and Management of Large Marine Ecosystems

EDITED BY
Kenneth Sherman,
Lewis M. Alexander, and Barry D. Gold

Routledge
Taylor & Francis Group
LONDON AND NEW YORK

First published 1991 by Westview Press, Inc.

Published 2018 by Routledge
52 Vanderbilt Avenue, New York, NY 10017
2 Park Square, Milton Park, Abingdon, Oxon OX14 4RN

Routledge is an imprint of the Taylor & Francis Group, an informa business

Library of Congress Cataloging-in-Publication Data
Food chains, yields, models, and management of large marine ecosystems / edited by Kenneth Sherman, Lewis M. Alexander, and Barry D. Gold.
p. cm.
Includes bibliographical references.
ISBN 0-8133-8386-2
1. Marine ecology. 2. Food chains (Ecology). 3. Marine productivity. 4. Marine resources—Management. I. Sherman, Kenneth, 1932- . II. Alexander, Lewis M., 1921- . III. Gold, Barry D.
QH541.5.S3F66 1991
574.5′3′09162—dc20 91-16534
CIP

ISBN 13: 978-0-367-01256-4 (hbk)
ISBN 13: 978-0-367-16243-6 (pbk)

Contents

Preface

This volume is the fourth in a series on Large Marine Ecosystems (LMEs) based on symposia held during the annual meetings of the American Association for the Advancement of Science (1984, 1987, 1988, 1990). In the volumes, emphasis is placed on LMEs as a unifying concept for conducting research, monitoring, and managing of regional units of ocean space around the globe. Much marine science is specialized into disciplines dealing with the biology, physics, and chemistry of the oceans. Specialists generally publish separately on the use and management of ocean resources. It is not often that the results of multidisciplinary scientific and management studies are brought together in a single volume, as they are here.

The LME approach is an effort to bring multidisciplinary marine studies to bear on global problems of resource sustainability by examining the causes of variability in productivity of those regions around the margins of the world oceans from which 95 percent of the annual yields of useable fisheries biomass is harvested. Emphasis is placed on the identification of the principal, secondary, and tertiary driving forces controlling the large-scale variability of biomass yields among LMEs. The LMEs adjacent to the continents are being stressed by overexploitation, pollution, and natural environmental perturbations. This volume focuses on the food chains and biomass yields of the U.S. Exclusive Economic Zone and adjacent waters, providing case studies of LMEs and offering new insights on the trophodynamics and management strategies for increasing the long-term sustainability of the living marine resources under the jurisdiction of the United States.

The volume represents the collective efforts of oceanographers, fishery biologists, ecologists, modellers, and managers to improve our understanding of LMEs and to ensure their continued productivity. By way of comparison, the volume includes a contribution from Dr. Vladimir Borisov of the Soviet Union, who underscores the need for improvement in the management of human-induced interventions changing the structure of an LME through excessive fishing mortality. He bases his arguments on the significant reduction of annual biomass yields from 2.8 million metric tons (mmt) in 1977 to 356,000 mt in 1987, during just one decade in waters of the Barents Sea ecosystem.

The theoretical contributions are aimed at linking food chain studies and sustainable yield models. It is clear that more effort is needed to make the appropriate linkages. However, the studies presented represent a basis for addressing more holistic research, monitoring, and management of LMEs around the globe.

The editors are indebted to the contributors for their willingness to take time out of busy schedules to develop the expert syntheses and reviews needed to move marine resources science and management forward. We are pleased to acknowledge the interest and financial support of NOAA's Office of Oceanography and Marine Assessment. We also gratefully acknowledge support for the Symposium provided by the National Marine Fisheries Service of NOAA, the U.S. Marine Mammal Commission, and the National Science Foundation, Ocean Science Division. We are especially indebted to Jennie Dunnington for her skill and energy in overseeing the editorial production of the final manuscript.

K. Sherman

L. M. Alexander
Narragansett, Rhode Island

B. D. Gold
Washington, D.C.

Contributors and Editors

LEWIS M. ALEXANDER, Department of Geography, University of Rhode Island, Kingston, RI

M. BASSON, Renewable Resources Assessment Group, Imperial College of Science, Technology, and Medicine, London, UK

JOHN R. BEDDINGTON, Renewable Resources Assessment Group, Imperial College of Science, Technology, and Medicine, London, UK

VLADIMIR M. BORISOV, All Union Research Institutes of Marine Fisheries and Oceanography (VNIRO), Moscow, USSR

JOAN A. BROWDER, NOAA, NMFS, Southeast Fisheries Center, Miami Laboratory, Miami, FL

BRADFORD E. BROWN, NOAA, NMFS, Southeast Fisheries Center, Miami Laboratory, Miami, FL

EDWARD B. COHEN, NOAA, NMFS, Northeast Fisheries Center, Woods Hole Laboratory, Woods Hole, MA

JEREMY S. COLLIE, Juneau Center for Fisheries and Ocean Sciences, University of Alaska, Juneau, AK

MICHAEL DAGG, Louisiana Universities Marine Consortium, Chauvin, LA

PAUL G. FALKOWSKI, Oceanographic and Atmospheric Sciences Division, Brookhaven National Laboratory, Upton, NY

BARRY D. GOLD, National Academy of Sciences, Washington, DC

CAROLE D. GOODYEAR, NOAA, NMFS, Southeast Fisheries Center, Miami Laboratory, Miami, FL

CHURCHILL GRIMES, NOAA, NMFS, Southeast Fisheries Center, Panama City Laboratory, Panama City, FL

S. LOHRENZ, Center for Marine Science, University of Southern Mississippi, Stennis Space Center, MS

MARC MANGEL, Zoology Department and Center for Population Biology, University of California, Davis, CA

B. McKEE, Louisiana Universities Marine Consortium, Chauvin, LA

MICHAEL M. MULLIN, Marine Life Research Group, Scripps Institution of Oceanography, La Jolla, CA

JOSEPH POWERS, NOAA, NMFS, Southeast Fisheries Center, Miami Laboratory, Miami, FL

ANDREW A. ROSENBERG, NOAA, NMFS, Northeast Fisheries Center, Woods Hole Laboratory, Woods Hole, MA

BRIAN J. ROTHSCHILD, University of Maryland Center for Environmental and Estuarine Studies, Chesapeake Biological Laboratory, Solomons, MD

KENNETH SHERMAN, NOAA, NMFS, Northeast Fisheries Center, Narragansett, RI

MICHAEL P. SISSENWINE, NOAA, NMFS, Silver Springs, MD

THEODORE SMAYDA, Graduate School of Oceanography, University of Rhode Island, Kingston, RI

R. TWILLEY, Department of Biology, University of Southwestern Louisiana, Lafayette, LA

WILLIAM WISEMAN, JR., Coastal Studies Institute, Louisiana State University, Baton Rouge, LA

JAMES A. YODER, Graduate School of Oceanography, University of Rhode Island, Kingston, RI

Kenneth Sherman

1. Sustainability of Resources in Large Marine Ecosystems

Sustainable Biosphere Initiative and LMEs

The long-term sustainability of biological resources in the sea requires the application of sound principles for the conservation and management of resources at risk from overexploitation, environmental stress, and global climate change. The need for improvements in how marine science supporting ocean management is conducted has recently been underscored by the Ecological Society of America in their "Sustainable Biosphere Initiative (SBI): An Ecological Research Agenda" (Lubchenko et al., 1991). The initiative attempts to define research priorities for ecology during the last decade of the 20th century, recognizing:

> the need to ameliorate the rapidly deteriorating state of the environment and to enhance its capacity to sustain the needs of the world's population. . . . Ecological knowledge and understanding are needed to detect and monitor changes, to evaluate consequences of a wide range of human activities, and to plan for the management of sustainable natural and human-dominated ecological systems.

The SBI proposes three research priorities: (1) global change, (2) biological diversity, and (3) sustainable ecological systems. The last named priority calls for a major new integrated program of research on the sustainability of ecological systems that is to be focused on understanding the underlying ecological processes in natural and human-dominated ecosystems in order to prescribe restoration and management strategies that will enhance the sustainability of the Earth's ecological systems. The Society argues that:

> current research efforts are inadequate for dealing with sustainable systems that involve multiple resources, multiple ecosystems, and large spatial scales. Moreover, much of the current research focuses on commodity-based management

> systems, with little attention paid to the sustainability of natural ecosystems whose goods and services currently lack a market value. Addressing the topic of sustainable ecological systems will require integration of social, physical, and biological science.

It is of interest to note that the very same concerns led to the convening of the first symposium on the variability and management of large marine ecosystems (LMEs) in 1984. Participants included marine scientists, geographers, economists, managers, lawyers, and diplomats concerned with the increasing stresses imposed on marine resources from overexploitation, pollution, and environmental perturbations. The response to the need for addressing research, monitoring, and mitigation actions on the scale of LMEs was positive and from 1984 through 1991, led to four subsequent symposia, one conference, one workshop, and three published volumes.

The LMEs are extensive areas of ocean space adjacent to the continents; they are of 200,000 km^2 (58,000 nm^2) or greater in size, characterized by distinct hydrographic regimes, submarine topography, productivity, and trophically-dependent populations (Sherman and Alexander, 1986). The ecological concept that critical processes controlling the structure and functioning of biological communities can best be addressed on a regional basis (Ricklefs, 1987; Levin, 1990; Graham et al., 1991) is consistent with the LME approach to research on living marine resources, the "health" of their habitats, and their management.

Human intervention and global change are sources of increasing variability in the natural productivity of the world ocean. Since the 1960s, the number of reports of changes in the structure and community dynamics of marine ecosystems attributed to the effects of excessive fishing mortality have been growing (Sherman et al., 1990a). Overfishing has caused multimillion metric ton biomass flips among the dominant pelagic components of the fish community off the northeastern United States (Sissenwine, 1986). The biomass flip, wherein a dominant species rapidly drops to a low level to be succeeded by another species, can generate cascading effects among other important components of the ecosystem, including marine birds (Powers and Brown, 1987), marine mammals, and zooplankton (Overholtz and Nicolas, 1979; Payne et al., 1990). Other sources of ecosystem perturbations to marine populations through human intervention are the incidental catches of marine mammals in netting during fishing operations and the growing impacts of pollution on biomass yields. Efforts to reduce stress and mortality on marine mammal by-catch are being pursued (Bonner, 1982; Loughlin and Nelson, 1986; Waring et al., 1990). Pollution problems at the continental margins of marine ecosystems that impact on natural productivity cycles, including eutrophication from high nitrogen and phosphorus effluent from estuaries, the presence of toxins in poorly treated sewage discharge, and loss of wetland nursery areas to coastal development are also being addressed (GESAMP, 1990). The growing awareness that biomass yields are being influenced by multiple but different driving forces in marine ecosystems around the globe has accelerated efforts to broaden research strategies to encompass the effects of food chain dynamics, environmental perturbations, and pollution on living marine resources from an ecosystem perspective. Mitigating actions to reduce stress on living resources in the oceans are required to ensure the long-term sustainability of biomass yields.

Historical Perspective of LMEs

In 1987, the global yield of marine fisheries was 80.5 million metric tons (mmt). Of this amount 4.2% (3.4 mmt) including catches of tunas, billfishes, and bonitos was attributed to biomass catches in the pelagic high-seas. The major biomass, constituting 95.8% of the annual yield, was caught within the geographic limits of 49 LMEs. The boundaries of the LMEs are depicted in Figure 1.1. A list of the LMEs is given in Table 1.1. Criteria used for defining the geographical limits of the LMEs included consideration of distinct bathymetry, hydrography, productivity, and trophically dependent populations. Several occupy semi-enclosed seas, such as the Black Sea, the Mediterranean Sea, and the Caribbean Sea. Some of these can be divided into domains, or subsystems, such as in the case of the Adriatic Sea, a subsystem of the Mediterranean Sea LME. In other LMEs geographic limits are defined by the scope of continental margins. Among these are the U.S. Northeast Continental Shelf, the East Greenland Sea, the Northwestern Australian Shelf. The seaward limit of the LMEs extend beyond the physical outer limits of the shelves, themselves, to include all or a portion of the continental slopes as well. Care was taken to limit the seaward boundaries to the areas affected by ocean currents, rather than relying simply on the 200-mile Exclusive Economic Zone or fisheries zone limits. Among the ocean current LMEs are the Humboldt Current, Canary Current, and Kuroshio Current. The LMEs that together produce approximately 95% of the annual global fisheries biomass yield are listed in Table 1.2.

For nearly 75 years, since the turn of the century, fishery scientists were preoccupied with single-species stock assessments, while during this same period biological oceanographers did not achieve any great success in predicting fish yield based on food chain studies. As a result, through the mid 1970s, the predictions of the

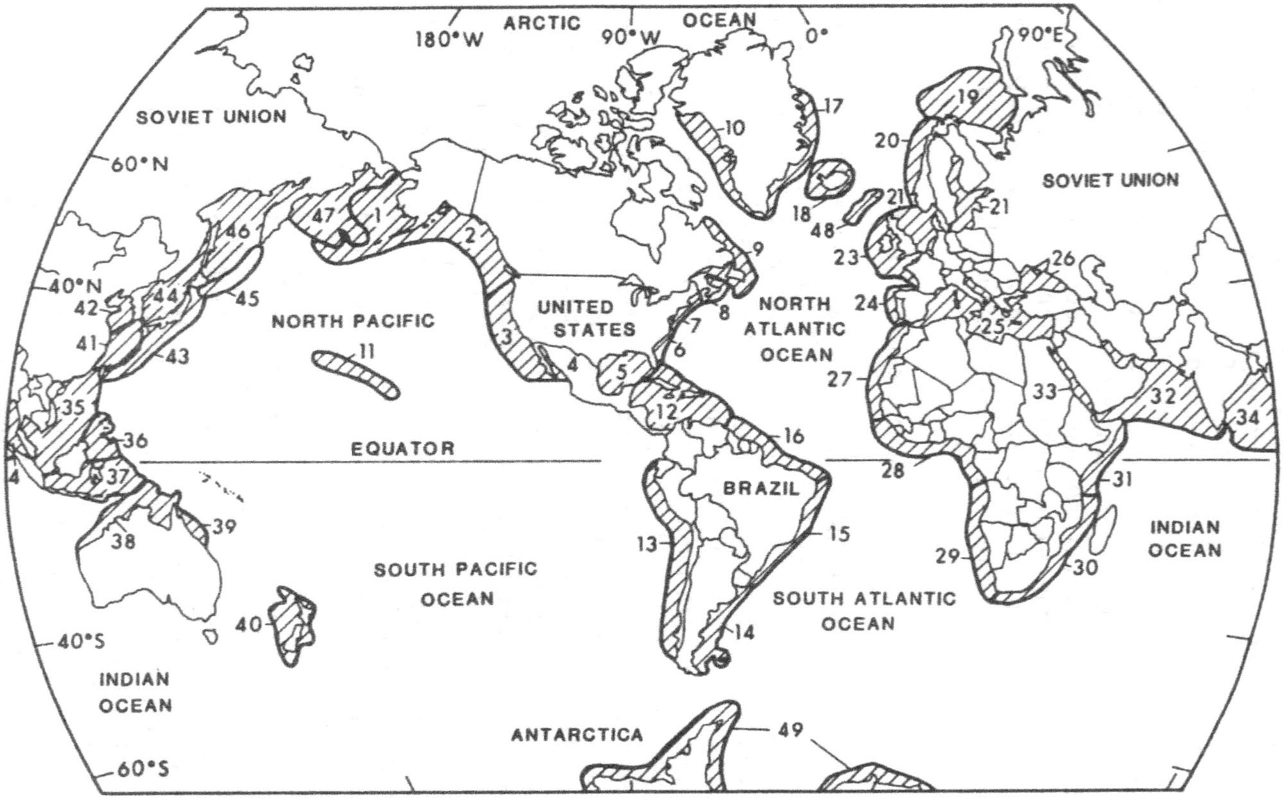

Figure 1.1. Boundaries of 49 LMEs.

Table 1.1. Number codes for the map of large marine ecosystems.

1. Eastern Bering Sea	25. Mediterranean Sea
2. Gulf of Alaska	26. Black Sea
3. California Current	27. Canary Current
4. Gulf of California	28. Guinea Current
5. Gulf of Mexico	29. Benguela Current
6. SE U.S. Continental Shelf	30. Agulhas Current
7. NE U.S. Continental Shelf	31. Somalia Coastal Current
8. Scotian Shelf	32. Arabian Sea
9. Newfoundland Shelf	33. Red Sea
10. West Greenland Shelf	34. Bay of Bengal
11. Insular Pacific--Hawaiian	35. South China Sea
12. Caribbean Sea	36. Sulu-Celebes Seas
13. Humboldt Current	37. Indonesian Seas
14. Patagonian Shelf	38. Northern Australian Shelf
15. Brazil Current	39. Great Barrier Reef
16. Northeast Brazil Shelf	40. New Zealand Shelf
17. East Greenland Shelf	41. East China Sea
18. Iceland Shelf	42. Yellow Sea
19. Barents Sea	43. Kuroshio Current
20. Norwegian Shelf	44. Sea of Japan
21. North Sea	45. Oyashio Current
22. Baltic Sea	46. Sea of Okhotsk
23. Celtic-Biscay Shelf	47. West Bering Sea
24. Iberian Coastal	48. Faroe Plateau
	49. Antarctic

levels of biomass yields for different regions of the world ocean were open to disagreement (Ryther, 1969; Alverson et al., 1970; Lasker, 1988). A milestone in fishery science was achieved in 1975 with the convening of a symposium by the International Council for the Exploration of the Sea that was focused on changes in the fish stocks of the North Sea and their causes. The symposium, which dealt with the North Sea as an ecosystem following the lead of Steele (1974), Cushing (1975), Andersen and Ursin (1977), and others, was prompted by a rather dramatic shift in the dominance of the finfish species of the North Sea from a balanced pelagic and demersal finfish community prior to 1960 to demersal domination from the mid-1960s through the mid-1970s. Although no consensus on cause and effect was reached by the participants, it was suggested by the convener (Hempel, 1978) that the previous studies of seven-and-a-half decades may have been too narrowly focused, and that future studies should take into consideration fish stocks, their competitors,

Table 1.2. Contributions by country and large marine ecosystem (LME) representing 95% of the annual global catch in 1987.

Country	Percentage of [1] world marine nominal catch	LMEs producing annual biomass yield	Cumulative percentages
Japan	14.43	Oyashio Current, Kuroshio Current; Sea of Okhotsk, Sea of Japan, Yellow Sea, East China Sea, W. Bering Sea, E. Bering Sea, and Scotia Sea	
USSR	12.63	Sea of Okhotsk, Barents Sea, Norwegian Shelf, W. Bering Sea, E. Bering Sea, and Scotia Sea	
USA	7.03	Northeast US Shelf, Southeast US Shelf, Gulf of Mexico, California Current, Gulf of Alaska, and E. Bering Sea	
China	6.72	W. Bering Sea, Yellow Sea, E. China Sea, and S. China Sea	
Chile	5.98	Humboldt Current	
Peru	5.65	Humboldt Current	50
Korea Republic	3.50	Yellow Sea, Sea of Japan, E. China Sea, and Kuroshio Current	

[1]Percentages based on fish catch statistics from FAO Yearbook, vol. 64, 1989.

Table 1.2. continued.

Country	Percentage of [1] world marine nominal catch	LMEs producing annual biomass yield	Cumulative percentages
Thailand	2.48	South China Sea, and Indonesian Seas	
Indonesia	2.45	Indonesian Seas	
Norway	2.40	Norwegian Shelf and Barents Sea	
India	2.09	Bay of Bengal and Arabian Sea	
Denmark	2.07	Baltic Sea and North Sea	
Iceland	2.02	Icelandic Shelf	
Korea D. P. Rep.	1.99	Sea of Japan and Yellow Sea	
Philippines	1.78	S. China Sea, Sulu-Celebes Sea	
Canada	1.75	Scotian Shelf, North-east U.S. Shelf, Newfoundland Shelf	75
Spain	1.69	Iberian Coastal Current and Canary Current	
Mexico	1.55	Gulf of California, Gulf of Mexico, and California Current	

[1]Percentages based on fish catch statistics from FAO Yearbook, vol. 64, 1989.

Table 1.2. continued.

Country	Percentage of [1] world marine nominal catch	LMEs producing annual biomass yield	Cumulative percentages
South Africa	1.12	Benguela Current and Agulhas Current	
France	1.00	North Sea, Biscay-Celtic Shelf, Mediterranean Sea	80
Ecuador	0.84	Humboldt Current	
UK-Scotland	0.82	North Sea	
Poland	0.80	Baltic Sea	
Viet Nam	0.77	South China Sea	
Malaysia	0.74	Gulf of Thailand, Andaman Sea, Indonesian Seas, and S. China Sea	
Brazil	0.72	Patagonian Shelf and Brazil Current	
Turkey	0.72	Black Sea, Mediterranean Sea	
Argentina	0.69	Patagonian Shelf	
Namibia	0.64	Benguela Current	
Italy	0.62	Mediterranean Sea	
Morocco	0.61	Canary Current	

[1]Percentages based on fish catch statistics from FAO Yearbook, vol. 64, 1989.

Table 1.2. continued.

Country	Percentage of[1] world marine nominal catch	LMEs producing annual biomass yield	Cumulative percentages
New Zealand	0.54	New Zealand Shelf Ecosystem	
Netherlands	0.53	North Sea	
Portugal	0.49	Iberian Shelf and Canary Current	
Faroe Islands	0.44	Faroe Plateau	90
Pakistan	0.42	Bay of Bengal	
Ghana	0.40	Gulf of Guinea	
Senegal	0.35	Gulf of Guinea and Canary Current	
Venezuela	0.34	Caribbean Sea	
Ireland	0.31	Biscay-Celtic Shelf	
UK Eng., Wales	0.30	North Sea	
Bangladesh	0.29	Bay of Bengal	
Hong Kong	0.28	S. China Sea	
Sweden	0.26	Baltic Sea	
Australia	0.25	N Australian Shelf and Great Barrier Reef	

[1]Percentages based on fish catch statistics from FAO Yearbook, vol. 64, 1989.

Table 1.2. continued.

Country	Percentage of [1] world marine nominal catch	LMEs producing annual biomass yield	Cumulative percentages
Cuba	0.25	Caribbean Sea	
Romania	0.25	Black Sea	
German D. Rep.	0.22	Baltic Sea and Scotia Sea	
Panama	0.21	California Current and Caribbean Sea	
Sri Lanka	0.19	Bay of Bengal	
Nigeria	0.18	Gulf of Guinea	
Uruguay	0.17	Patagonian Shelf	
Finland	0.16	Baltic Sea	95

[1]Percentages based on fish catch statistics from FAO Yearbook, vol. 64, 1989.

predators and prey, and interactions of the fish stocks with their environments, the fisheries, and pollution from an ecosystems perspective.

Perturbations and Driving Forces in LMEs

There is a growing awareness of the utility of a more holistic ecosystems approach to resource management among marine scientists, geographers, economists, government representatives, and lawyers (Byrne, 1986; Christy, 1986; Alexander, 1989; Belsky, 1989; Crawford et al., 1989; Morgan, 1989; Prescott, 1989). Effective management from an ecosystems perspective will be contingent on the identification of the major, secondary, and tertiary driving forces causing large-scale changes in biomass yields. Management of species responding to strong environmental signals will be enhanced by improving the understanding of the physical factors forcing biological changes, whereas in other LMEs when the prime driving force is predation, either by natural predators or by human predation expressed as excessive fishing mortalities, options can be explored for implementing adaptive management strategies. Mitigation actions are required to ensure that the "pollution" of the coastal zone of LMEs is reduced and does not become a principal driving force in any LME. Concerns remain regarding the socioeconomic and political difficulties in management across national boundaries as in the case of the Sea of Japan ecosystem where the fishery resources are shared by five countries (Morgan, 1988), or the North Sea ecosystem, or the 38 nations sharing the resources of the Caribbean Sea ecosystem.

Management of LMEs

Several LMEs are presently being subjected to ecosystem management. Among these are: the Yellow Sea ecosystem, where the principal effort is underway by the Peoples Republic of China (Tang, 1989); the multispecies fisheries of the Benguela Current ecosystem under the management of the government of South Africa (Crawford et al., 1989); the Great Barrier Reef ecosystem (Bradbury and Mundy, 1989; Kelleher, 1990), the Northwest Australian Continental Shelf ecosystem (Sainsbury, 1988) under management of the state and federal governments of Australia; the Antarctic marine ecosystem under the Commission for the Conservation of Antarctic Marine Living Resources (CCAMLR) and its 21-nation membership (Scully et al., 1986; Sherman and Ryan, 1988). Scientists and managers of the states of Oregon and California have developed a plan for research, monitoring, and management of marine resources of the Northern California Current ecosystem (Bottom et al., 1989).

Comparative LME Studies

Although development of research, monitoring, and management strategies for LMEs is an evolving process (Morgan, 1988; Alexander, 1989), sufficient progress has been made to allow for useful comparisons of the principal forces driving variability in fisheries biomass yields among selected LMEs around the globe. The large-scale spatial dimensions of LMEs precludes a strictly controlled experimental approach to their study. However, as regional ecosystems, they are amenable to the comparative method of science as described by Mayr (1982) and practiced by Bakun (1986; 1990) and Bakun and Parish (1980; 1990). Since 1984, thirty case studies investigating the major causes of large-scale perturbations in biomass yields of LMEs have been completed (Sherman and Alexander 1986, 1989; Sherman et al., 1990a, and this volume). Changes in the ocean climate of the northern North Atlantic during the late 1960s and early 1970s have been considered by several marine scientists as the dominant cause of change in food chain structure and fisheries biomass yields of three LMEs--the Norwegian Shelf ecosystem (Ellertsen et al., 1990), the Barents Sea ecosystem (Skjoldal and Rey, 1989; Borisov, this volume), and the West Greenland Sea ecosystem (Hovgaard and Buch, 1989). Excessive fishing mortality was a secondary factor in the decline of the cod and capelin stocks of these LMEs.

In the North Sea ecosystem the large-scale changes in the species composition of the catches from a pelagic dominance to a demersal species dominance during the 1960s and 1970s have been attributed to both changing hydrographic conditions and overexploitation. However, none of the arguments can be considered more than speculative at this time, pending the analysis of more recent information (Hempel, 1978; Daan, 1986; Zijlstra, 1988). Further to the south, the variability in biomass yields of pelagic species of the Iberian Shelf ecosystem have been attributed to shifts in water masses and thermal structure (Wyatt and Perez-Gandaras, 1989). In the Benguela Current ecosystem, long-term variability in fisheries biomass yield and marine bird populations are the result of changes in the oceanographic regime (Crawford et al., 1989).

Around the Pacific rim, the large-scale changes in fisheries biomass yields reported for the increases in clupeoid abundance have been related to changes in productivity of the Oyashio Current ecosystem (Minoda, 1989), Kuroshio Current ecosystem (Terazaki, 1989), and the Humboldt Current ecosystem (Canon, 1986, Alheit and Bernal, in press). The importance of secondary effects of fishing on the recovery of the depressed anchoveta population of the Humboldt Current ecosystem off the coast of Peru has been stressed recently by Pauly (1989). Although of lesser magnitude, the long-term shift in abundance levels between sardines and anchovies within the California Current ecosystem is considered the result, primarily, of natural environmental change and secondarily, from intensive fishing, rather than from any density-dependent competition between the two species (MacCall, 1986). Changes in biomass yields of two other Pacific rim LMEs have been the result of overexploitation in the Gulf of Thailand ecosystem (Piyakarnchana, 1989) and the Yellow Sea ecosystem (Tang, 1989).

The importance of a natural predator, the crown-of-thorns starfish, driving an ecosystem is supported by the study of Australian scientists reporting on changes in the community structure of the Great Barrier Reef ecosystem (Bradbury and Mundy, 1989). The

Banda Sea ecosystem located northwest of Australia remains relatively healthy, with a potential of supporting a long-term yield of pelagic fish at a level higher than the average annual yield of 30,000 metric tons, presently caught (Zijlstra and Baars, 1990).

Food Chains and Yields of LMEs

In the present volume, the contributors extend the information base on LMEs by providing information on the food chains of large marine ecosystems off the coasts of the United States and the Soviet Union, including the U.S. Northeast Shelf ecosystem, U.S. Southeast Shelf ecosystem, the Gulf of Mexico ecosystem, the California Current ecosystem, and the Barents Sea ecosystem. In the latter part of the volume, the focus is on the use of food chain information and fisheries yield models and their advantages and disadvantages in supporting management strategies from an ecosystems perspective.

In the first part of the volume, the difficulties in linking primary production to fish production are described by P. Falkowski. The nature of the spring bloom in the southern part of the Northeast Shelf ecosystem was examined using new buoy technology to produce a fine-scale characterization of energy flow at the lower end of the food chain. In the southern part of the Northeast Shelf ecosystem, the classical single event of a large-scale spring phytoplankton bloom was not observed. The spring bloom consisted of a series of short pulses of phytoplankton of approximately 3- to 5-day duration related to advective processes. On the short-time scale, coupling of phytoplankton and zooplankton showed a good deal of variability. Whereas integration of data sets into longer time-scale segments of 30 days or greater, phytoplankton and zooplankton are highly coupled. The need for additional studies aimed at the integration of plankton food chain dynamics into physical circulation models is recommended by Falkowski.

Further south, along the eastern coast of the United States, in the region designated as the Southeast Continental Shelf ecosystem (NOAA, 1988), J. Yoder observed similar short-term pulses of primary production coupled with nutrient enrichment of the inner margin of the ecosystem from estuarine runoff, whereas the outer margin of the ecosystem is dependent on upwellings from the nutrient rich bottom waters associated with Gulf Stream frontal processes. These short-term events are more important to plankton production than any seasonal bloom phenomenon. The important fisheries supported by the plankton production of the ecosystem includes a sizeable shrimp fishery and "live-bottom" habitats. The importance of physical processes for advecting menhaden larvae from oligotrophic waters of the Gulf Stream to enriched and highly productive waters of near-coast and estuarine habits is considered by J. Yoder as an important consideration in the recruitment of menhaden year-classes.

These studies of the Northeast Shelf and Southeast Shelf ecosystems emphasize the important continuous propagation of short-term pulses in primary productivity in contrast to large amplitude seasonal pulses of the kind described in the earlier literature for Georges Bank (Riley, 1947). The probability of a match between the availability of planktonic prey for fish larvae and their time of first-feeding is very likely enhanced by the short-term-pulse propagation. Miss-matches between larvae and the high amplitude seasonal blooms of phytoplankton and swarming of zooplankton prey have been considered by Cushing (1975, 1988) as a principal cause of recruitment failure of cod in the North Sea ecosystem. In an effort to improve our understanding of the linkages between ocean physics, primary production, and fish production, the International Council for the Exploration of the Sea has endorsed a comparative study of cod dynamics in marine ecosystems across the northern North Atlantic Ocean. Initial efforts will be directed to the coupling processes at the lower end of the food chain including the nature of linkages between phytoplankton, zooplankton, and larval cod fish (ICES, 1990).

Continuing around the Exclusive Economic Zone of the United States, the study of food chains of the northern area of the Gulf of Mexico ecosystem by M. Dagg et al. emphasizes the importance of the inflow of Mississippi River water on the basic productivity of the Gulf of Mexico ecosystem. The high phytoplankton production of the system is attributed to the introduction of high concentrations of dissolved nutrients during the annual production cycle, reaching an annual seasonal high during spring flood of the Mississippi, which replenishes nitrates along a broad band of the coastal habitat, westerly around the Gulf as far as west Texas and Mexico. Little evidence of nutrient enrichment was found in the vicinity of the slope/shelf front. The principal physical characteristic of the Gulf is the presence of fronts at the river plumes, between mid-shelf and inner-shelf, and the outer-shelf and open waters of the Gulf ecosystem. The phytoplankton and zooplankton support a high biomass of fish species, including the menhaden *Brevoortia patronus* which accounts for the highest volume fishing in the United States. Preliminary evidence suggests that the abundance of estuarine habitat and concentrating processes of the fronts enhance growth and survival of larval stages of fish. Although hypoxia, fueled by copepod fecal pellets, is observed during the summer months the spatial and temporal impact is apparently limited largely to the waters off Louisiana.

Two contributions describe the results of analyses that incorporate food chain dynamics to fisheries biomass yields. In the first of these, M. Sissenwine and E. Cohen review the "carrying-capacity" of the Northeast Continental Shelf ecosystem, based on productivity information on plankton, and estimated energy transfer efficiencies up the food chain to fish. They conclude that within the fish community of the ecosystem, large fish consume a large quantity of small fish, with nearly 70% of the annual fish production serving as prey for the fish component of the ecosystem. Therefore, the major determinant of long-term sustainability of fisheries biomass yield is interspecific predation, that could be controlled to maximize yield through selective fishing mortality directed to high biomass but low-valued elasmobranch species (e.g., spiny dogfish and skates) that have increased in abundance during the past two decades. However, they emphasize the need to develop and validate carefully constructed hypotheses to ensure success in any adaptive management of elasmobranchs designed to enhance recovery of depleted higher-valued gadoid and flounder species.

In the second case, B. Brown et al. provide initial estimates of the energetic requirements of the fish community within the Gulf of Mexico ecosystem. They caution that any expansion of fisheries for coastal pelagics will need to consider the impact of the reduction of the small coastal pelagics on higher trophic level fish dependent on the coastal pelagics as a food base. They also emphasize the importance as prey of the small fish discarded at sea as by-catch by the shrimp fisheries, and argue that in relation to long-term sustainability of the Gulf of Mexico fish stocks, that shrimp by-catch should be significantly reduced. In both studies it is clear that strategies for management of fisheries stocks would be greatly enhanced by a better understanding of the recruitment process of fish stocks. The results of M. Mullin's study of spatial and temporal scales among phytoplankton and zooplankton components within the California Current ecosystem occupying a significant area of EEZ of the west coast, are consistent with the need for a better understanding of the linkage between plankton and the fish recruitment process in LMEs. In his presentation, M. Mullin emphasizes the gaps in understanding of secondary production, especially the need to improve estimates of herbivores, protozoa, and zooplankton smaller than 0.5 mm. With this information, it should be possible to examine spatio-temporal linkages between secondary productivity and larval fish growth, survival, and subsequent year-class recruitment.

In the latter part of the volume, the studies progress from food chain dynamics to an analysis of a severely stressed LME, the Barents Sea, and several studies that examine the theoretical basis for management of fisheries biomass yield. The concluding chapter underscores the problems of ecosystem instability at the very base of the food chain, the phytoplankton and the consequences of noxious blooms on marine food chains in LMEs.

In the Barents Sea ecosystem, V. Borisov reports that the fisheries biomass yield decreased from a maximum of 2.8 mmt in 1977 to 350,000 mt in 1987. Borisov reports that the ecosystem is influenced by its location in a boundary region between the cool waters of the subarctic and the relatively warmer waters of the Atlantic. This boundary condition has had a measurable impact on the recruitment success of cod (Ellertsen et al., 1990). In addition, the heavy fishing mortality has reduced the stocks of cod and other principal commercial species including herring, capelin, polar cod, haddock, redfish, Greenland halibut, and other species. The author emphasizes the need for a reduction in fishing effort and vigorously applied enforcement of other management measures designed to reduce fishing effort on spawning stocks, and allow escapement of immature fish by regulating the size of mesh in the trawls. The drastic decline in the fish component of the ecosystem has resulted in an apparent increase in the abundance of euphausiids. The influence on the reduction of zooplanktivores on the natural productivity cycle of zooplankton is not clear. However, what is clear is the "overfished" state of the Barents Sea. A similar "overfished state" has been reported for the Northeast Shelf ecosystem of the U.S. by M. Sissenwine and E. Cohen (this volume) who have proposed consideration of adaptive management of the fish component in the ecosystem in an effort to restore depleted gadoid and flounder stocks and enhance long-term sustainability of fisheries yields. Whether the Barents Sea ecosystem can be restored to an earlier state of higher sustained fisheries yield is an open question that is deserving of attention over the next decade. Based on the recovery of mackerel and herring stocks of the Northeast Shelf ecosystem following a decade of fishing-effort reduction, it would appear that reduction of excessive fishing mortality can result in a measurable enhancement of recruitment.

The theoretical paper by A. Rosenberg et al. underscores the importance of fish-stock biomass levels and multispecies predator-prey interactions in relation to equilibrium yields on long-term sustainability levels for cod, capelin, and herring of the Barents Sea ecosystem, and krill and icefish in the Antarctic marine ecosystem. The capelin, herring, and cod represent different trophic levels in the Barents Sea food chain, with latter species longer-lived than the other two, and the herring longer-lived than the relatively fast-growing, but short-lived capelin. The confidence limits about their long-term average recruitment levels decreases from capelin to cod. The results of the modelling effort suggests that the risk to the sustainability of Barents Sea stocks, in terms of spawning stock biomass depression, increases more rapidly with harvest rate for predation higher in the food chain than for species at the lower end of the food chain, where the probability of strong year classes is higher. In the Antarctic ecosystem the probability of recovery of depleted icefish stocks is confounded by the krill fishery which, in the vicinity of South Georgia, carries the potential for significant by-catch of juveniles and, hence, represents an added risk to spawning stock biomass recovery.

An extension of the multispecies yield models is given in the paper by J. Collie who reviews the theory of adaptive management as it applies to large marine ecosystems, and is presently practiced in at least one LME (Sainsbury, 1988). The application of adaptive management is described for six stocks of yellowtail flounder of the U.S. Northeast Shelf ecosystem and the Scotian Shelf ecosystem off Canada, where harvest rates were simulated to optimize sustained yields over a 50-yr period. In the simulation, it was not possible to distinguish the correct model if a low productivity level was used as the initial input as the stock did not respond enough to distinguish the correct model. The simulations for the Eastern Bering Sea included information on trophic interactions among yellowfin sole (*Limanda aspera*), Pacific cod (*Gadus macrocephalus*), arrowtooth flounder (*Atheresthes stomias*), and sablefish (*Anoplopoma fimbria*). The groundfish species appear to have developed in different areas of the Eastern Bering Sea as principal habitat with pollock (*Theragra chalcogramma*) occurring over the outer shelf and oceanic domains where they prey on copepods, euphausiids, and juvenile pollock. Pacific cod occupy the middle and outer shelf and prey on shrimp, crabs, juvenile pollock, and flatfish. Whereas yellowfin sole inhabits the coastal and mid-shelf domain and feeds on bivalves and benthic invertebrates. Most of the diet overlap among the species is with juvenile pollock; they also serve as principal prey of fur seals, kittiwakes, and murres. In addition, cannibalism is a major biological regulator of pollock recruitment. Among the "experimental" options are to fish down the pollock to intermediate levels and monitor recruitment (heavy fishing mortality may contribute to this "experiment"); or compare recruitment between the Eastern Bering Sea stock and the Gulf of Alaska stock which is at a relatively low abundance level to test the effects of high and lower stock size on recruitment of pollock. The other species that could be considered for adaptive management are the cod and tanner crab. Although these species are presently under a management regime of the North Pacific Fisheries Management Council, Collie maintains that the catch quotas are determined without benefit of any "adaptive management" protocol, based on management alternatives across a set of hypotheses, and that considerations of long-term sustainability of the stocks could benefit from adoption of an adaptive management strategy that would consider alternative hypotheses against which to evaluate harvesting policies.

The paper by M. Mangel takes exception to the unstated

assumptions of fisheries yield models, particularly those on the one-way linkages between a stock and the ecosystem. Mangel demonstrates the impact of fish stocks on the ecosystem, using case studies (e.g., MacCall, 1986, Walters, 1986; Paine, 1988; Bradbury and Mundy, 1989; Smith et al., 1989) and argues that yield models should take this observation into account. He cites the work of Paine (1984) as a basis for encouraging the study of the role of ecosystem disturbance on stock-size and sustained fisheries biomass yields. Mangel recognizes the difficulty in conducting "disturbance" experiments in large marine ecosystems as does Collie (this volume) in proposing adaptive management strategies. However, Mangel goes on to state that "natural disturbance is continually providing perturbations which, if we are sufficiently clever, may take the place of experimental perturbations." He then provides a recommended protocol for quantifying changes in LMEs and predicting their outcomes on the basis of multispecies predator-prey interactions that includes effects of resource density, consumer density, predator density, on predator and resource population growth rates to elucidate the interaction strength of LME food chains. He concludes his arguments by recommending the development of a new generation of ecological models that link predator-prey strength with ecosystem feedback processes.

The chapter by B. Rothschild underscores the need for understanding the relationship between the population-dynamics process and the physical environment. Rothschild argues that efforts to link fish-stock variability to the physical environment have been difficult to discover, largely because environmental interactions have been treated as statistical noise by some fishery scientists, and that overfishing is the main cause of the collapse of fish stocks. The importance of high-dimensionality of ecosystems and the difficulty in selecting the best relationship to account for species variability is discussed. Rothschild considers physical events in the North Atlantic as measured by the kinetics of winds, storms, and cool water mass advection "coupled" with biological feedback in the form of northward movement of benthos, spread of intertidal organisms during the warm decades of the 1900s; the cooler periods were linked to the decline in zooplankton and phytoplankton in the Northeast Atlantic and declines in the catch of sardines within the Iberian Shelf ecosystem off Portugal. Other examples of the effects of kinetics on the growth of phytoplankton and larval sardines are given. Based on these observations and a consideration of turbulence as a function of energy dissipation rate and wind strength, Rothschild concludes that turbulence is a critical physical characteristic of marine ecosystems and that careful measurements of biological and physical linkages in the context of the kinetics of turbulence in relation to predator-prey contact rates at the lower end of the food chain can provide important insights to cause and effect relationships in the sustainability of biomass yields in the ocean. The linkage between small scale events at the predator-prey interface of larval fish and their zooplankton prey and basin-wide changes in ocean climate expressed in wind and water mass terms, when considered along with human interventions, provides a unifying approach to the biological and environmental processes important to the sustainable production of fisheries yields in LMEs.

Following the kinetics linkage of Rothschild, is the study by T. Smayda that is focussed on the growing frequency and extent of noxious phytoplankton blooms and the associated stress they impose on LMEs. The frequency and extent of red, green, yellow, brown, or white water discolorations from phytoplankton production are treated by Smayda collectively as red-tide or noxious bloom events. The losses of economically viable biomass to bloom events is significant, amounting to millions to tens of millions of dollars annually. The bloom condition has been reported from LMEs of the Indo-Pacific, the North Sea ecosystem, the U.S. Northeast Shelf ecosystem, the Black Sea ecosystem, the Baltic Sea ecosystem, and the Sea of Japan ecosystem. In some regions the bloom was accompanied by elevated nitrogen and phosphorus and reductions in ionic ratios with silica, tending toward a niche structure that favors flagellate production over diatoms. The importance of dinoflagellates as prey of larval fish is reviewed, including incidents where larval feeding on toxic dinoflagellates can lead to mortality. Ichthyotoxins produced by dinoflagellates are also harmful to juvenile and adult fish. Toxic dinoflagellates have also been implicated in the deaths of humpback whales and fur seals. Growth and recruitment of zooplankton can be impaired by toxic dinoflagellates. Smayda provides examples of the impact of toxic or nuisance phytoplankton bloom events on ecosystem trophodynamics either via direct or indirect trophic transfer or by anoxic die-offs to benthic organisms caused by bloom events.

Smayda argues that the increased pulse-like occurrence of toxic phytoplankton bloom events following the classical spring bloom in north temperate coastal waters "represent a major change in the phytoplankton dynamics of coastal waters." Should this trend persist, it is possible that recruitment of some fish species could be adversely impacted as postulated by Gosselin et al. (1989). Future research and monitoring strategies for large marine ecosystems should consider the arguments of Smayda that suggest noxious phytoplankton blooms may be influencing the structure and dynamics of marine ecosystems around the globe, and design a program of monitoring that will aid in quantifying the global extent of bloom events as part of an effort to mitigate their adverse impacts.

The future is encouraging. Based largely on the deliberations held during the October 1990 LME Conference in Monaco, the concept of linking LME research, monitoring, and management projects between a developed country and a developing country has taken the form of action in a planned symposium on "The Status and Future of Large Marine Ecosystems of the Indian Ocean" being organized jointly by Kenya and Belgium to be convened in Mombasa, Kenya, in 1992. Other regional LME symposia are being considered for the Yellow Sea ecosystem and the Bay of Bengal ecosystem.

The present symposium volume is designed principally to provide a synthesis on the present state-of-the-art for linking food chain dynamics studies with fisheries yield, population models, and management strategies under consideration within the United States.

Although variations in time and space of constituent elements of LMEs are demanding in time and effort to monitor, assess, and predict, recent technical advances have been made and applied to improve monitoring strategies for LMEs. Interventions by humans can have a stabilizing or destabilizing influence on the natural variability and cause or accelerate large-scale shifts in ecosystems. Therefore, it is important to include information in fish catches in any long-term LME monitoring study (Wise, 1986) and some means to measure changes in the productivity levels of LMEs. Can information on the cause and effect of these shifts provide insights that can be used to improve management strategies? We are entering a time of uncertainty about the effects of global change on human and environmental stresses on living marine resources. The consequences of additional stress on the structure and function of ecosystems sustaining the global fisheries are poorly understood. The present array of single species, and multispecies models need to be augmented to consider variability from spatial and temporal effects on species interactions and environmental conditions. Two methods that have proven successful for monitoring change in LMEs are regularly scheduled trawling, environmental, and plankton surveys. Using a stratified random sampling design, the large-scale changes in the fisheries of the North Sea and the Northeast Continental Shelf of the United States have been successfully analyzed for several decades (Azarovitz and Grosslein, 1987). The surveys have been conducted by relatively large research vessels. However, standardized sampling procedures, when deployed from small calibrated trawlers, can provide important information on fish stocks and their environment. The fish catch provides biological samples for stomach analyses, data for clarifying and quantifying multispecies trophic relationships; samples can be used for age and growth, fecundity, and size comparisons (ICES, 1991). Samples of trawl-caught fish can also be used to monitor the effect of gross pathological conditions that may be associated with coastal pollution. The need for both biological and environmental monitoring in the North Sea ecosystem has been emphasized following the Symposium on Long-Term Changes in the

Fish Stocks of the North Sea ecosystem (Hempel, 1978). In this regard, physical measurements can be made from small trawlers or ships-of-opportunity, using readily available and relatively inexpensive systems for measuring temperature and salinity of the water column, including the salinity and temperature sensors placed on trawl cables. Standard logs for weather observations, important in detecting global change, are an important component of the data-collecting effort. The monitoring of changes in fish stocks is ongoing in LMEs across the North Atlantic basin, including the Northeast U.S. Shelf, the Canadian Scotian Shelf, the Newfoundland Shelf, and on the Greenland Shelf, Icelandic Shelf, Norwegian Shelf, Barents Sea Shelf, and the North Sea.

The plankton of LMEs can be measured at a relatively low cost by deploying continuous plankton recorder (CPR) systems from commercial vessels of opportunity (Glover, 1967). The advanced plankton recorders can be fitted with sensors for temperature, salinity, chlorophyll, nitrate/nitrite, light, petroleum hydrocarbons, bioluminescence, zooplankton, and ichthyoplankton (Aiken, 1981; Williams and Aiken, 1990), providing the means to monitor changes in species composition, and dominance, and long-term changes in the physical and nutrient characteristics of the LME, as well as longer term changes relating to the biofeedback of the plankton to the stress of climate change (Colebrook, 1986; Dickson et al., 1988; Jossi and Smith, 1990; Sherman et al., 1990b). Plankton monitoring using the CPR system is at present expanding in the North Atlantic (UNESCO, 1990), the Mediterranean, and in 1992, new routes are being planned for the Western Indian Ocean, off east Africa, across the Somalia Current ecosystem, off west Africa in the Gulf of Guinea ecosystem, in the Aegean and Adriatic subsystems of the Mediterranean, and in the Yellow Sea ecosystem.

The temporal and spatial scales influencing important processes in biological production in the sea have been the topic of a number of studies. The selection of scale in any study is related to the process under investigation. An excellent treatment of this topic can be found in Steele (1988). He indicates that in relation to general ecology, the best known work in fish population dynamics is represented by the early studies of Schaefer (1954) and Beverton and Holt (1957). Evolution to more holistic ecological models was introduced in the energy flow approach of Steele (1965) following the early pioneering approach of Lindeman (1942). However, as noted by Steele they are unsuitable for consideration of temporal or spatial variability in the ocean. The concept of large marine ecosystems (LMEs) defines the unit of study on the order of thousands of kilometers in scale, with regard to fish and fisheries yields. In this approach, large-scale climatic or environmental changes are examined in relation to multidecadal fisheries-yield patterns of marine ecosystems. Changes in the fish communities of LMEs can trigger a cascade effect involving higher trophic levels of marine mammal and bird populations, and lower trophic levels of phytoplankton and zooplankton, and the economies dependent on the resources of the ecosystems. A list of reports describing the effects of biological and physical perturbations on the fisheries biomass yields of 30 large marine ecosystems is given in Table 1.3. The questions generally posed by these investigations are not dissimilar from those posed a few years ago by Beddington (1984):

> There are a number of scientific questions which are central to the rational management of marine communities, but all revolve around the question of sustainability.
>
> What levels of mortality imposed by a fishery will permit a sustainable yield? Are there levels below which a fish population will not recover? Can judicious manipulation of the catch composition of the fishery alter the potential of the community to produce yields of a particular type, e.g., high value species? Can a community be depleted to a level where its potential for producing a harvestable resource is reduced?
>
> With the exception of the first question, these questions and others like them are rarely explicitly addressed in the scientific bodies of the various fisheries' organizations. Instead, such bodies concentrate on the estimation of stock abundance and the calculation of allowable catch levels, although often implicit in the advice given by these bodies to management are a set of beliefs about the answers to such questions.

Given the increasing number of responsibilities of government agencies for: (1) managing fisheries, (2) mitigating pollution, (3) reducing environmental stress, and (4) restoration of lost habitat, it is not surprising that interest is growing to pursue resource management problems from an ecosystem perspective.

The topic of change and persistence in marine communities and the need for multispecies and ecosystem perspectives in fishery management relate to the reports of changing states of marine ecosystems (Sugihara et al., 1984). Collapses of the Pacific sardine in the California Current ecosystem, the pilchard in the Benguela Current ecosystem, and the anchovy in the Humboldt Current ecosystem, are but a few examples of cascading effects on other ecosystem components including marine birds (MacCall, 1986; Croxall, 1987; Burger, 1988; Crawford et al., 1989).

Table 1.3. List of 30 large marine ecosystems and subsystems for which syntheses relating to principal, secondary, or tertiary driving forces controlling variability in biomass yields have been completed for inclusion in LME volumes through February 1991.

Large marine ecosystem	Volume no.*	Authors
U.S. Northeast Continental Shelf	1	M. Sissenwine
	4	P. Falkowski
U.S. Southeast Continental Shelf	4	J. Yoder
Gulf of Mexico	2	W. Richards and M. McGowan
	4	B. Brown et al.
California Current	1	A. MacCall
	4	M. Mullin
	5	D. Bottom
Eastern Bering Shelf	1	L. Incze and J. Schumacher
West Greenland Shelf	3	H. Hovgaard and E. Buch
Norwegian Sea	3	B. Ellertsen et al.
Barents Sea	2	H. Skjoldal and F. Rey
	4	V. Borisov
North Sea	1	N. Daan
Baltic Sea	1&5	G. Kullenberg
Iberian Coastal	2	T. Wyatt and G. Perez-Gandaras
Mediterranean-Adriatic Sea	5	G. Bombace
Canary Current	5	C. Bas
Gulf of Guinea	5	D. Binet and E. Marchal
Benguela Current	2	R. Crawford et al.
Patagonian Shelf	5	A. Bakun
Caribbean Sea	3	W. Richards and J. Bohnsack
South China Sea-Gulf of Thailand	2	T. Piyakarnchana
Yellow Sea	2	Q. Tang
Sea of Okhotsk	5	V. Kusnetsov et al.
Humboldt Current	5	J. Alheit and P. Bernal
Indonesia Seas-Banda Sea	3	J. Zijlstra and M. Baars
Bay of Bengal	5	S. Dwivedi
Antarctic Marine	1&5	R. Scully et al.
Weddell Sea	3	G. Hempel
Kuroshio Current	2	M. Terazaki
Oyashio Current	2	T. Minoda
Great Barrier Reef	2	R. Bradbury and C. Mundy
	5	G. Kelleher
Gulf of California	5	L. Mee
South China Sea	5	D. Pauly and V. Christensen

Sherman [1991]. Reprinted by permission.
*Footnotes on next page.

Marine Resources and LMEs

As the trend for the management of living marine resources moves from single species to multispecies assemblages, it becomes increasingly important to encompass entire ecosystems as management units. This approach will ensure that management measures designed to optimize the natural productivity of target species will also include consideration for related competitor/predator populations and their environments.

A systems approach to the management of LMEs is depicted in Table 1.4. The system allows for the LMEs to serve as the link between local events (e.g., fishing, pollution, storms) occurring on the daily-to-seasonal temporal scale and their effects on living marine resources and the more ubiquitous global effects of climate changes on the multidecadal time-scale. The regional and temporal focus of season to decade is consistent with the evolved spawning and feeding migrations of the fishes--the keystone species of most large marine ecosystems. These migrations are seasonal and occur over hundreds to thousands of kilometers within the unique physical and biological characteristics of the regional LME to which they have adapted. As the fisheries represent most of the useable biomass yield of the LMEs and fish populations consist of several age classes, it follows that measures of variability in growth, recruitment, and mortality should be conducted over multiyear-time-scales. This is necessary in order to interpret the effects of environmental, biological, and fishing effects on changing abundance levels of the year class to the populations of the species constituting the fish community, their predators and prey, and physical environment.

Notes to Table 1.3.

Vol. 1, Variability and Management of Large Marine Ecosystems, Edited by K. Sherman and L. M. Alexander, AAAS Selected Symposium 99, Westview Press, Boulder, CO, 1986.

Vol. 2, Biomass Yields and Geography of Large Marine Ecosystems, Edited by K. Sherman and L. M. Alexander, AAAS Selected Symposium 111, Westview Press, Boulder, CO, 1989.

Vol. 3, Large Marine Ecosystems: Patterns, Processes, and Yields, Edited by K. Sherman, L. M. Alexander, and B. D. Gold. AAAS Symposium, Am. Assoc. Advancement Sci., Washington, D.C. 1990.

Vol. 4, Food Chains, Yields, Models, and Management of Large Marine Ecosystems, Edited by K. Sherman, L. M. Alexander, and B. D. Gold, AAAS Symposium, Westview Press, Boulder, CO. (in press)

Vol. 5, Stress, Mitigation, and Sustainability of Large Marine Ecosystems. Edited by K. Sherman, L. M. Alexander, and B. D. Gold. Am. Assoc. Advancement Sci., Washington, D.C. (in press).

Table 1.4. Key spatial and temporal scales and principal elements of a systems approach to the research and management of large marine ecosystems.

1. Spatial-Temporal Scales

Spatial	Temporal	Unit
1.1 **Global** (World Ocean)	Millennia-decadal	Pelagic biogeographic
1.2 **Regional** (Exclusive Economic Zones)	Decadal-seasonal	Large marine ecosystems
1.3 **Local**	Seasonal-daily	Subsystems

2. Research Elements
   - 2.1 Spawning strategies
   - 2.2 Feeding strategies
   - 2.3 Productivity, trophodynamics
   - 2.4 Stock fluctuations/recruitment/mortality
   - 2.5 Natural variability (hydrography, currents, water masses, weather)
   - 2.6 Human perturbations (fishing, waste disposal, petrogenic hydrocarbon impacts, aerosol contaminants, eutrophication effects)
3. Management Elements--Options and Advice--International, National, Local
   - 3.1 Bioenvironmental and socioeconomic models
   - 3.2 Management to optimize fisheries yields
4. Feedback Loop
   - 4.1 Evaluation of ecosystem status
   - 4.2 Evaluation of fisheries status
   - 4.3 Evaluation of management practices

Sherman [1991]. Reprinted by permission.

Consideration of the naturally occurring environmental events and the human-induced perturbations affecting demography of the populations within the ecosystem is necessary. Based on a firm, scientific understanding of the principal causes of variability in abundance and with due consideration to socioeconomic needs, management options can be considered for implementation from an ecosystems perspective. The final element in the systems approach

Table 1.5. Selected hypotheses concerning variability in biomass yields of large marine ecosystems. Note that references can be found in Table 1.3.

Ecosystem	Predominant variables	Hypothesis
Oyashio Current Kuroshio Current California Current Humboldt Current Benguela Current Iberian Coastal	Density-independent natural environmental perturbations	<u>Clupeoid Population Increases:</u> Predominant variables influencing changes in biomass of clupeoids are major increases in water-column productivity resulting from shifts in the direction and flow velocities of the currents and changes in upwelling within the ecosystem.
Yellow Sea U. S. Northeast Continental Shelf Gulf of Thailand	Density-dependent predation	<u>Declines in Fish Stocks</u>: Precipitous decline in biomass of fish stocks is the result of excessive fishing mortality, reducing the probability of reproductive success. Losses in biomass are attributed to excesses of human predation expressed as overfishing.
Great Barrier Reef	Density-dependent predation	<u>Change in Ecosystem Structure</u>: The extreme predation pressure of crown-of-thorns starfish has disrupted normal food chain linkage between benthic primary production and the fish component of the reef ecosystem.
East Greenland Sea Barents Sea Norwegian Sea	Density-independent natural environmental perturbations	<u>Shifts in Abundance of Fish Stock Biomass</u>: Major shifts in the levels of fish stock biomass within the ecosystems are attributed to large-scale environmental changes in water movements and temperature structure.
Baltic Sea	Density-independent pollution	<u>Changes in Ecosystem Productivity Levels</u>: The apparent increases in productivity levels are attributed to the effects of nitrate enrichment resulting from elevated levels of agricultural contaminant inputs from the bordering land masses.

Table 1.5 continued.

Ecosystem	Predominant variables	Hypothesis
Antarctic marine	Density-dependent perturbations	Status of Krill Stocks: Annual natural production cycle of krill is in balance with food requirements of dependent predator populations. Surplus production is available to support economically significant yields, but sustainable level of fishing effort is unknown.
	Density-independent natural environmental perturbations	Shifts in Abundance in Krill Biomass: Major shifts in abundance levels of krill biomass within the ecosystem are attributed to large-scale changes in water movements and productivity.

Sherman [1991]. Reprinted by permission.

is the feedback loop that allows for evaluation of the effects of management actions at the fisheries level (single species, multispecies) and the ecosystem level, with regard to the concept of resource maintenance and sustained yield. It will be necessary to conduct supportive research on the processes controlling sustained productivity of LMEs. Within several of the LMEs, important hypotheses concerned with the growing impacts of pollution, overexploitation, and environmental changes on sustained biomass yields are under investigation (Table 1.5). By comparing the results of research among the different systems, it should be possible to accelerate an understanding of how the systems respond and recover from stress; the comparisons should allow for narrowing the context of unresolved problems and capitalizing on research efforts underway in the different ecosystems.

References

Aiken, J. 1981. The Undulating Oceanographic Recorder Mark 2. J. Plankton Res. 3:551-560.

Alexander, L. M. 1989. Large marine ecosystems as global management units. Pp. 339-344. *In* K. Sherman and L. M. Alexander (Eds.) Biomass yields and geography of large marine ecosystems. AAAS Selected Symposium 111. Westview Press, Inc., Boulder. 493 Pp.

Andersen, K. P., and E. Ursin. 1977. A multispecies extension to the Beverton and Holt Theory of Fishing with accounts of phosphorus circulation and primary production. Medd. Dan. Fish.-Havunders. N.S. 7:319-345.

Azarovitz, T. R., and M. D. Grosslein. 1987. Fishes and squids. Pp. 315-346. *In* R. H. Backus (Ed.) Georges Bank, MIT Press, Cambridge, MA. 593 Pp.

Bakun, A. 1986. Definition of environmental variability affecting biological processes in large marine ecosystems. Pp. 89-108. *In* K. Sherman and L. M. Alexander (Eds.) Variability and management of large marine ecosystems. AAAS Selected Symposium 99. Westview Press, Inc., Boulder.

Bakun, A. 1990. Global climate change and intensification of coastal ocean upwelling. Science 247:198-201.

Bakun, A., and R. H. Parrish. 1980. Environmental inputs to fishery population models for eastern boundary current regions. Pp. 67-104. *In* G. D. Sharp (Ed.) Workshop on the effects of environmental variation on the survival of larval pelagic fishes. IOC Workshop Report 28, UNESCO, Paris.

Bakun, A., and R. H. Parrish. 1990. Comparative studies of coastal pelagic fish reproductive habitats: The Brazilian sardine (*Sardinella aurita*). J. Cons. int. Explor. Mer 46:269-283.

Beddington, J. R. 1984. The response of multispecies systems to perturbations. Pp. 209-255. *In* R. M. May (Ed.) Exploitation of marine communities. Springer-Verlag, Berlin. 366 Pp.

Belsky, M. H. 1989. The ecosystem model mandate for a comprehensive United States ocean policy and Law of the Sea. San Diego L. Rev. 26(3):417-495.

Beverton, R.J.H., and S. J. Holt. 1957. On the dynamics of exploited fish populations. Fish. Invest. Minist. Agric. Fish. Food (G. B.) Ser.II 19:1-533.

Bonner, W. N. 1982. Seals and man, a study of interactions. Univ. Washington Press, Seattle, 170 Pp.

Borisov, V. This volume. The state of the main commercial species of fish in the changeable Barents Sea ecosystem.

Bottom, D. L., K. K. Jones, J. D. Rodgers, and R. F. Brown. 1989. Management of living resources: A research plan for the Washington and Oregon continental margin. National Coastal Resources Research & Development Institute, Newport, OR. NCRI-T-89-004, 80 Pp.

Bradbury, R. H., and C. N. Mundy. 1989. Large-scale shifts in biomass of the Great Barrier Reef ecosystem. Pp. 143-167. *In* K. Sherman and L. M. Alexander (Eds.) Biomass yields and geography of large marine ecosystems. AAAS Selected Symposium 111. Westview Press, Inc., Boulder. 493 Pp.

Burger, J. 1988. Interactions of marine birds with other marine vertebrates in marine environments. Pp. 3-28. *In* J. Burger (Ed.) Seabirds and other marine vertebrates. Columbia Univ. Press, New York. 339 Pp.

Byrne, J. 1986. Large marine ecosystems and the future of ocean studies. Pp. 299-308. *In* K. Sherman and L. M. Alexander (Eds.) Variability and management of large marine ecosystems. AAAS Selected Symposium 99. Westview Press, Inc., Boulder. 319 Pp.

Canon, J. R. 1986. Variabilidad ambiental en relacion con la pesqueria neritica pelagica de la zona Norte de Chile. Pp. 195-205 *In* P. Arana (Ed.) La Pesca en Chile. Escuela de Ciencias del Mar, Facultad de Recursos Naturales, Universidad Catolica de Valparaiso, Chile.

Christy, F. T., Jr. 1986. Can large marine ecosystems be managed for optimum yields? Pp. 263-267. *In* K. Sherman and L. M. Alexander (Eds.) Variability and management of large marine ecosystems. AAAS Selected Symposium 99. Westview Press, Inc., Boulder. 319 Pp.

Colebrook, J. M. 1986. Environmental influences on long-term variability in marine plankton. Hydrobiologia 142:309-325.

Collie, J. S. This volume. Adaptive strategies for management of fisheries resources in large marine ecosystems.

Crawford, R.J.M., L. V. Shannon, and P. A. Shelton. 1989. Characteristics and management of the Benguela as a large marine ecosystem. Pp. 169-219. *In* K. Sherman and L. M. Alexander (Eds.) Biomass yields and geography of large marine ecosystems. AAAS Selected Symposium 111. Westview Press, Inc., Boulder. 493 Pp.

Croxall, J. P. (Editor). 1987. Seabirds: feeding ecology and role in marine ecosystems. Cambridge Univ. Press, Cambridge, UK. 408 Pp.

Cushing, D. H. 1975. Marine ecology and fisheries. Cambridge Univ. Press, London. 278 Pp.

Cushing, D. H. 1988. Review of: The Peruvian anchoveta and its upwelling ecosystem: Three decades of change. J. Cons. int. Explor. Mer 44:297-300.

Daan, N. 1986. Results of recent time-series observations for monitoring trends in large marine ecosystems with a focus on the North Sea. Pp. 145-174. *In* K. Sherman and L. M. Alexander (Eds.) Variability and management of large marine ecosystems. AAAS Selected Symposium 99. Westview Press, Inc., Boulder.

Dickson, R. R., P. M. Kelly, J. M. Colebrook, W. S. Wooster, and D. H. Cushing. 1988. North winds and production in the eastern North Atlantic. J. Plankton Res. 10:151-169.

Ellertsen, B., P. Fossum, P. Solemdal, S. Sundby, and S. Tilseth. 1990. Environmental influence on recruitment and biomass yields in the Norwegian Sea ecosystem. Pp. 19-35. *In* K. Sherman and L. M. Alexander (Eds.) Large marine ecosystems: Patterns, processes, and yields. Am. Assoc. Advancement Sci., Washington, DC.

FAO [Food and Agriculture Organization of the UN]. 1989. FAO yearbook. Fishery statistics: catches and landings. Vol. 64 (for 1987). FAO, Rome.

GESAMP [Group of Experts on the Scientific Aspects of Marine Pollution]. 1990. The state of the marine environment. UNEP Regional Seas Reports and Studies No. 115. Nairobi.

Glover, R. S. 1967. The continuous plankton recorder survey of the North Atlantic. Symp. Zool. Soc. Lond. 19:189-210.

Gosselin, S., L. Fortier, and J. A. Gagne. 1989. Vulnerability of marine fish larvae to the toxic dinoflagellate *Protogonyaulax tamarensis*. Mar. Ecol. Prog. Ser. 57:1-10.

Graham, R. L., C. T. Hunsaker, R. V. O'Neill, and B. L. Jackson. 1991. Ecological risk assessment at the regional scale. Ecol. Applications 1:196-206.

Hempel, G. (Editor). 1978. Symposium on North Sea fish stocks--Recent changes and their causes. Rapp. P.-v. Reun. Cons. int. Explor. Mer 172:449 Pp.

Hovgaard, H., and E. Buch. 1990. Fluctuation in the cod biomass of the West Greenland Sea ecosystem in relation to climate. Pp. 36-43. *In* K. Sherman, L. M. Alexander and B. D. Gold (Eds.) Large marine ecosystems: Patterns, processes and yields. AAAS Publ., Washington, DC. 242 Pp.

ICES [International Council for the Exploration of the Sea]. 1990. Report of the ICES Study Group on Cod Stock Fluctuations. Cod and climate change (CCC), framework for the study of global ocean ecosystem dynamics. ICES C.M. 1990/G:50 Ref. C+L. 29 Pp.

ICES. 1991. Report of the Multispecies Working Group. ICES C.M. 1991/Assess:7.

Jossi, J. W., and D. E. Smith. 1990. Continuous plankton records: Massachusetts to Cape Sable, N.S., and New York to the Gulf Stream, 1989. NAFO Ser. Doc. 90/66:1-11.

Kelleher, G. In press. Sustainable development of the Great Barrier Reef as a large marine ecosystem. *In* K. Sherman, L. M. Alexander, and B. D. Gold (Eds.) Stress, mitigation, and sustainability of large marine ecosystems. Am. Assoc. Advancement Sci., Washington, DC.

Levin, S. A. 1990. Physical and biological scales, and modelling of predator-prey interactions in large marine ecosystems. Pp. 179-187. *In* K. Sherman, L. M. Alexander, and B. D. Gold (Eds.) Large marine ecosystems: Patterns, processes, and yields. AAAS Publ., Washington, DC. 242 Pp.

Lindeman, R. L. 1942. The trophic dynamic aspect of ecology. Ecology 23:399-418.

Loughlin, T. R., and R. Nelson, Jr. 1986. Incidental mortality of northern sea lions in the Shelikof Strait, Alaska. Mar. Mammal Sci. 1:14-33.

Lubchenko et al. 1991. The sustainable biosphere initiative: An ecological research agenda. A report from the Ecological Society of America. Ecology 72:371-412.

MacCall, A. D. 1986. Changes in the biomass of the California Current system. Pp. 33-54. *In* K. Sherman and L. M. Alexander (Eds.) Variability and management of large marine ecosystems. AAAS Selected Symposium 99. Westview Press, Inc., Boulder. 319 Pp.

Mays, E. 1982. The growth of biological thought. Harvard Univ. Press, Cambridge, MA. 9974 Pp.

Minoda, T. 1989. Oceanographic and biomass changes in the Oyashio Current ecosystem. Pp. 67-93 *In* K. Sherman and L. M. Alexander (Eds.) Biomass yields and geography of large marine ecosystems. AAAS Selected Symposium 111, Westview Press, Inc., Boulder. 493 Pp.

Morgan, J. R. 1988. Large marine ecosystems: an emerging concept of regional management. Environment 29(10):4-9 & 26-34.

Morgan, J. R. 1989. Large marine ecosystems in the Pacific Ocean. Pp. 377-394. *In* K. Sherman and L. M. Alexander (Eds.) Biomass yields and geography of large marine ecosystems. AAAS Selected Symposium 111. Westview Press, Inc., Boulder. 493 Pp.

NOAA [National Oceanic and Atmospheric Administration]. 1988. Fishery resource programs. Folio Map No. 7, A national atlas: Health and use of coastal waters, United States of America. U.S. Dept. of Commer., NOAA, Nat. Ocean Serv., Office of Oceanography and Marine Assessment, Washington, DC.

Overholtz, W. J., and J. R. Nicolas. 1979. Apparent feeding by the fin whale *Balaenoptera physalus*, and humpback whale, *Megoptera novaeangliae*, on the American sand lance, *Ammodytes americanus*, in the Northwest Atlantic. Fish. Bull., U.S. 77:285-287.

Paine, R. T. 1984. Some approaches to modeling multispecies systems. Pp. 191-207. *In* R. M. May (Ed.) Exploitation of marine communities. Report of the Dahlem Workshop on Exploitation of Marine Communities. Springer-Verlag, Berlin.

Paine, R. T. 1988. Food webs: Road maps of interactions or grist for theoretical development? Ecology 69:1648-1654.

Pauly, D. 1989. Interactions and dynamics of the Peruvian Upwelling system: A postscript. Pp. 404-407. *In* D. Pauly, P. Muck, J. Mendo, and I. Tsukayama (Eds.) The Peruvian upwelling ecosystem: Dynamics and interactions. Proceedings of the Workshop on Models for Yield Prediction in the Peruvian Ecosystem, 24-28 August 1987, Callao, Peru. IMARPE, GTZ, and ICLARM, Manila, Philippines. PROCOPA Contrib. No. 92 and ICLARM Contrib. No. 409. 438 Pp.

Payne, P. M., D. N. Wiley, S. B. Young, S. Pittman, P. J. Clapham, and J. W. Jossi. 1990. Recent fluctuations in the abundance of baleen whales in the southern Gulf of Maine in Relation to changes in selected prey. Fish. Bull., U.S. 88:687-696.

Piyakarnchana, T. 1989. Yield dynamics as an index of biomass shifts in the Gulf of Thailand ecosystems. Pp. 95-142 *In* K. Sherman and L. M. Alexander (Eds.) Biomass yields and geography of large marine ecosystems. AAAS Selected Symposium 111, Westview Press, Inc., Boulder. 493 Pp.

Powers, K. D., and R.G.B. Brown. 1987. Seabirds. Chapter 34. Pp. 359-371. *In* R. H. Backus (Ed.) Georges Bank. MIT Press, Cambridge, MA. 593 Pp.

Prescott, J.R.V. 1989. The political division of large marine ecosystems in the Atlantic Ocean and some associated seas. Pp. 395-442. *In* K. Sherman and L. M. Alexander (Eds.) Biomass yields and geography of large marine ecosystems. AAAS Selected Symposium 111. Westview Press, Inc., Boulder. 493 Pp.

Ricklefs, R. E. 1987. Community diversity relative roles of local and regional processes. Science 235(4785):167-171.

Riley, G. A. 1947. Seasonal fluctuations of the phytoplankton populations in New England coastal waters. J. Mar. Res. 6:114-125.

Sainsbury, K. J. 1988. The ecological basis of multispecies fisheries, and management of a dermersal fishery in tropical Australia. Pp. 349-382. *In* J. A. Gulland (Ed.) Fish Population Dynamics, 2nd Ed., John Wiley & Sons, New York. 422 Pp.

Schaefer, M. B. 1954. Some aspects of the dynamics of populations important to the management of the commercial marine fisheries. Bull. Inter-Am. Trop. Tuna Comm. 1:27-56.

Scully, R. T., W. Y. Brown, and B. S. Manheim. 1986. The Convention for the Conservation of Antarctic Marine Living Resources: A model for large marine ecosystem management. Pp. 281-286. *In* K. Sherman and L. M. Alexander (Eds.) Variability and management of large marine ecosystems. AAAS Selected Symposium 99. Westview Press, Inc., Boulder. 319 Pp.

Sherman, K. [1991]. The large marine ecosystem concept: A research and management strategy for living marine resources. Ecol. Applications 1(4):(in press).
Sherman, K., and L. M. Alexander (Editors). 1986. Variability and management of large marine ecosystems. AAAS Selected Symposium 99. Westview Press, Inc., Boulder. 319 Pp.
Sherman, K., and L. M. Alexander. 1989. Biomass yields and geography of large marine ecosystems. AAAS Selected Symposium 111. Westview Press, Inc., Boulder. 493 Pp.
Sherman, K., L. M. Alexander, and B. D. Gold (Editors). 1990a. Large marine ecosystems: Patterns, processes and yields. AAAS Publ., Washington, DC. 242 Pp.
Sherman, K., J. Jossi, and J. Goulet. 1990b. Comparative stability of zooplankton communities of the Northeast U.S. Shelf ecosystem and the North Sea ecosystem in relation to climatic variability. ICES C.M. 1990/L:23.
Sherman, K., and A. F. Ryan. 1988. Antarctic marine living resources. Oceanus 31(2):59-63.
Sissenwine, M. P. 1986. Perturbation of a predator-controlled continental shelf ecosystem. Pp. 55-85. *In* K. Sherman and L. M. Alexander (Eds.) Variability and management of large marine ecosystems. AAAS Selected Symposium 99, Westview Press, Inc., Boulder. 319 Pp.
Sissenwine, M. P., and E. B. Cohen. This volume. Resource productivity and fisheries management of the northeast shelf ecosystem.
Skjoldal, H. R., and F. Rey. 1989. Pelagic production and variability of the Barents Sea ecosystem. Pp. 241-286. *In* K. Sherman and L. M. Alexander (Eds.) Biomass yields and geography of large marine ecosystems. Westview Press, Inc., Boulder.
Smith, P. E., M. D. Ohman, and L. E. Eber. 1989. Analysis of the patterns of distribution of zooplankton aggregations from an acoustic Doppler current profiler. CalCOFI Rep. 30:89-103.
Steele, J. H. 1965. Some problems in the study of marine resources. Spec. Publ. Int. Comm. Northw. Atlan. Fish. 6:463-476.
Steele, J. H. 1974. The structure of marine ecosystems. Harvard Univ. Press, Cambridge, MA.
Steele, J. H. 1988. Scale selection for biodynamic theories. Pp. 513-526. *In* B. J. Rothschild (Ed.) Toward a theory on biological-physical interactions in the World Ocean, NATO ASI Series C: Mathematical and Physical Sciences, Vol. 239. Kluwer Academic Publishers, Dordrecht. 650 Pp.
Sugihara, G., S. Garcia, J. A. Gulland, J. H. Lawton, H. Maske, R. T. Paine, T. Platt, E. Rachor, B. J. Rothschild, E. A. Ursin, B.F.K. Zeitzschel. 1984. Ecosystem dynamics: group report. Pp. 130-

153. *In* R. M. May (Ed.) Exploitation of marine communities. Springer-Verlag, Berlin. 366 Pp.

Tang, Q. 1989. Changes in the biomass of the Yellow Sea ecosystems. Pp. 7-35. *In* K. Sherman and L. M. Alexander (Eds.) Biomass yields and geography of large marine ecosystems. AAAS Selected Symposium 111. Westview Press, Inc., Boulder. 493 Pp.

Terazaki, M. 1989. Recent Large-Scale Changes in the Biomass of the Kuroshio Current Ecosystem. Pp. 37-65 *In* K. Sherman and L. M. Alexander (Eds.) Biomass yields and geography of large marine ecosystems. AAAS Selected Symposium 111. Westview Press, Inc., Boulder.

UNESCO [United Nations Education, Scientific and Cultural Organization]. 1990. UNEP-IOC-WMO meeting of experts on long-term global monitoring system of coastal and near-shore phenomena related to climate change. Intergovernmental Oceanographic Commission Reports of Meetings of Experts and Equivalent Bodies. Paris, 10-14 December 1990.

Waring, G. T., P. M. Payne, B. L. Parry, and J. R. Nicolas. 1990. Incidental take of marine mammals in foreign fishery activities off the northeast United States, 1977-88. Fish. Bull., U.S. 88:347-360.

Williams, R., and J. Aiken. 1990. Optical measurements from underwater towed vehicles deployed from ships-of-opportunity in the North Sea. *In* Environment and pollution measurement sensors and systems, 14-15 March 1990, The Hague, The Netherlands. SPIE 1269:186-194.

Wise, J. P. 1986. Fisheries statistics--Boring stuff until you need them for monitoring. Oceans '86 Conference Proceedings, Vol. 3 Monitoring Strategies Symposium. Mar. Tech. Soc., Washington, DC.

Wyatt, T., and G. Perez-Gandaras. 1989. Biomass changes in the Iberian ecosystem. Pp. 221-239. *In* K. Sherman and L. M. Alexander (Eds.) Biomass yields and geography of large marine ecosystems. AAAS Selected Symposium 111. Westview Press, Inc., Boulder. 493 Pp.

Zijlstra, J. J. 1988. The North Sea Ecosystem. Pp. 231-278 *In* H. Postma and J. J. Zijlstra (Eds.) Ecosystems of the world 27. Continental Shelves. Elsevier, Amsterdam.

Zijlstra, J. J., and M. A. Baars. 1990. Productivity and fisheries potential of the Banda Sea ecosystem. Pp. 54-65. *In* K. Sherman, L. M. Alexander, and B. D. Gold (Eds.) Large marine ecosystems: Patterns, processes, and yields. Am. Assoc. Advancement Sci., Washington, DC.

Paul G. Falkowski

2. A Carbon Budget for the Northeast Continental Shelf Ecosystem: Results of the Shelf Edge Exchange Process Studies

Abstract

Classically, phytoplankton losses have been attributed primarily to zooplankton grazing. Within the past decade there has been increased evidence that microzooplankton and bacteria also mediate the oxidation of phytoplankton, especially in the oligotrophic central ocean gyres. On the continental margins, however, export of shelf phytoplankton to the adjacent ocean basins may also occur. The Shelf Edge Exchange Processes (SEEP) program is a multidisciplinary effort aimed at quantifying the various fates of phytoplankton of the Northeast Continental Shelf ecosystem of the United States. The results of the first SEEP field program, conducted off the coast of Long Island, New York and Martha's Vineyard, Massachusetts, in 1984, suggested that about 35% of the spring phytoplankton production was consumed by herbivorous zooplankton, 45% was oxidized on the shelf by microbes and the benthos, and the remainder was possibly exported to the interior of the adjacent ocean basin and slope sediments. In the second SEEP program, conducted off the Delmarva Peninsula from March 1988 to May 1989, phytoplankton and zooplankton biomass, current speed and direction, incident irradiance, soluble oxygen, as well as temperature and salinity were continuously measured at 10 moorings by state-of-the-art instrumentation. The results revealed that a single pulse of phytoplankton, constituting a bloom, does not occur in the spring, that phytoplankton and zooplankton biomass are highly coherent, and that on a time scale of 2 to 10 days, physical processes dominate the dynamics between primary and secondary producers. The results imply that understanding food chain dynamics requires a much higher temporal resolution of components than has heretofore been available, and that on the continental shelf, fisheries are more likely to be limited by recruitment and mortality than food availability.

Introduction

A paradigm in biological oceanography is that in temperate coastal and continental shelf regions of the world oceans there is a spring phytoplankton bloom (Riley, 1946, Harvey, 1957, Harris, 1980). On shelves, when insolation increases and ample nutrients are available, integrated water column primary production can exceed respiratory and grazing losses, leading to a rapid increase in phytoplankton biomass (Sverdrup, 1953; Steele, 1974). That paradigm implies that primary production is made available to other trophic levels in large pulses, lasting about a month (Harris, 1980). The efficiency of the coupling of the spring bloom to secondary production is often assumed to be a critical factor in the production, yield, recruitment, and possibly species composition of fish (Cushing, 1962). Marine ecologists have attempted to heuristically model the timing and magnitude of spring blooms, the transfer of primary production to herbivorous zooplankton grazers, and extrapolate from secondary production to potential fish yields (Steele, 1974; Cushing, 1975; Walsh, 1981). If, for some reason, a bloom is not efficiently transferred to pelagic herbivores, presumably the primary production would either be utilized by benthic fauna, buried in the sediment, exported to the deep ocean basins, and/or oxidized on the shelf by microbes and protozoa.

Our recent observations for the Northeast Continental Shelf ecosystem of the United States suggest that the spring bloom paradigm, consisting of a broad phytoplankton biomass and primary production maximum followed by a peak of zooplankton abundance and grazing, greatly oversimplifies food chain dynamics on a archetypical temperate continental shelf. Patches of phytoplankton forced by physical processes operating on a time scale of days-to-weeks are observed rather than a gradual increase of phytoplankton abundance.

Here I will examine the problems of understanding the coupling between primary and secondary producers within the context of the Shelf Edge Exchange Process (SEEP) program; a multidisciplinary study which was conducted within the United States Northeast Shelf ecosystem.

A Description of the Northeast Continental Shelf

The Northeast Continental Shelf ecosystem of the United States extends from the Gulf of Maine to Cape Hatteras (Sherman et al., 1988) (Figure 2.1), a distance of about 1300 km. The shelf-slope break occurs between the 100- and 200-m isobaths, and the shelf is

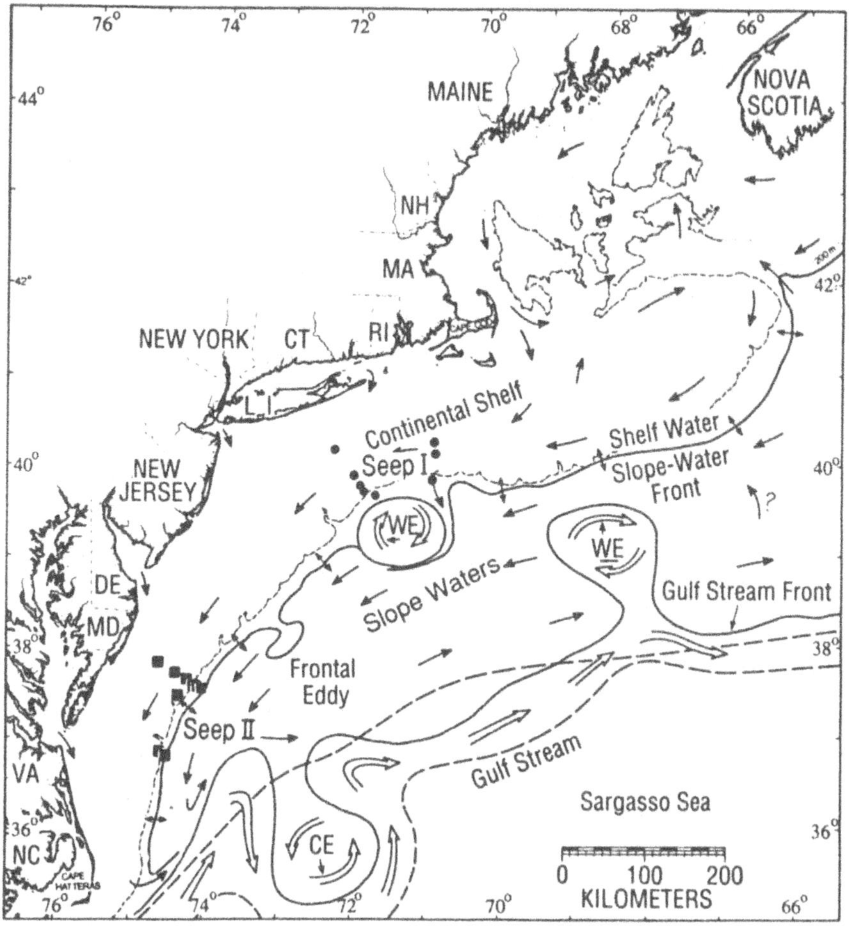

Figure 2.1. Generalized flow diagram for the northeast continental shelf of the United States. The shelf-slope break is shown as the dashed line corresponding to the 200-m isobath. Water enters the shelf from the northeast from the Gulf of Maine and the Nova Scotia region. The major component of flow is to the southwest at ca. 5 cm/s. From October to May, a front (see Figure 2.2) demarks the boundary between cold, fresher shelf waters and warm saltier slope waters. Exchange of fluid across the shelf-slope front is primarily related to wind events. Warm core rings (WE), which entrain waters from the Gulf Stream, occasionally meander onto the shelf. As well, slope and shelf waters are occasionally entrained in cold core eddies (CE) which can be advected in the Gulf Stream. The circulation in the slope is thought to be characterized by a cyclonic gyre. The positions of SEEP I moorings are shown as filled circles, while the positions of SEEP II moorings are shown as filled squares.

widest in the northeast, becoming increasingly narrower to the southwest.

The circulation on the shelf is relatively complicated. Shelf waters are formed in Labrador and Nova Scotia, and move southwest into the Gulf of Maine and around both sides of Georges Bank entering the shelf at Nantucket Shoals. An along-shore pressure gradient, which is not locally forced by wind stress, maintains a predominant flow field along the shelf to the southwest at about 5 cm/s. At Cape Hatteras shelf waters exit and are thought to be entrained with warmer, saltier slope waters, in an anticlockwise gyre. The slope-sea gyre is separated from the open waters of the central Northwest Atlantic by the Gulf Stream. From time to time warm core eddies pinch off from the Gulf Stream and wander onto the shelf. As well, cold shelf, and mixtures of shelf and slope, waters may become entrained and encapsulated by Gulf Stream waters, forming cold core rings. Warm and cold core rings are one means of transporting fluid between the shelf the North Atlantic central basin.

Three major estuaries, the Hudson, Delaware and Chesapeake, drain into the shelf, freshening the waters overall by about 1%. From October through mid-May the fresher, colder shelf waters are separated from the warmer, saltier slope waters by a front (Figure 2.2). The foot of the front wanders between the 60- and 90-m isobaths, while the surface manifestation of the front may extend from ca. the 60-m isobath seaward to over the 2000-m isobath, depending on wind forcing, offshore pressure gradients and intrusions of warm core rings which have broken off the Gulf Stream. The front is an impediment to advective mixing of slope and shelf waters, thus effectively isolating shelf waters from the Northwest Atlantic central basin. While rings may facilitate fluid transport, the volume transported by rings is small (on the order of 3 to 5%) compared with the volume of shelf water.

The Origin of the Export Hypothesis

In the mid-1970s the Oceanographic Sciences Division at Brookhaven National Laboratory began studying the coupling between primary and secondary production on the Northeast Shelf off the coast of Long Island. That work indicated that annual primary production was approximately 300 g C m^{-2} y^{-1} (Malone et al., 1982), of which about 40% appeared to be produced between March and April. Curiously, however, zooplankton biomass, assessed primarily by monthly net tows (Judkins et al., 1980), was low in April and May, did not peak until June and July, i.e., one to two months after the bloom had dissipated. Moreover, a few direct measurements of zooplankton grazing (not measured during the spring), suggested that

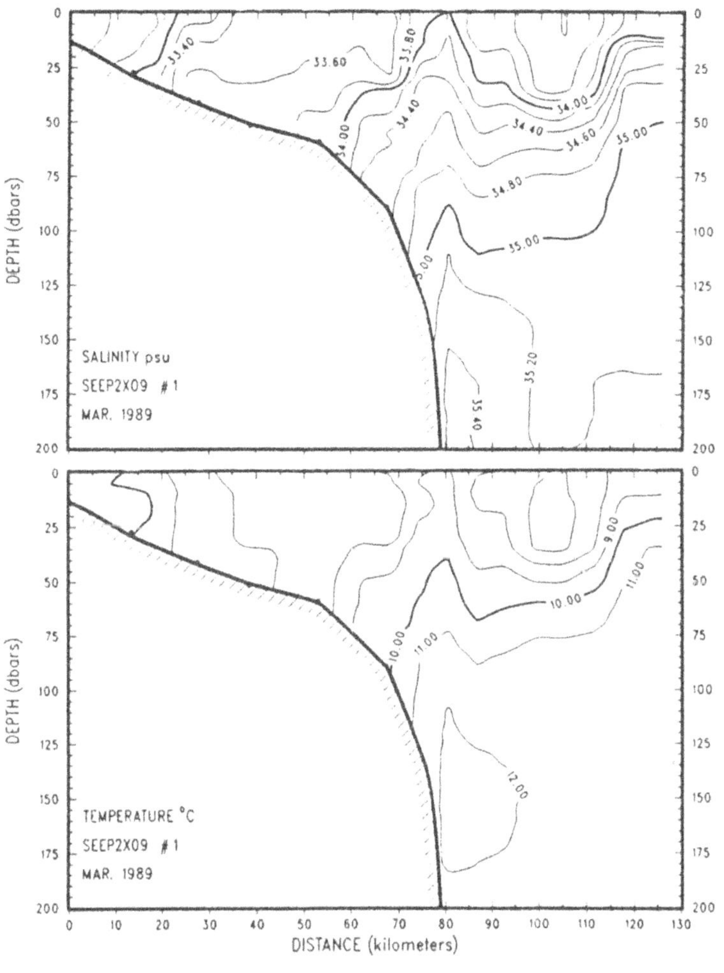

Figure 2.2. A section across the northeast continental shelf in March 1989 east of the Delmarva Peninsula, near the SEEP II moorings (see Figure 2.1) showing the changes in salinity (upper), and temperature and the relative position of the shelf-slope front (lower).

herbivores could only account for about 35% of the daily primary production. It thus appeared that the spring phytoplankton bloom was not terminated as a direct result of zooplankton grazing pressure, and that the transfer of primary production to zooplankton was less efficient than predicted by accepted models of marine food chains (Steele, 1974). Where did all the "excess" phytoplankton carbon go?

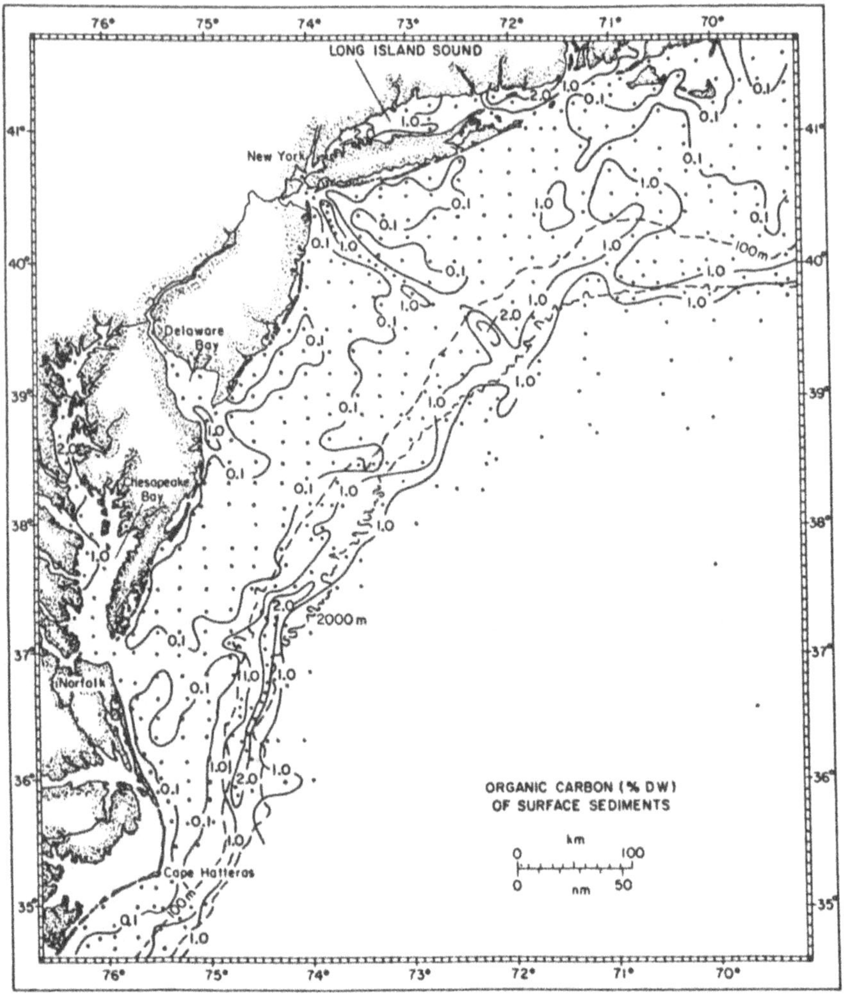

Figure 2.3. The distribution of organic carbon in the surface sediments of the northeast continental shelf and slope. The shelf sediments are mostly coarse-grained relict sands, containing low concentrations of organic carbon. In contrast, the continental slope sediments are finer grained and contain 1 to 2% organic carbon. The data were derived from Hathaway (1971).

One possibility is burial in the sediments. With the exception of the "mud patch" south of Cape Cod, the sediments of the shelf are course and sandy, containing only about 0.1% organic carbon by weight (Figure 2.3). In contrast however, the sediments on the continental slope are fine grained and contain between 1% and 2% organic carbon. Might the spring phytoplankton bloom be exported from the shelf to the slope depocenter and the interior of the central Northwest Atlantic basin?

The so-called export hypothesis, proposed by Walsh et al. (1981), was supported by an attempt to construct a carbon budget for the shelf. Based on the calculated annual primary production for the shelf, measurements of benthic respiration (see review by Rowe et al., 1988), and data provided by the National Marine Fisheries Service on terminal yields of fin and shell fish, Walsh et al. (1981) suggested that as much as 90% of the spring phytoplankton bloom, and 65% of the annual production could not be accounted for within the shelf ecosystem and therefore must be exported to the slope sea.

It is difficult to conceive of how phytoplankton could be easily transported from the shelf to the slope across the front. Two possibilities emerge. In the spring the phytoplankton is dominated by large, netplankton diatoms which have relatively high sinking rates. During periods between wind events (which occur on average every 5 days in the spring) the phytoplankton tend to sink, forming large near-bottom, nepheloid layers (Falkowski et al., 1988). It was proposed that during an off-shore wind event the phytoplankton are resuspended and move via Ekmann transport over the surface of the front. As, during this type of wind regime the front would be leaning offshore, phytoplankton would be able to sink through the front to the upper slope. This mechanism would allow the movement of particles through the front without the concomitant transport of water. Alternatively, phytoplankton could be transported at the base of the front bottom by the effective scouring of the bottom as the base of the front sweeps back and forth between the 60- and 90-m isobath.

The SEEP Program

The Shelf Edge Exchange Processes (SEEP) program was developed to examine if and how particles, such as phytoplankton, were transported across the shelf-slope front. The first field experiment, conducted off the coasts of Long Island and Massachusetts occurred between February and May 1984. The second field experiment occurred off the Delmarva Peninsula (Figure 2.1) between February 1988 and May 1989. Both field programs utilized moored instrumentation, especially moored fluorometers. The fluorometers, designed at BNL specifically for the SEEP program, indirectly measure phytoplankton biomass by measuring in vivo chlorophyll fluorescence. The instruments internally record chlorophyll fluorescence every 15 to 30 min for 3 to 6 mo. Additionally they measure temperature and two other optional sensor attributes (e.g., oxygen, beam attenuation, and/or insolation).

The results of the first SEEP field experiment were rather surprising. First, the moored fluorometer records failed to reveal the occurrence of a spring bloom, although the moorings were deployed at a time which typically bracketed the bloom period based on shipboard observations (Falkowski et al., 1988). Secondly, near bottom instruments on the shelf recorded frontal movements at the 80-m isobath, based on temperature excursions. However, when these occurred, chlorophyll on the slope side of the front was consistently much lower than that on the shelf side (Walsh et al., 1988). Moreover, when fluorescence records were combined with simultaneously collected current meter records, the calculated export of phytoplankton carbon was extremely small off the coast of Long Island, and there actually was an import of phytoplankton carbon from the slope to the shelf off the coast of Massachusetts.

Cross-shelf sections of chlorophyll distributions (Figure 2.4) during SEEP I revealed a build up of phytoplankton in the mid-shelf near the bottom though the early spring. At the beginning of February chlorophyll on the shelf was generally low. Only a few weeks later, it increased by about a factor of two, and the hint of a near bottom maximum, containing about 3 μg Chl/l, was found at the shallowest station. By early March the mid-shelf stations had 10 μg Chl/l and by mid-April that had increased to 25 μg/l. These data reveal that chlorophyll accumulated on the shelf near the bottom, but provided little evidence for export of phytoplankton.

A two-dimensional model of the time and space-dependent changes in oxygen concentrations measured during the first SEEP field program suggested that about 70 to 75% of the primary production produced on the shelf was also oxidized on the shelf during in the spring (Falkowski et al., 1988). Based on these and other measurements, we concluded that in mid to late winter, when production is low, carbon which is not grazed by zooplankton and which does not appear to be oxidized, amounts to about 25 to 40% of the average daily production. In the early spring, when primary production is high, only about 15% of the phytoplankton carbon produced on the shelf is available for export, the rest is oxidized on the shelf. Thus the first SEEP experiment suggested that the export hypothesis was not supported.

The second SEEP field experiment lasted 15 mo and bracketed two spring periods. In addition to moored fluorometers, we also added an acoustic Doppler current profiler to the moored instrumentation. That instrument simultaneously recorded the acoustic back-scatter signal strength, as well as the Doppler shift of the acoustic signal. From empirical calibration, Flagg and Smith (1989) were able to relate the acoustic target strength to zooplankton biomass, and moreover, could resolve the vertical distribution of zooplankton biomass throughout the 15-mo deployment (Figure 2.5). Thus, for the first

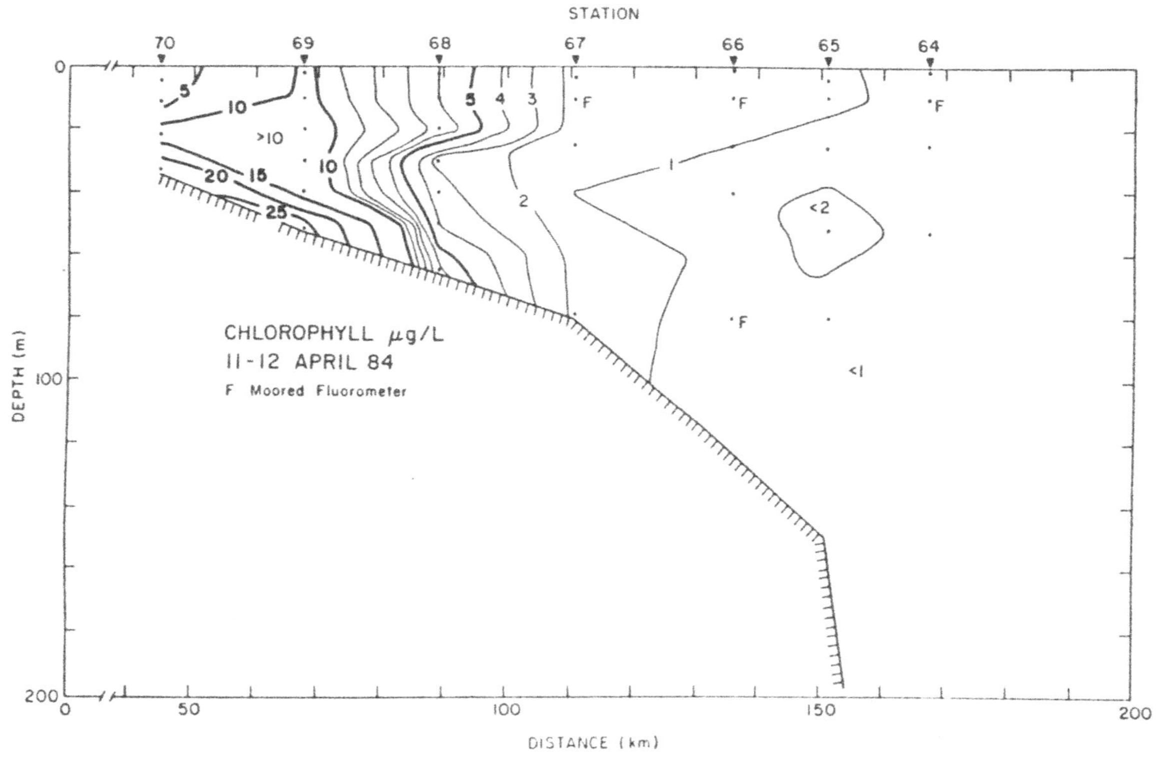

Figure 2.4. A cross-shelf section of chlorophyll *a* obtained during the SEEP I experiment south of Long Island, New York, showing high concentrations of phytoplankton in the middle of the shelf near the bottom. The near-bottom nepheloid layers appear to largely decompose on the shelf after the water column stratifies in mid-May.

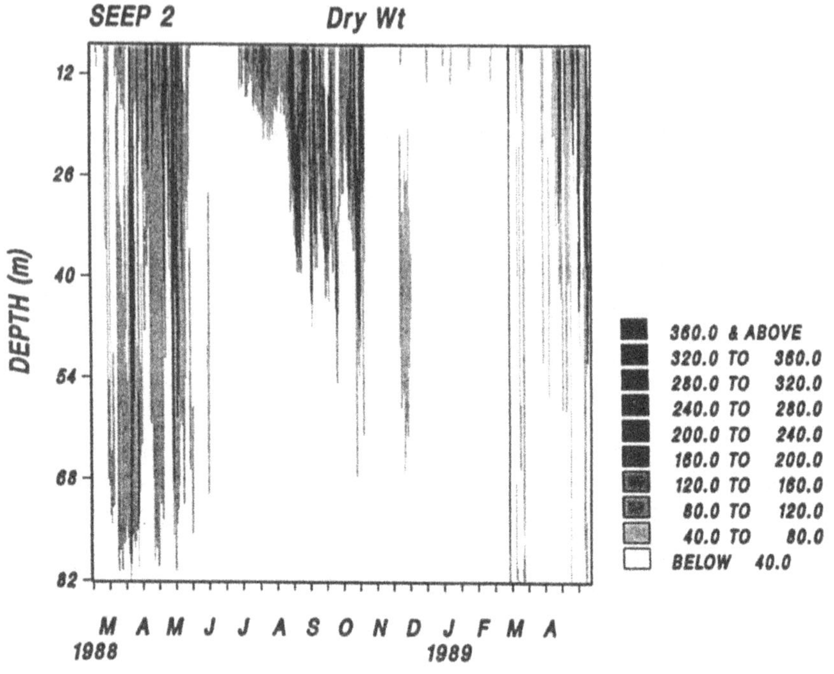

Figure 2.5. A 15-mo time series of the depth-resolved zooplankton biomass distributions (in mg/m^3) derived from the acoustic back-scatter intensity from an acoustic Doppler current profiler (Flagg and Smith, 1989). The instrument, moored in 90 m of water at the shelf slope break, off the Delmarva Peninsula in the SEEP II program (see Figure 2.1), internally recorded the acoustic back-scatter every 30 min.

time, oceanographers were able to resolve time dependent changes in phytoplankton biomass (from three moored fluorometers), zooplankton biomass (from the ADCP) and current speed and direction (from the ADCP) for a 15-mo period. The results of that time series indicate a highly correlated relationship between phytoplankton and zooplankton biomass (Figure 2.6) on the shelf.

The fluorometer records did not reveal a specific spring bloom during either 1988 or 1989. Instead, one can observe short pulses of phytoplankton, lasting 3 to 5 days, associated with advective processes. Furthermore, the relationship between phytoplankton and zooplankton does not appear to follow the paradigm for temperate shelf marine food chain dynamics. Zooplankton biomass does not increase shortly after a pulse of phytoplankton, and at times a pulse of zooplankton precedes by a few weeks a pulse of phytoplankton. Empirical orthogonal analysis of these records suggests that at short

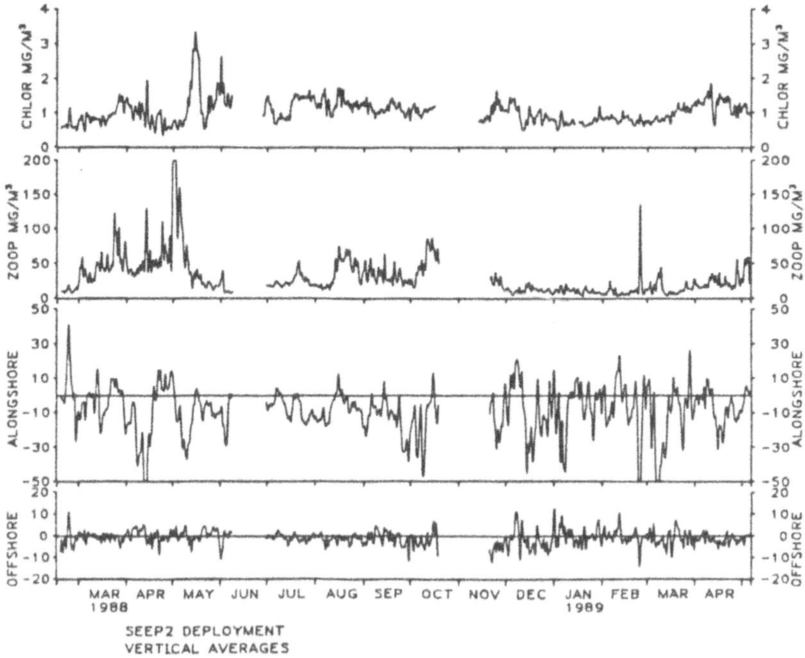

Figure 2.6. A 15-mo time series showing the average water column chlorophyll *a* concentration (derived from in-vivo fluorescence records), zooplankton biomass (from acoustic back-scatter) and the vertically averaged alongshore and cross-shelf components of current velocity (from the Doppler shift of the acoustic signals). Note the relative coherence between phytoplankton and zooplankton biomass and the relative lack of a distinctive spring bloom. Most of the variance in both phytoplankton and zooplankton on short temporal scales is associated with advective processes.

time scale of 5 to 10 days, both phytoplankton and zooplankton are forced by advective processes. On a time scale of a month or so, however, phytoplankton and zooplankton are highly coupled, and a simple grazing model predicts that during the spring up to 80% of the net daily primary production is grazed by herbivorous zooplankton (Flagg and Wirick, pers. commun.).

A Shelf Carbon Budget

Based on the SEEP field measurements we can construct a carbon budget for the northeast shelf for the early steps in the marine food chain. Approximately 60 to 65% of the average annual primary production is consumed by herbivorous macrozooplankton. An additional 20% is consumed by the benthos. In this regard, benthic bacteria in the upper 0.5 cm of the sediment and at the sediment-water interface, are especially important in the oxidation of phytodetritus. Approximately 5% of the average daily carbon fixed is consumed by pelagic bacteria and microzooplankton and an additional 2% is consumed by phytophagous fish such as menhaden. The remaining 10 to 15% of the production which is not accounted for in this budget may be exported to the adjacent slope sea. It should be pointed out that only about 1% of the carbon available for export is caught in sediment traps on the continental slope, suggesting that if 10% of the shelf production is exported, most of it (i.e., 90%) is oxidized within the water column at the ocean margin.

Conclusions and Implications for Fisheries

The SEEP experiments have fundamentally changed the way oceanographers think about marine food chains in temperate waters. While the measurements suggest that the carbon budgets are much more closely balanced than the export hypothesis suggested, the moored observations suggest that the concept of a single, major spring phytoplankton bloom followed by a zooplankton bloom is grossly oversimplified. There is little evidence of a specific spring bloom in the southern part of the Northeast Shelf ecosystem. On short time scales (<10 days) physical processes mask biological ones. In order to decipher the true biological processes from the fluid motion, it will be necessary to couple three-dimensional, time dependent circulation models with biological process models. At present, such models require a state of the art supercomputer, and are still somewhat crude in their resolution (Hofmann, 1988).

It is difficult to directly extrapolate the results of the SEEP experiments to fisheries, however, it would appear that tight coupling between phytoplankton and zooplankton on time scales of a month would intuitively provide a more temporally buffered food supply to higher trophic levels than the pulse mode predicted by the spring bloom paradigm. In principle, yields should be more sensitive to recruitment and mortality than food supply. It should be pointed out however, that the vast majority of anthropogenic pollutants enter higher levels of the marine food chain by adsorption on the surfaces of particles and subsequent ingestion. As the first point of entry of pollutants is via phytoplankton, and approximately 65% of the phytoplankton on the shelf are consumed by zooplankton, there is a high probability that increased anthropogenic insult to coastal waters will result in increased health risk to humans. It is not reasonable to assume that anthropogenic pollutants will quickly be advected to the central ocean basins, thereby reducing their potential effects on human health.

The SEEP data illustrate the difficulty of applying food chain dynamic models to calculate the transfer of primary production to zooplankton, let alone to higher levels of the marine food chain. The principle difficulty is to integrate planktonic food chain dynamics into physical circulation models. Should fisheries managers give up hope of ever being able to apply food chain dynamics to predict yields, recruitment and populations? Clearly, marine ecologists are far from that elusive goal. However, in striving to reach it, dynamic ocean processes are becoming increasingly understood. Thus, understanding fisheries inevitably requires an understanding of how marine food chains work, yet pragmatic management may not.

Acknowledgments

This research was conducted under contract DE-AC02-76CH00016 by the US Dept. of Energy, Office of Health and Environmental Research. All opinions and interpretations are the author's, and do not represent official policy and/or opinion of any agency of the United States government or Brookhaven National Laboratory. I thank my colleagues in the Oceanographic and Atmospheric Sciences Division for discussions, comments and criticisms.

References

Cushing, D. H. 1962. Production of a pelagic fishery in the sea. Fish. Invest. London, Ser. 2:18. 103 Pp.

Cushing, D. H. 1975. Marine ecology and fisheries. Cambridge Univ. Press, London. 278 Pp.

Falkowski, P. G., C. N. Flagg, G. T. Rowe, S. L. Smith, T. E. Whitledge, and C. D. Wirick. 1988. The fate of a spring phytoplankton bloom: oxidation or export? Cont. Shelf Res. 8:457-484.

Flagg, C. N., and S. L. Smith. 1989. On the use of the acoustic Doppler current profiler to measure zooplankton abundance. Deep-Sea Res. 36:455-474.

Harris, G. P. 1980. Temporal and spatial scales in phytoplankton ecology. Mechanisms, methods, models and management. Can. J. Fish. Aquat. Sci. 37:877-900.

Harvey, H. W. 1957. The Chemistry and Fertility of Sea Water. 2nd ed. Cambridge Univ. Press. 234 Pp.

Hathaway, J. C. 1971. Woods Hole Oceanographic Institution, Data File, Continental Margin program, Atlantic Coast of the United States. Vol. 2, WHOI Technical Rep. 71-75.

Hofmann, E. E. 1988. Plankton dynamics on the outer southeastern U.S. continental shelf. Part III: A coupled physical-biological model. J. Mar. Res. 46:919-946.

Judkins, D. C., C. D. Wirick, and W. E. Esaias. 1980. Composition, abundance, and distribution of zooplankton in the New York Bight, September 1974-September 1975. Fish. Bull., U.S. 77:669-684.

Malone, T. C., T. S. Hopkins, P. G. Falkowski, and T. E. Whitledge. 1983. Production and transport of phytoplankton biomass over the continental shelf of the New York Bight. Cont. Shelf. Res. 1:305-337.

Riley, G. A. 1946. Factors controlling phytoplankton populations on Georges Bank. J. Mar. Res. 6:54-73.

Rowe, G. T, R. Theroux, W. Phoel, H. Quinby, R. Wilke, D. Koschoreck, T. Whitledge, P. Falkowski, and C. Fray. 1988. Benthic carbon budgets for the continental shelf south of New England. Cont. Shelf Res. 8:511-527.

Sherman, K., M. Grosslein, D. Mountain, D. Busch, J. O'Reilly, and R. Theroux. 1988. The continental shelf ecosystem off the northeast coast of the United States. Pp. 279-337. *In* H. Postma and J. J. Zijlstra (Eds.) Ecosystems of the world 27, Continental Shelves. Elsevier, Amsterdam.

Steele, J. H. 1974. The Structure of Marine Ecosystems. Harvard Univ. Press., Cambridge, MA. 128 Pp.

Sverdrup, H. U. 1953. On conditions for the vernal blooming of phytoplankton. J. Cons. int. Explor. Mer 18:287-295.

Walsh, J. J. 1981. A carbon budget for overfishing off Peru. Nature 290:300-304.

Walsh, J. J., G. T. Rowe, R. L. Iverson, and C. P. McRoy. 1981. Biological export of shelf carbon is a sink of the global CO_2 cycle. Nature 291:196-201.

James A. Yoder

3. Warm-Temperate Food Chains of the Southeast Shelf Ecosystem

Abstract

Food chain dynamics in waters off the southeast United States within the Southeast Shelf ecosystem are principally controlled by oceanographic processes occurring at the shelf break and at the coastal boundary. At the shelf break, upwelling along the Gulf Stream front is the dominant source of new plant nutrients sustaining plankton production. Upwelling dynamics determine the temporal and spatial attributes of plankton production on the outer shelf.

On the inner southeastern shelf, a coastal salinity front delineates the boundary between a 10-20-km wide zone of productive coastal waters from waters just seaward of the front which have very low productivity. Plankton dynamics of coastal waters shoreward of the front are unusual in that primary production is very high throughout the year, whereas concentrations of inorganic nitrogen (NH_4, NO_3 + NO_2) are quite low. New N sources, such as rivers, provide only a small fraction of N required to sustain the observed rates of production. These observations imply that recycled N is the most important source on the inner shelf, and that photosynthesis and respiration are very tightly coupled throughout the year.

The Setting

Geography

The Southeastern Shelf ecosystem extends from the tip of Florida to Cape Hatteras, North Carolina. This review will focus on the South Atlantic Bight (SAB) or that portion of the southeastern shelf between Cape Canaveral, Florida, and Cape Hatteras (Figure 3.1). The width of the shelf in the SAB varies between 50-200 km and the area between the coast and the 60-m isobath (shelf break) is 90,600 km^2 (Atkinson and Menzel, 1985).

For dynamical, as well as biological reasons, the shelf area is

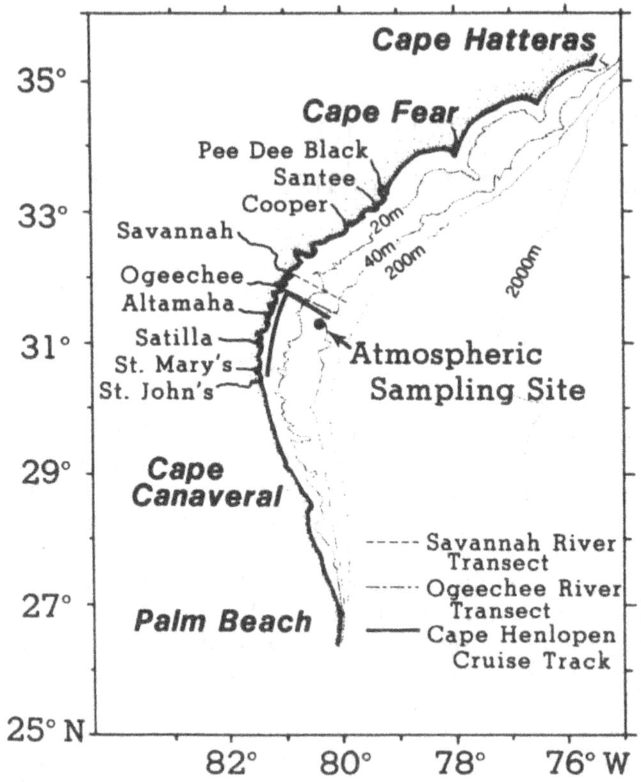

Figure 3.1. Bathymetric chart of South Atlantic Bight (SAB) showing capes, general bathymetry, and cities.

subdivided based on isobath range as the inner (0-20-m isobaths), middle (20-40-m isobaths) and outer (40-m shelf break) shelf zones (e.g., Bishop et al., 1980). In the central portion of the SAB off Georgia, each zone is approximately 40 km wide (Figure 3.1). With some exceptions (to be noted), along-shelf gradients in biological, physical and chemical properties and processes are minor compared with changes occurring across the shelf.

Physical Oceanography

Three characteristics of shelf circulation and hydrography are of particular significance to biological oceanographic processes on the southeastern shelf. Plankton dynamics will be discussed in relation to these three important characteristics.

The location of the Gulf Stream cyclonic front at the shelf break is the most important physical oceanographic characteristic of the southeastern shelf. Secondly, fresh water enters the SAB as a line source owing to the numerous small rivers along the coasts of Georgia, South Carolina, and North Carolina. The rivers discharge into small estuaries surrounded by extensive salt marshes (ca. 32,500 km^2) (Turner, 1981). Mean annual river runoff is 84 km^3 (into a total shelf water volume of ca. 2400 km^3) generally peaking in March (Atkinson and Menzel, 1985). Finally, there is no nutrient-rich water mass resident on the southeastern shelf, in part because shelf waters are relatively shallow (shelf break occurs ca. 60-m isobath) and stratification is relatively weak throughout the year (except during upwelling/intrusion events to be discussed) (Atkinson, 1985). Inorganic nitrogen (NO_3, NO_3, NH_4) rarely accumulates in resident shelf waters to concentrations exceeding 1.0 micromolar, even within near-bottom waters (except during upwelling/intrusion events to be discussed). One important implication of low nutrient concentrations in resident shelf waters is that storm-induced vertical mixing does not enrich the euphotic zone with significant concentrations of inorganic nitrogen and other major plant nutrients. This is a major difference from conditions that occur within shelf waters north of Cape Hatteras (e.g. Walsh et al., 1978) and is the principal reason why the Southeastern Shelf ecosystem lacks a spring phytoplankton bloom (Bishop et al., 1980; Yoder, 1985).

The most important process affecting the flux of new nutrients to southeastern shelf waters is Gulf Stream-induced upwelling of North Atlantic Central Water (NACW) at the shelf break (Atkinson, 1985). Upwelling is caused by frontal eddies and other processes occurring along the cyclonic Gulf Stream front. Upwelling processes occur throughout the year with a characteristic time scale of ca. 1 event per 10 d (Lee et al., 1981; Lee and Atkinson, 1983).

The fate of upwelled NACW depends on the density of shelf waters and on wind direction and strength (Atkinson, 1977). During winter and early spring, outer shelf waters are relatively cold (<20°C) and are of comparable density to upwelled NACW. As a result, upwelled NACW does not penetrate beyond the outer shelf. During summer, shelf waters are 28°C and are significantly less dense than upwelled NACW. Under these hydrographic conditions, and with prevailing winds that are generally from the southeast, upwelled NACW moves shoreward as a bottom intrusion (Atkinson et al., 1987) (Figure 3.2).

Plankton production and distributions on the inner shelf off Georgia and South Carolina are very much influenced by coastal circulation dynamics. During much of the year, low salinity coastal waters are constrained near the coast by a coastal density front which acts as a dynamic barrier to across-shelf exchange of both dissolved and particulate in-water constituents (Blanton, 1981; Blanton and Atkinson, 1983) (Figure 3.3). New nutrients (from rivers), and

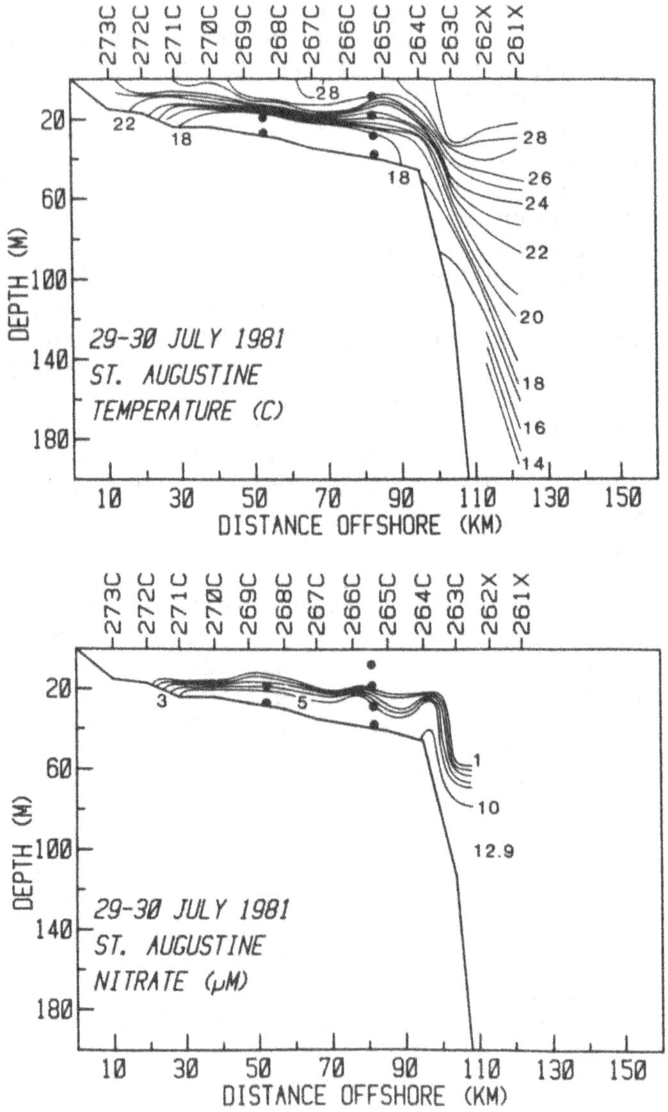

Figure 3.2. Temperature and nitrate cross-shelf sections through a subsurface intrusion of NACW. From Atkinson et al. (1987).

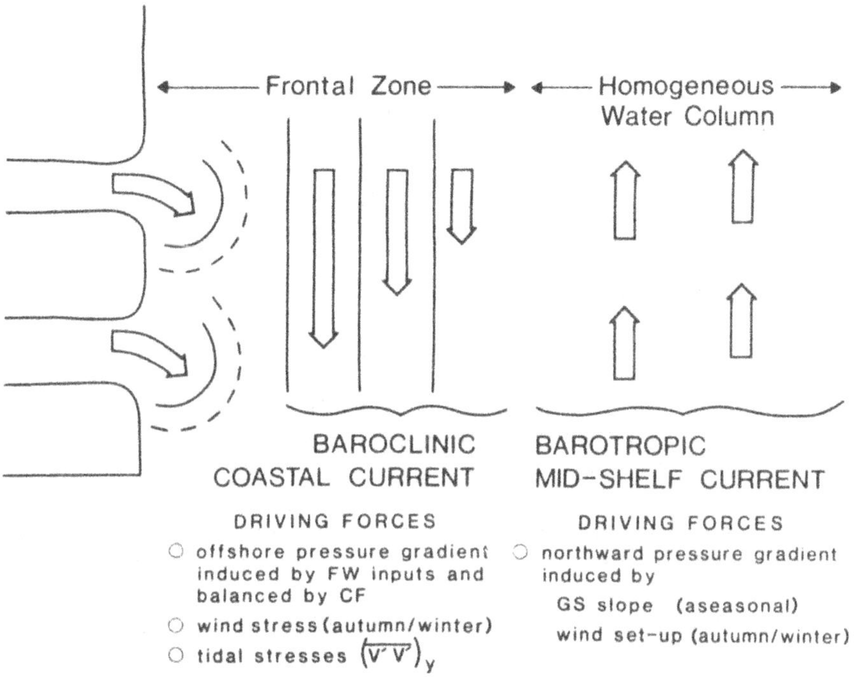

Figure 3.3. Schematic of a baroclinic coastal current in a frontal zone bounded by the coast. FW = freshwater, CF = Coriolis force, GS = Gulf Stream. From Blanton (1981).

nutrients recycled between the benthos and water column, are trapped on the inner shelf within the low salinity waters shoreward of the frontal zone.

In summary, the conceptual model of plankton dynamics of the Southeastern Shelf ecosystem begins with the physical dynamics and attributes of two frontal systems: the Gulf Stream front near the shelf break and a coastal salinity front on the inner shelf. Both frontal systems affect the flux rates and residence times of plant nutrients. In the relatively shallow and well-illuminated (because of the latitude) waters of the southeastern shelf, plankton dynamics are closely coupled to nutrient flux. This is an important distinction from shelf waters north of Cape Hatteras, where seasonal changes in incident irradiance, as well as stratification/destratification of shelf waters, strongly influence plankton dynamics.

Plankton Dynamics

The dynamics of plankton production in southeastern shelf waters was reviewed previously (Paffenhofer, 1985; Pomeroy, 1985; Yoder, 1985). One of the main conclusions was that phytoplankton and bacterioplankton production of the middle and outer shelf is principally controlled by the upwelling/intrusion events associated with Gulf Stream frontal processes. Secondly, upwelling/intrusion events, rather than seasonal plankton blooms, are most important to plankton production on the outer/middle southeastern shelf.

Winter/Spring Upwelling

The effects of upwelling during winter and spring on near-surface chlorophyll distribution are illustrated in Figure 3.4. High chlorophyll concentrations (>4 mg m^{-3}) are within the upwelled core of NACW of a Gulf Stream frontal eddy surrounded by near-surface Gulf Stream waters and shelf waters having relatively low pigment concentrations (<0.5 mg m^{-3}) (Figure 3.4). Primary production during this event was as high as 6.0 g C m^{-2} d^{-1} within upwelled waters which were dominated by centric diatoms (Yoder et al., 1981).

Current meter records were used to estimate the frequency of Gulf Stream-induced upwelling events and thus estimate their contribution to seasonal (ca. November-April) outer shelf primary production. Results of these analyses showed primary production of the outer shelf (Figure 3.1) to be 175 g C m^{-2} 6 mo^{-1} of which at least 50% is new production (Yoder et al., 1983).

Bacterial populations also respond to upwelling events, and their abundance tracks temporal changes in phytoplankton productivity (Pomeroy, 1985). Bacterial numbers range from 10^5 to 10^6 per ml in middle and outer shelf waters with maximum concentrations of $5x10^6$ ml^{-1} associated with phytoplankton blooms in upwelled waters (Pomeroy et al., 1983; Pomeroy, 1985).

The fate of phytoplankton biomass synthesized during winter/spring upwelling on the outer shelf is difficult to ascertain. Evidence suggests that large populations of copepods generally do not develop during these months, but under some hydrographic conditions, blooms of pelagic tunicates occur within upwelled waters that become stranded on the middle/outer shelf (Deibel, 1985). Strong northerly currents occur on the outer shelf due to the presence of the Gulf Stream near the shelf break. Numerical models of biological/physical processes during the winter/spring season show that upwelled waters are advected north, and probably off the shelf, within days to weeks following upwelling events (Hofmann, 1988; Hofmann and Ambler, 1988). Thus there is inadequate time for

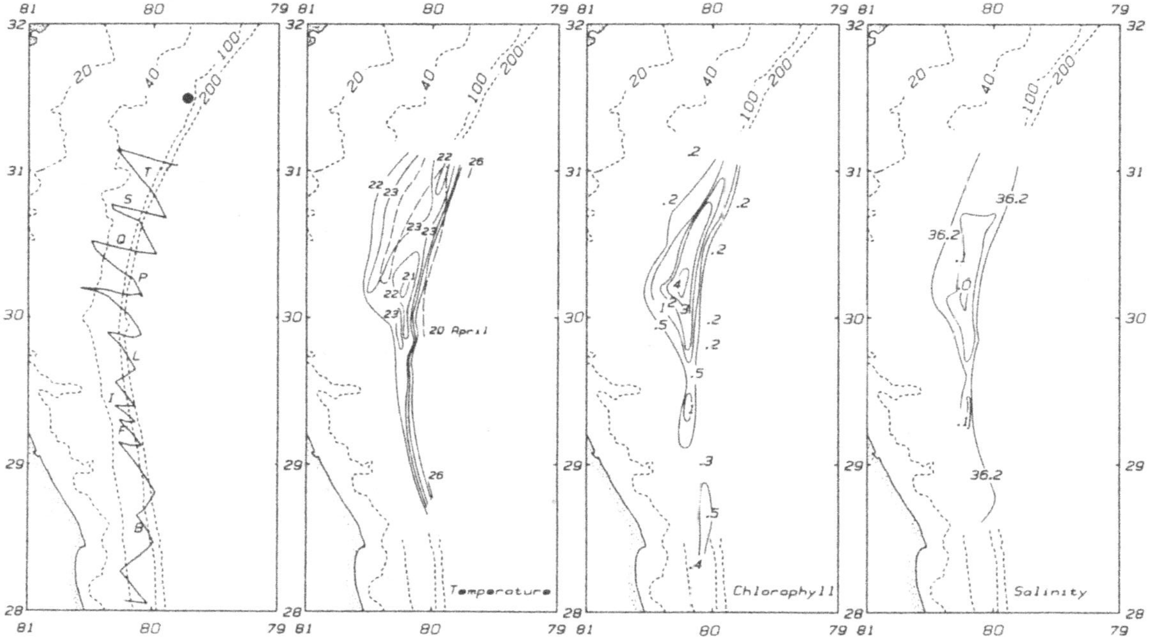

Figure 3.4. Cruise track and surface temperature, Chl *a* (mg m^{-3}) and salinity (ppt) as determined by continuous shipboard measurements 20-22 April 1979. Dashed line in temperature frame indicates Gulf Stream surface thermal front as determined by satellite on 20 April. From Yoder et al. (1981).

copepod blooms to develop in response to enhanced levels of phytoplankton/bacterioplankton production within upwelled waters. The models suggest that much (most) of the phytoplankton production that occurs during upwelling is advected north and than off the shelf near Cape Hatteras. During winter, as much as 20% of the new production occurring within upwelled waters may advect (cascade) off the shelf and down the slope (Yoder and Ishimaru, 1989).

Summer Intrusions

Summer in the SAB is loosely defined as the months encompassing May through October. During summer, near-surface waters on the outer shelf generally exceed 24°C (Atkinson et al., 1983). From May through August, prevailing winds are from the southeast (i.e., intrusion-favorable) and switch to the northeast in September and October (Weber and Blanton, 1980). As expected from these wind and hydrographic conditions, subsurface intrusions of NACW are most commonly observed from May to August.

The most complete study of subsurface intrusions occurred during 1981 (Yoder et al., 1985, and see Progress in Oceanography, Vol 19:221-441). From 20 July through 13 August, 1981, several upwelling events occurred at the shelf break triggering the flow of nutrient-rich NACW across the shelf as a subsurface intrusion (Figure 3.5). As intruded NACW moved onto the middle shelf, subsurface NO_3 concentrations increased from undetectable levels on 20 July to ca. 5.0 micromolar on 23 July. A phytoplankton bloom developed within days to weeks of intruded waters reaching the middle shelf (Figure 3.6). Beginning on 20 July, Chl *a* steadily increased and peaked at ca. 75 mg m^{-2} on 3 August with all of the increase occurring within NACW occupying the lower 15 m of the 30 m-deep water column (Figure 3.6). Peak primary production exceeded 3 g C m^{-2} d^{-1}, and NO_3-N uptake measurements showed that new production was more than 90% of total primary production at the peak of the bloom (Yoder et al., 1985).

With respect to regional plankton dynamics, the 1981 results and studies of other summer intrusions in the SAB (e.g., Dunstan and Atkinson, 1976; Yoder et al., 1983; Paffenhofer, 1983; Pomeroy et al., 1983; Paffenhofer et al., 1984) show:

1. Large intrusions result in summer rates of middle shelf primary production exceeding 150 g C m^{-2}. Mean rates of daily primary production (~1.9 g C m^{-1} d^{-1}) are equivalent to typical values reported for seasonal phytoplankton blooms at temperate and higher latitudes. Intrusions studied in summer, 1981, covered an area of approximately 10^4 km^2 (Yoder et al., 1985).

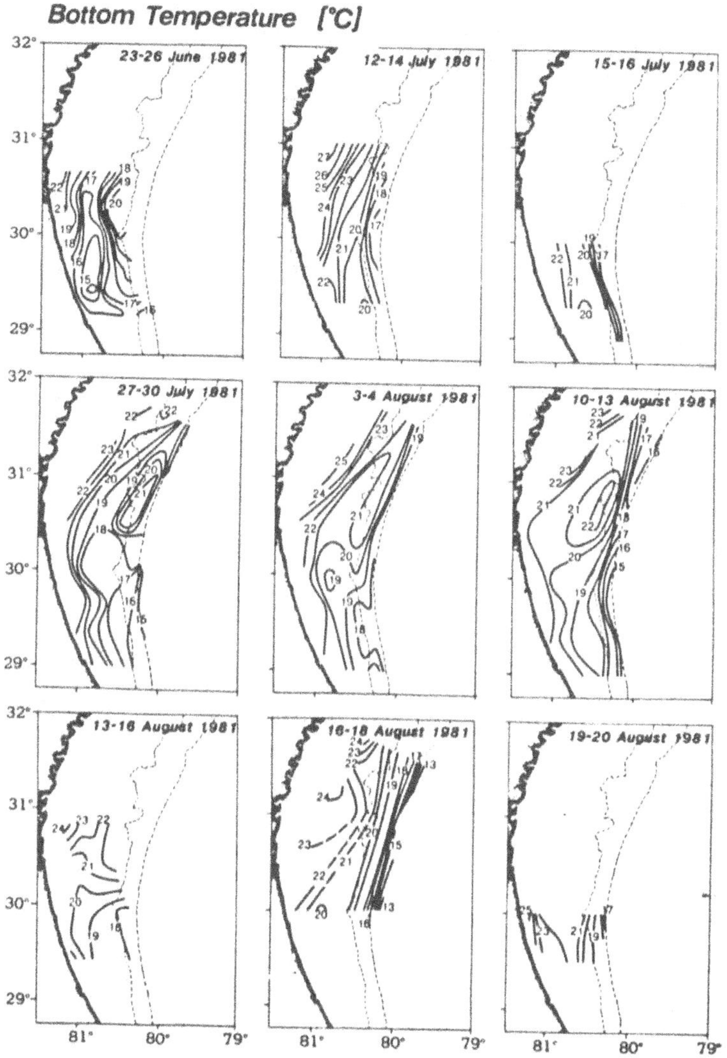

Figure 3.5. Bottom temperature distributions on the southeastern shelf during summer, 1981, showing intrusions of cold (<24°C) of NACW. From Atkinson et al. (1987).

2. Interannual differences in the size and effect of subsurface intrusions can be inferred from field studies conducted from 1977-1981, but characteristic time scales for large intrusions (e.g., 1981) having dramatic effects on water column productivity are not known (Yoder et al., 1985).

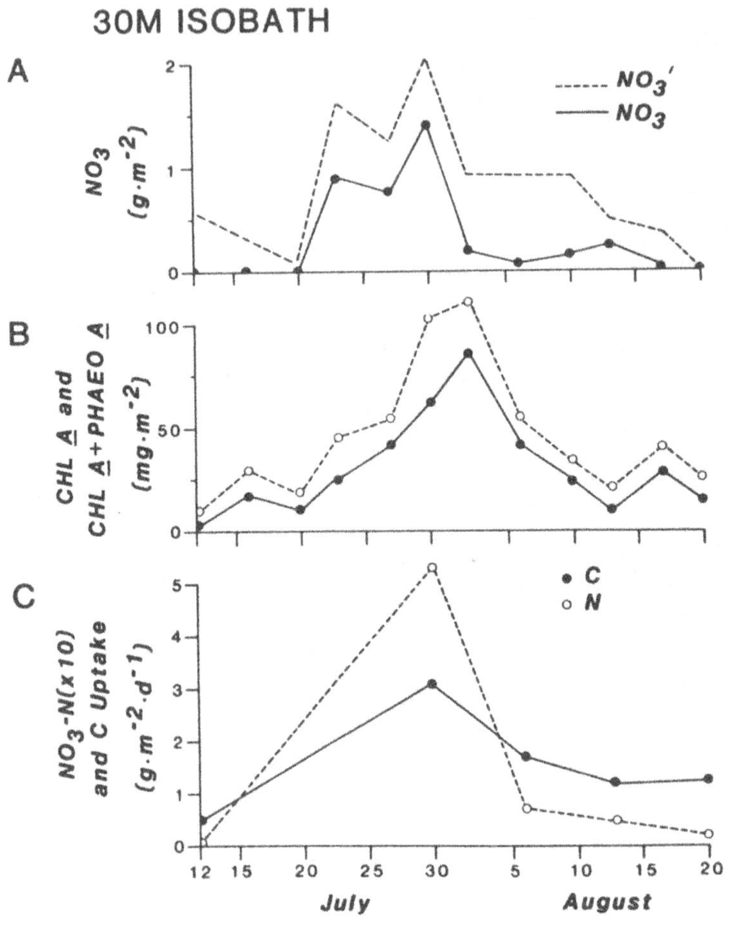

Figure 3.6. Middle shelf (30- to 35-m isobath) time series at 30°N during summer, 1981. NO_3* is initial NO_3 (water column integral) as calculated from observed vertical temperature distribution and a NO_3 vs. temperature relation of newly upwelled NACW. NO_3* - NO_3 indicates the amount of NO_3 taken up by phytoplankton. From Yoder et al. (1985).

3. Peak bacterioplankton biomass occurs in the near-surface waters overlying phytoplankton-rich intruded waters (Pomeroy et al., 1983). During summer, 1981,free-living bacteria exceed 10^6 ml^{-1} within and above intruded waters, which is 10-fold higher than typical concentrations on the middle and outer shelf.

4. Zooplankton populations develop rapidly within intrusions. Copepods such as *Temora turbinata* reproduce within phytoplankton-rich (>1 mg Chl *a* m^{-3}) waters of intrusions to form patches exceeding >1000 copepodids and adults per cubic meter (Paffenhofer et al., 1987a). Salps and doliolids are also abundant and during summer, 1981, increased in numbers by 25-fold over a period of 10 d (~50% increase per day) (Paffenhofer and Lee, 1987).

5. Intrusions of NACW are the most important process affecting summer plankton productivity of the middle and outer shelf of the SAB and of the inner shelf off northern Florida (Paffenhofer et al., 1987b). Intrusions do not penetrate to the inner shelf off Georgia and South Carolina. As a result, plankton dynamics there are very different off northeast Florida.

Inner Shelf Off Georgia and South Carolina

Concentrations of Chl *a*, suspended particles, and dissolved Si and P are very high off the Georgia and South Carolina coasts but decrease dramatically across the near-shore salinity front located a few km from the coast (Figure 3.7; Bishop et al., 1980; Yoder, 1985). In contrast to Si and P, concentrations of dissolved inorganic nitrogen (DIN = NO_3 + NO_2 + NH_4) are relatively low (<2 micromolar) on both sides of the front. This pattern persists throughout the year, although peak Chl *a*, Si and P concentrations are highest during summer (Yoder, 1985). The location of the strongest density and chlorophyll *a* gradients are closely related and move back and forth with tide a distance of more than 5 km (Yoder, unpublished). Bottom topography appears to control the position of the strongest surface gradients in density and Chl *a* (Yoder et al., 1987). Shoreward of the front, inner shelf waters are very turbid, having suspended particle concentrations as high as $2x10^5$ mg m^{-3} (Oertel and Dunstan, 1981). Euphotic depth (depth of the 1% isolume) is as shallow as 1 m (Yoder and Bishop, 1985).

Near-shore primary production is highest during summer (ca. 3 g C m^{-2} d^{-1}), but exceeds 1 g C m^{-2} d^{-1} even during winter (Thomas, 1966). Centric diatoms dominate the biomass of coastal phytoplankton (Bishop et al., 1980; Jacobsen et al., 1983; Verity and Yoder, unpublished). Common species include *Skeletonema costatum*, *Asterionella japonica*, and various species of *Chaetoceros* and *Rhizosolenia*. In a recent study, phytoplankton cells retained by an 8 micromolar mesh filter contributed more than 50% of Chl *a*, primary production and total phytoplankton nitrogen demand in summer. During winter, cells >8 micromolar represented more than 80% of these same 3 measures of phytoplankton productivity.

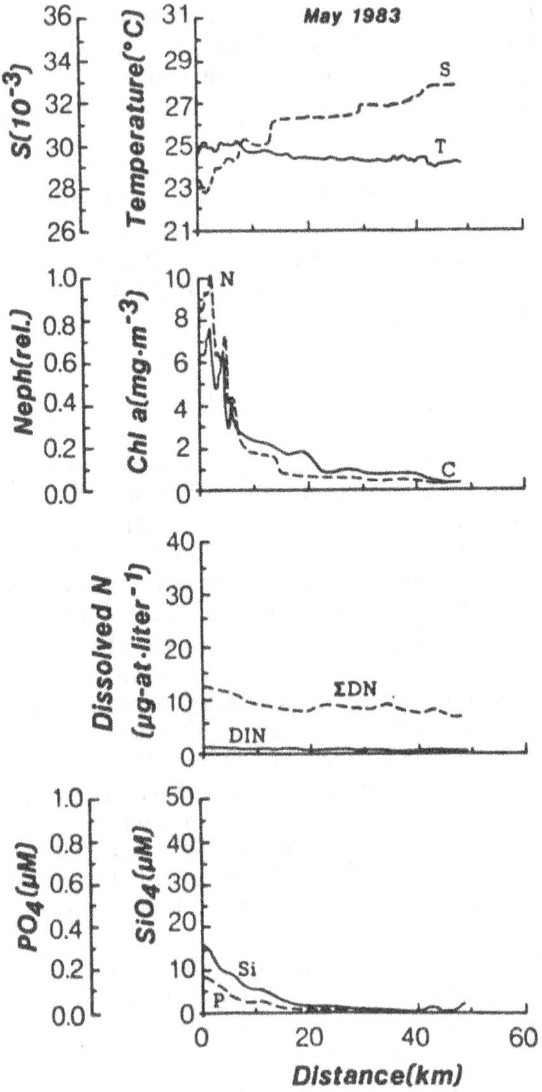

Figure 3.7. Surface hydrographic, nutrient and Chl *a* concentrations along a cross-shelf transect starting at coast off Wassaw Island, Georgia. The 20-m isobath (outer limit of the inner shelf) is located about 40 km offshore along this transect (Yoder, unpublished).

DIN concentrations do not vary seasonally suggesting that N is utilized at approximately the same rate at which it is introduced to near-shore waters and that the supply of N limits phytoplankton production (Bishop et al., 1984). Rivers and estuaries introduce N and other nutrients to Georgia coastal waters, but export of DIN from estuaries to coastal waters is balanced by import of phytoplankton and other sources of particulate N (Dame et al., 1986) suggesting a tightly coupled plankton system encompassing the estuaries and near-shore waters shoreward of the coastal front.

High phytoplankton biomass concentrations and specific rates of primary production, such as occur in Georgia/South Carolina coastal waters, are generally associated with coastal upwelling ecosystems where new N is the dominant source sustaining phytoplankton primary production (Eppley and Peterson, 1979). Mass balance calculations indicate that recycled N is the principal source and may account for more than 90% of the N required by inner shelf phytoplankton within coastal waters off Georgia and South Carolina (Haines, 1974; Yoder, 1985). These calculations have yet to be confirmed with experimental or observational evidence (e.g., Hanson and Robertson, 1988).

Recent unpublished studies show the importance of protozoans as grazers of phytoplankton and bacterioplankton and as nutrient recyclers. Summer abundances of heterotrophic nannoplankton (10^4 cells ml^{-1}) and ciliates (60 ml^{-1}) on the inner Georgia shelf are among the highest recorded for non-polluted waters. In a recent study most of the bacterial production was ingested by heterotrophic nannoplankton, and most of the nannoplankton production was grazed by ciliates. Based on published metabolic rates and conversion efficiencies, ciliate and heterotrophic nannoplankton contribution to rates of regenerated nitrogen is significant.

Little is known of the distribution and production of macrozooplankton on the inner shelf off Georgia and South Carolina. *Acartia tonsa* is the dominant species in Georgia estuaries and presumably on the inner shelf as well (Paffenhofer and Stearns, 1988).

Some Relations Between Plankton Dynamics and Fisheries

With respect to demersal fishes, most of the open shelf is relatively unproductive, and commercially important species such as snapper and grouper are associated with live-bottom habitats which are extensively distributed along and across the shelf (Barans and Burrell, 1976; Miller and Richards, 1980). On the inner shelf, the shrimp fishery is most important accounting for approximately 10%

of total tonnage landed and 30% of the value of all South Atlantic region fisheries in 1983 and 1984 (U.S. Department of Commerce, 1985). In the South Atlantic region, most shrimp are caught within 5 km of the coast (U.S. Department of Commerce 1985), and thus are a component of the productive inner shelf ecosystem.

Larval stages of other species of fish apparently depend upon outer shelf plankton blooms for nutrition during the critical early stages of development (Yoder, 1983). Migratory species known to spawn on the outer southeastern shelf include Atlantic menhaden (Nelson et al., 1977; Nicholson, 1978), bluefish (Kendall and Walford, 1979), chub mackerel (Berrien, 1978), and short-finned squid (Rowell et al., 1985). In the case of short-finned squid and Atlantic menhaden, spawning is maximal within upwelled NACW along the Gulf Stream front (Rowell et al., 1985; Checkley et al., 1988).

Spawning near the Gulf Stream front may have other advantages for Atlantic menhaden and bluefish. Recent studies (Checkley et al., 1988) show that menhaden spawn near the Gulf Stream front during winter in part because the water is warm and because of a unique density-driven circulation that transports larval menhaden towards the coast (Checkley et al., 1988). To survive, larval menhaden must reach southeastern estuaries where they develop as juveniles. In the case of bluefish, adults migrate from north of Cape Hatteras into southeastern shelf waters to spawn on the outer shelf during late winter and spring. In contrast to Atlantic menhaden, bluefish larvae do not develop into juveniles or adults in the southeast but need to reach estuaries north of Cape Hatteras (Kendall and Walford, 1979). The Gulf Stream provides an obvious mechanism for northerly transport of larval bluefish.

Conclusions

European investigators such D. H. Cushing and J. H. Steele developed much of the original theory relating plankton dynamics to fisheries. Their conclusions were based on studies of the North Sea and other temperate seas where seasonal plankton blooms, and in particular the spring bloom, dominate plankton production cycles. In contrast to temperate seas, the seasonal signal of the Southeastern Shelf ecosystem is very much suppressed. On the inner shelf, efficient nutrient recycling sustains plankton production at a high level throughout the year (although summer rates are higher than winter which results in some degree of seasonal changes). On the middle and outer shelf, upwelling/intrusion of nutrient-rich NACW causes short-lived phytoplankton blooms. These events are the most important signal in plankton production. Thus, the important time scales associated with plankton production of the Southeastern Shelf ecosystem are "events" (days-week) for the middle and outer shelf and "relatively constant" for the inner shelf. Strong seasonal pulses of plankton production are not very important. A spring bloom is virtually absent. For each individual large marine ecosystem, biomass, yield models and management strategies for fisheries must recognize and account for the unique time scales associated with plankton production.

References

Atkinson, L. P. 1977. Modes of Gulf Stream intrusion into the South Atlantic Bight shelf waters. Geophys. Res. Letters 7:583-586.

Atkinson, L. P. 1985. Hydrography and nutrients of the southeastern U.S. continental shelf. Pp. 93-103. *In* L. P. Atkinson, D. W. Menzel, and K. A. Bush, (Eds.) Oceanography of the southeastern United States continental shelf. Coast. Estuar. Sci. Vol. 2, AGU, Washington, DC.

Atkinson, L. P., and D. W. Menzel. 1985. Introduction: Oceanography of the southeastern United States continental shelf. Pp. 1-9. *In* L. P. Atkinson, D. W. Menzel, and K. A. Bush, (Eds.) Oceanography of the southeastern United States continental shelf. Coast. Estuar. Sci. Vol. 2, AGU, Washington, DC.

Atkinson, L. P., et al. 1983. Climatology of the southeastern United States continental shelf waters. J. Geophys. Res. 88: 4705-4718.

Atkinson, L. P., et al. 1987. Summer Upwelling on the southeastern continental shelf of the U.S.A. during 1981. Hydrographic Observations. Prog. Oceanogr. 19:231-266.

Barans, C. A., and V. G. Burrell, Jr. 1976. Preliminary findings of trawlings on the continental shelf off the southeastern United States during four seasons (1973-1975). South Carolina Marine Resources Center, Tech. Rep. No. 13.

Berrien, P. L. 1978. Eggs and larvae of *Scomber scombrus* and *Scomber japonicus* in continental shelf waters between Massachusetts and Florida. Fish. Bull., U.S. 76:95-115.

Bishop, S. S., K. A. Emmanuele, and J. A. Yoder. 1984. Nutrient limitation of phytoplankton growth in Georgia nearshore waters. Estuaries 7:506-512.

Bishop, S. S., J. A. Yoder, and G.-A. Paffenhofer. 1980. Phytoplankton and nutrient variability along a cross-shelf transect off Savannah, Georgia, U.S.A. Estuar. Coast. Mar. Sci. 11:359-368.

Blanton, J. O. 1981. Ocean currents along a nearshore frontal zone on the continental shelf of the southeastern United States. J. Phys. Oceangr. 11:1627-1637.

Blanton, J. O., and L. P. Atkinson. 1983. Transport and fate of river discharge on the continental shelf of the southeastern United States. J. Geophys Res. 88:4730-4738.

Checkley, D. M., et al. 1988. Winter storm effects on the spawning and larval drift of a pelagic fish. Nature 335:346-348.

Dame, R., and others. 1986. The outwelling hypothesis and North Inlet, South Carolina. Mar. Ecol. Prog. Ser. 33:217-229.

Deibel, D. 1985. Blooms of the pelagic tunicate, *Dolioletta gegenbauri*: Are they associated with Gulf Stream frontal eddies? J. Mar. Res. 43:211-236.

Dunstan, W. M., and L. P. Atkinson. 1976. Sources of new nitrogen for the south Atlantic Bight. Pp. 69-78. *In* Estuarine processes, Vol. 1, editor, Academic Press, New York.

Eppley, R. W., and B. C. Peterson. 1979. Particulate organic matter flux and planktonic new production in the deep ocean. Nature 282:677-680.

Haines, E. B. 1974. Processes affecting production in Georgia coastal waters. Ph.D. dissertation, Duke Univ., Durham, NC. 118 Pp.

Hanson, R. B., and C. Y. Robertson. 1988. Spring recycling rates of ammonium in turbid continental shelf waters off the southeastern United States. Cont. Shelf Res. 8:49-68.

Hofmann, E. E. 1988. Plankton dynamics on the outer southeastern U.S. continental shelf, Part III: A coupled physical-biological model. J. Mar. Res. 46:919-946.

Hofmann, E. E., and J. W. Ambler. 1988. Plankton dynamics on the outer southeastern U.S. continental shelf, Part II: A time dependent biological model. J. Mar. Res. 46:883-913.

Jacobsen, T. R., L. R. Pomeroy, and J. O. Blanton. 1983. Autotrophic and heterotrophic abundance and activity associated with a nearshore front off the Georgia coast, U.S.A. Estuarine, Coastal, Shelf Sci. 17:509-520.

Kendall, A. W., and L. A. Walford. 1979. Sources and distribution of bluefish, *Pomatomus salatrix*, larvae and juveniles off the east coast of the United States. Fish. Bull., U.S. 77:213-227.

Lee, T. N., and L. P. Atkinson. 1983. Low-frequency current and temperature variability from Gulf Stream frontal eddies and atmospheric forcing along the southeast U.S. outer continental shelf. J. Geophys. Res. 88:4617-4632.

Lee, T. N., L. P. Atkinson, and R. Legeckis. 1981. Observations of a Gulf Stream frontal eddy on the Georgia continental shelf, April 1977. Deep-Sea Res. 28A:347-378.

Miller, G. C., and W. J. Richards. 1980. Reef fish habitat, faunal assemblages, and factors determining distributions in the South Atlantic Bight. Proceedings, 32nd Annual Gulf and Caribbean Fisheries Institute, Miami Beach, FL. Pp. 114-130.

Nelson, W. R., M. C. Ingham, and W. E. Schaaf. 1977. Larval transport and year-class strength of Atlantic menhaden, *Brevoortia tyrannus*. Fish. Bull., U.S. 75:23-41.

Nicholson, W. R. 1978. Movements and population structure of Atlantic menhaden indicated by tag returns. Estuaries 1:141-150.

Oertel, G. F., and W. M. Dunstan. 1981. Suspended-sediment distribution and certain aspects of phytoplankton production off Georgia, U.S.A. Mar. Geol. 40:171-197.

Paffenhofer, G.-A. 1983. Vertical zooplankton distribution on the northeastern Florida shelf and its relation to temperature and food abundance. J. Plank. Res. 5:15-33.

Paffenhofer, G.-A. 1985. The abundance and distribution of zooplankton on the southeastern shelf of the United States. Pp. 104-117. *In* L. P. Atkinson, D. W. Menzel, and K. A. Bush (Eds.) Oceanography of the southeastern United States continental shelf. Coast. Estuar. Sci. Vol. 2, AGU, Washington, DC.

Paffenhofer, G.-A. and T. N. Lee. 1987. Development and persistence of patches of Thaliacea. *In* A. I. L. Payne, J. A. Gulland, and K. H. Brink (Eds.) The Benguela and comparable ecosystems. S. Afr. J. mar. Sci. 5:305-318.

Paffenhofer, G.-A., and D. E. Stearns. 1988. Why is *Acartia tonsa* (Copepoda: Calanoida) restricted to nearshore environments? Mar. Ecol. Prog. Ser. 42:33-38.

Paffenhofer, G.-A., B. K. Sherman, and T. N. Lee. 1987a. Summer Upwelling on the southeastern continental shelf of the U.S.A. during 1981. Abundance, distribution and patch formation of zooplankton. Prog. Oceanogr. 19:403-436.

Paffenhofer, G.-A., B. T. Wester, and W. D. Nicholas. 1984. Zooplankton abundance in relation to state and type of intrusions onto the southeastern United States shelf during summer. J. Mar. Res. 42:995-1017.

Paffenhofer, G.-A., et al. 1987b. Summer Upwelling on the southeastern continental shelf of the U.S.A. during 1981. Summary and conclusions. Prog. Oceanogr. 19:437-441.

Pomeroy, L. R., et al. 1983. Microbial distribution and abundance in response to physical and biological processes on the continental shelf of southeastern U.S.A. Cont. Shelf Res. 2:1-20.

Pomeroy, L. R. 1985. The microbial food web of the southeastern U.S. continental shelf. Pp. 118-129. *In* L. P. Atkinson, D. W. Menzel, and K. A. Bush, (Eds.) Oceanography of the southeastern United States Continental Shelf. Coast. Estuar. Sci. Vol. 2, AGU, Washington, DC.

Rowell, T. W., R. W. Trites, and E. G. Dawe. 1985. Distribution of short-finned squid (*Illex illecebrosus*) larvae and juveniles in relation to the Gulf Stream frontal zone between Florida and Cape Hatteras. NAFO Sci. Coun. Studies 9:77-92.

Thomas, J. P. 1966. Influence of the Altamaha River on primary production beyond the mouth of the river, M.S. Thesis, Univ. of Georgia, Athens, 88 Pp.

Turner, R. E. 1981. Plankton productivity and the distribution of fishes on the southeastern U.S. continental shelf. Science 214:353-354.

United States Department of Commerce. 1985. Fisheries of the United States, 1984. Current Fishery Statistics No. 8360. Washington, DC. 121 Pp.

Walsh, J. J., et al. 1978. Wind events and food chain dynamics within the New York Bight. Limnol. Oceanogr. 23:659-683.

Weber, A. H., and J. O. Blanton. 1980. Monthly mean wind fields for the South Atlantic Bight. J. Phys. Oceanogr. 10:1256-1263.

Yoder, J. A. 1983. Statistical analysis of the distribution of fish eggs and larvae on the southeastern U.S. continental shelf with comments on oceanographic processes that may affect larval survival. Estuar. Coast. Shelf Sci. 17:637-650.

Yoder, J.A. 1985. Environmental control of phytoplankton production on the southeastern U.S. continental shelf. Pp. 93-103. *In* L. P. Atkinson, D. W. Menzel, and K. A. Bush, (Eds.) Oceanography of the southeastern United States continental shelf. Coast. Estuar. Sci. Vol. 2, AGU, Washington, DC.

Yoder, J. A., and S. S. Bishop. 1985. Effects of mixing-induced irradiance fluctuations on photosynthesis of natural assemblages of coastal phytoplankton. Mar. Biol. 90:87-93.

Yoder, J. A., and T. Ishimaru. 1989. Phytoplankton advection off the southeastern United States continental shelf. Cont. Shelf Res. 9:547-553.

Yoder, J. A., et al. 1981. Role of Gulf Stream frontal eddies in forming phytoplankton patches on the outer southeastern shelf. Limnol. Oceanogr. 26:1103-1110.

Yoder, J. A., et al. 1983. Effect of upwelling on phytoplankton productivity of the outer southeastern United States continental shelf. Cont. Shelf. Res. 1:385-404.

Yoder, J. A., et al. 1985. Phytoplankton dynamics within Gulf Stream intrusions on the southeastern United States continental shelf during summer 1981. Cont. Shelf Res. 4:611-635.

Yoder, J. A., et al. 1987. Spatial scales in CZCS-chlorophyll imagery of the southeastern U.S. continental shelf. Limnol. Oceanogr. 32:929-941.

M. Dagg, C. Grimes, S. Lohrenz,
B. McKee, R. Twilley, and W. Wiseman, Jr.

4. Continental Shelf Food Chains of the Northern Gulf of Mexico

Abstract

Biological productivity in the northern Gulf is significantly affected by the Mississippi River. The freshwater discharge (577 km^3 yr^{-1}, approx 10% of the volume of water on the shelf) contains high concentrations of dissolved nutrients (100-150 μmol NO_3 l^{-1}). Flow is primarily constrained by prevailing winds to the continental shelf west of the Mississippi Delta. River plumes are regions of high phytoplankton stock (>30 g Chl l^{-1}) and production (5 g C m^{-2} d^{-1}), high copepod stocks (nauplius concentrations >1000 l^{-1}) and high ichthyoplankton stocks (larval concentrations >50 m^{-3}). The high temperature of shelf waters assures high physiological rates, implying high rates of trophic transfer and high turnover rates. The primary fate of phytoplankton production is grazing by macrozooplankton and microzooplankton. However, sinking of phytoplankton and other organic material fuels the annual development of a band of hypoxic water along the Louisiana coast. Fisheries production is high; the northern Gulf supports the largest volume fishery in the United States, the Gulf menhaden, *Brevoortia patronus*. The Loop Current in its northernmost position affects shelf processes to the east of the Delta. Anticyclonic rings derived from the Loop Current occasionally impact on the Louisiana shelf west of the Delta but usually drift over to the western Gulf resulting in exchange of oceanic and shelf water off Texas.

Introduction

The morphology of the shelves of the northern Gulf of Mexico is relatively simple (Figure 4.1). DeSoto Canyon is the easternmost boundary of the region. The shelf break occurs near the 100-m isobath which lies 100-150 km offshore east of the Mississippi River

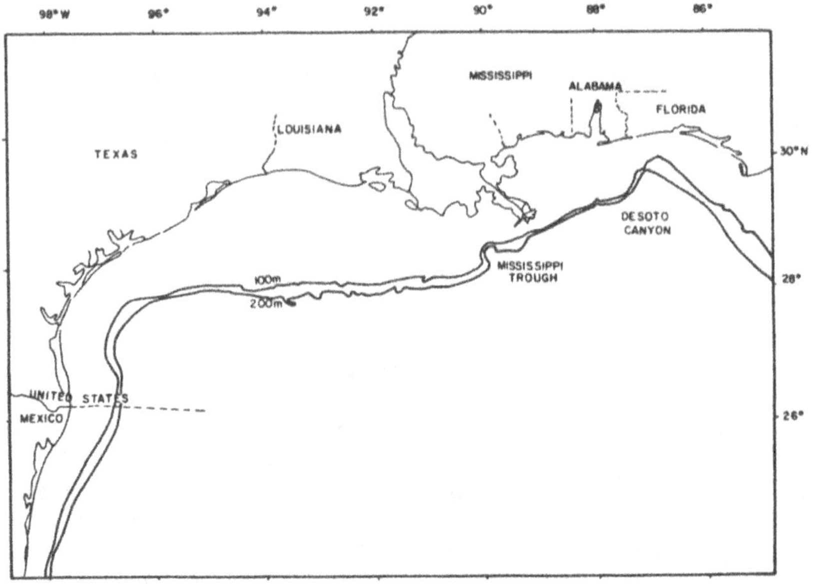

Figure 4.1. The northern Gulf of Mexico with bathymetry simplified to emphasize the broad shallow continental shelf of the region.

Delta. The Delta has built out nearly to the shelf break causing the shelf to narrow abruptly as one approaches the Delta. West of the Delta, a small canyon-like feature, the Mississippi Trough, cuts across the shelf and may represent a region of preferential off-shelf transport of particulate material. Further west, near 92°W, the shelf broadens to a width of approximately 200 km and then narrows again westward. Offshore of the Texas coast, it turns gently southward as one approaches the Mexican border.

The northern Gulf of Mexico is highly productive biologically. Portions of the shelf affected by the Mississippi River discharge have long been noted as regions of high phytoplankton stocks and productivity (Riley, 1937; El-Sayed, 1972). Phytoplankton production rates in excess of 5 g C m^{-2} d^{-1} have recently been reported in Mississippi River plume waters (Lohrenz et al., 1990; Ortner et al., submitted). There is evidence that zooplankton productivity is also high in the northern Gulf of Mexico (Dagg et al., 1987; Ortner et al., 1989) and fisheries production is very high. For example, the gulf menhaden, *Brevoortia patronus*, supports the largest volume fishery in the United States (Warlen, 1988).

Several factors suggest that the Mississippi River system is the ultimate source of much of the biological productivity on the Louisiana/Texas shelf: the river discharge is large and much of it remains on the broad shallow shelf for several months; river waters contain high concentrations of dissolved nutrients; and the open Gulf of Mexico does not appear to be a significant source of shelf nitrogen, although little is known about shelf/slope exchanges in this region.

The purpose of this paper is to present an initial characterization of the shelf environment of the northern Gulf of Mexico, and to indicate some of the processes that support the high biological production of this region.

Physical Regime

The dominant forcings of the shelf circulation are the strong runoff signal due to the Mississippi River system and the seasonally varying local meteorology. The Mississippi River system runoff averages 577 $km^3\ yr^{-1}$. Much of this occurs during the spring flood, and average daily discharge during the fall is only about 30% of the springtime high. Less than one third of the discharge flows eastward onto the Mississippi/Alabama shelf; most of the water discharging through the Mississippi River delta flows westward. About one third of the total system discharge enters the Gulf through the Atchafalaya River system and Atchafalaya Bay.

The summer breezes tend to be weak and southerly or southeasterly with characteristic time scales of the order of a few weeks. Occasional tropical storms and hurricanes represent significant disturbances to the area, particularly where they strike directly. During the fall, the summer winds weaken and the influence of cold air outbreaks from the north and northwest are felt on time scales of three to ten days. Latent heat loss is extensive and wave action over the shallow inner shelf is severe. Wind-driven set-up of a meter or more causes strong flooding of the adjacent wetlands and significant estuarine-shelf exchange (Schroeder and Wiseman, 1986). The role of these cold air outbreaks is dominant throughout the winter and early spring.

The relative strengths and timings of these weather and runoff variations influence the circulation and stratification of the shelf waters. These effects are better understood for the West Louisiana/Texas shelf than for the Louisiana/Mississippi/Alabama shelf. Therefore, this article primarily focuses on the shelf west of the birdfoot delta.

Due to buoyancy, water emanating from the major passes of the birdfoot delta of the Mississippi River rapidly separates from the bottom. It forms a well-defined plume with strong frontal boundaries which are regions of significant shear and convergence. The halocline beneath the plume breaks down through wind-mixing and shear instability (Wiseman et al., 1976a; Wiseman et al., 1976b). Local tides modulate the flow from the mouth of the river thereby affecting the entire plume structure and its associated fronts. The low salinity water emanating from the multiple river mouths and crevasses in evanescent plumes merges to form an identifiable flow of low salinity water which is traceable from the region of the Delta and which attaches to the coast within a few tens of kilometers downstream (Wiseman et al., 1982). A low salinity coastal boundary layer results and flows westward along the coast (Cochrane and Kelly, 1986). Variations in the velocity of this flow are driven by the winds on time scales of the synoptic weather patterns. The salinity of these waters is lowered by the effluent from the small coastal rivers, bayous and bays and by the major influx of river water from Atchafalaya Bay. The waters of the coastal boundary layer exchange efficiently with the local estuarine waters and are important to recruitment processes (Shaw et al., 1985). They are separated from the mid-shelf waters by a strong salinity front. Cross-frontal exchange processes are important to the transport of larvae and nutrients, but none of the mechanisms involved have been studied in detail.

Normal Fickian diffusion is probably weak in such a strongly stratified situation. Satellite imagery has shown examples of effluent plumes from estuaries traversing the front during cold air outbreaks. While such a process might take low salinity waters into the inner and mid-shelf regions, it appears to be a uni-directional process. Hydrographic surveys occasionally suggest meso-scale eddies forming along the front and possibly separating from it (Figure 4.2). During cold air outbreaks, the isopycnals in the front are more nearly vertical than during more weakly forced situations. It is not clear whether the altered density pattern is an advective response to wind forcing or whether the pycnocline breaks down due to mechanical mixing. The latter possibility would cause an efficient two-way exchange of shelf and coastal boundary layer waters. Finally, there exist regions of enhanced vertical mixing along the inner shelf such as the shoals immediately offshore of the Atchafalaya Bay (Chuang and Wiseman, 1983). It is possible that the majority of exchange across the front occurs in these regions. Cross-frontal exchange is an important process and clearly is worthy of further study.

Following the spring runoff maximum, the coastal boundary layer can be traced as far as the Mexican border and recent observations suggest that it remains an identifiable feature well into Mexican waters (Figure 4.2). In the late spring and early summer, winds along the Mexican and south Texas coast become upwelling favorable while those over the Louisiana coast remain downwelling favorable. (Indeed, weak upwelling along the south Texas coast

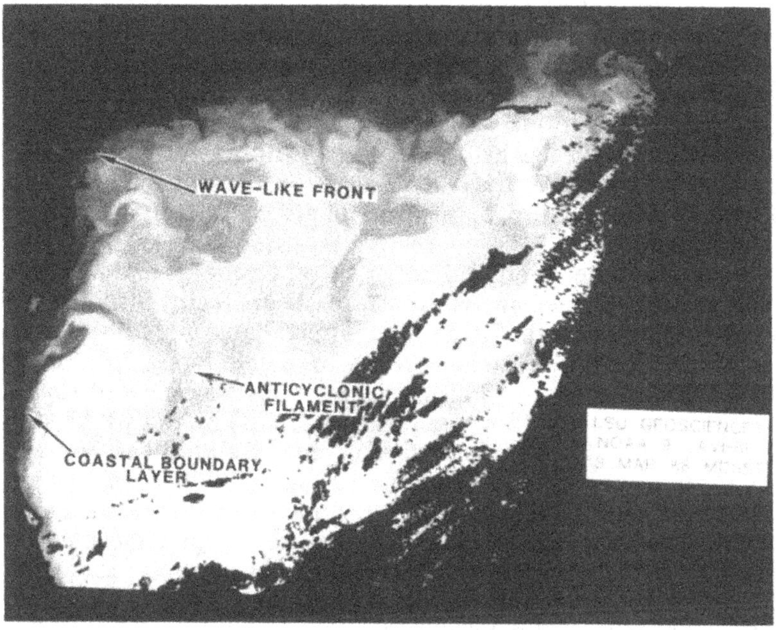

Figure 4.2. AVHRR image of the northwestern Gulf of Mexico for 3 March 1988. Land and clouds have been masked in black. Colder waters are represented by darker colors. Note: (1) the penetration of a cold coastal boundary layer into Mexican waters, (2) wave-like features on the frontal boundary of the coastal boundary layer in Texas waters, and (3) shelf break exchange as cold water is drawn offshore and rotates around an anticyclone in the western Gulf. (Courtesy of L. J. Rouse, Jr.).

appears to have been observed in hydrographic data [Cochrane and Kelly, 1986]). More importantly, the coastal boundary layer in this region changes flow direction. Waters return upcoast to the north and east along the Texas coast. A convergence zone develops and the low salinity waters flowing westward are turned offshore. This convergence zone migrates northeastward until mid-summer. Later, as the summer wind systems weaken, the southward flowing coastal boundary layer redevelops all along the Texas coast.

Under the weak summer winds, a strong halocline isolates the deeper nearshore waters from direct air-sea exchange. A region of severe near-bottom hypoxia develops along much of the Louisiana coast most years (Rabalais et al., in press). During winter storms, strong mixing vertically homogenizes the water column and reaerates the lower layer. A strong salinity front, though, still isolates the inner shelf waters from the mid-shelf waters. Between cold air outbreaks, when winds are weaker, a halocline may develop but it disappears with the arrival of the next cold front.

Wind forcing is normally a strong control on shelf circulation. Over the Texas/Louisiana shelf, though, direct wind forcing appears to leave a large portion of the subtidal current variance unexplained. Empirical orthogonal function analysis of existing data suggests that about half the subtidal variability is directly driven by the local winds. It is not clear what drives the remaining variability. Subtidal variance of the current field is large and seasonal mean currents estimated from direct measurements are often statistically insignificant.

Over the outer shelf and upper slope, Cochrane and Kelly (1986) posit a quasi-permanent, upcoast (eastward) current. The small number of direct current measurements available all indicate eastward flow in this region. Furthermore, the dynamic topography at the sea surface with respect to 70 db generally indicates eastward flow. The dynamics responsible for generating such a flow remain an enigma. The dynamic low which Cochrane and Kelly identify over the mid-shelf results in part from air-sea interaction occurring during winter cold air outbreaks. Cold, dry polar continental air flows out over the shelf and massive latent heat fluxes to the atmosphere occur (Nowlin and Parker, 1974). The salinity structure is such that the cooling produces denser water at mid-shelf than is found either inshore or further offshore. Some of this dense water sinks and runs downslope carrying with it dissolved material and material resuspended by wave-bottom interaction during the cold air outbreaks (McGrail and Carnes, 1983). The remainder comes into geostrophic equilibrium and contributes to the low in dynamic topography found over the shelf (Cochrane and Kelly, 1986).

Shelf break exchange is an important process but the studies necessary to fully document and explain it in this region have not yet been carried out. Offshelf flow, either into Mexican waters or into the deep Gulf, required to balance the freshwater budget of the shelf has been estimated (Dinnel and Wiseman, 1986) although the processes involved were not indicated. As mentioned above, along the south Texas shelf, the winds are upwelling favorable during much of the late spring, summer, and early fall. Other processes, though, appear to contribute as well.

Rings which break off from the Loop Current in the northern Gulf of Mexico typically migrate in a southwesterly direction and enter the western Gulf. Some, in contrast, move directly westward through the northern Gulf and interact with the slope west of the Delta. These anticyclonic rings are often associated with smaller cyclonic features. The circulation of the cyclones and anti-cyclones are observed, on AVHRR and CZCS imagery, to entrain chlorophyll laden shelf waters off the shelf (T. Leming, personal communication) (Figure 4.2). Concurrent surveys indicate that these waters also have increased ichthyoplankton concentrations. A final shelf break

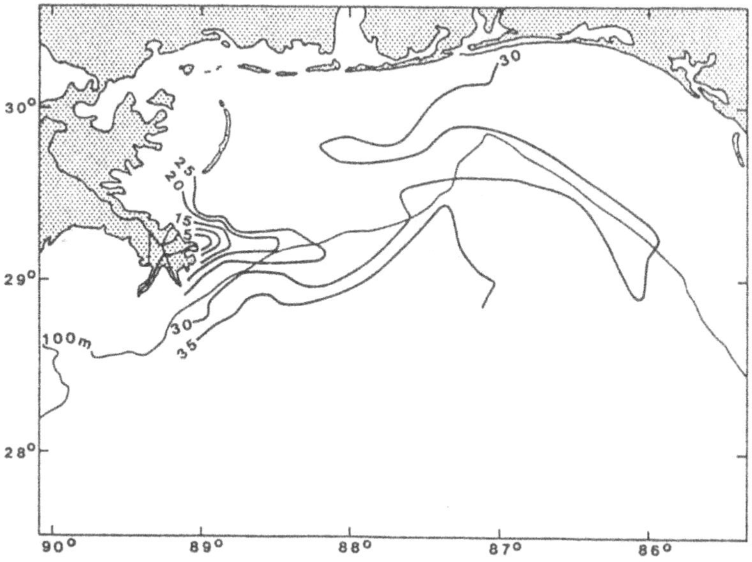

Figure 4.3. Sea surface salinity pattern east of the Mississippi River delta for 29 June-3 July 1964. Note the tongue of low salinity water characterized by the 30 $^{o}/_{oo}$ isohaline, which flows along the shelf break. (Redrawn from Drennen, 1968).

exchange process, mentioned above, occurs during winter months when heavy water forms at mid-shelf, sinks, and flows seaward (McGrail and Carnes, 1983).

East of the Mississippi River Delta, the shelf circulation is again strongly influenced by the Mississippi River discharge as well as by the fresh water discharge of the Mobile, Pascagoula and Pearl Rivers. A strong halocline is often present across the shelf. Mississippi River waters are usually found to flow eastward along the shelf break as a distinct low-salinity surface feature (Figure 4.3). Inner shelf, low-salinity waters flow westward, but are strongly wind-driven and respond to the seasonal and subtidal wind changes (Dinnel, 1988; Chuang et al., 1982).

To our knowledge, there is no current meter data from the mid-shelf region east of the Delta in the public domain. The U.S. Minerals Management Service is presently sponsoring a field program to collect such data. When it is available, it should enhance our knowledge of the circulation in this region.

Shelf waters are often entrained into the offshore by deep Loop Current intrusions into the northeastern Gulf (Schroeder et al., 1987; Huh et al., 1981). Dinnel (1988) has also hypothesized the existence of two regions of preferential offshore transport, one into the northwest portion of DeSoto Canyon and the other at approximately 88°W. Both of these regions are identified from mean hydrography and may only exist in a statistical sense. The role that such offshore transport of nutrient laden Mississippi discharge may play in the trophic dynamics of the open Gulf has recently been discussed (Walsh et al., in press).

Another important shelf break exchange process in the region east of the delta is wind-driven during the winter months. Southerly and southeasterly winds force water into the corner region formed by the Delta and the Mississippi coastline. As these winds relax, the water flows southward along the Delta and the Chandeleur Islands and crosses the shelf break immediately east of the Delta (Hart and Murray, 1978; Schroeder et al., 1985).

There is a potential for communication across the Delta front. Long-term current meter measurements in front of the Delta do not suggest a statistically significant mean flow except during periods when the Loop Current may be deeply intruded into the Gulf (Wiseman and Dinnel, 1988). Nevertheless, the subtidal oscillations in the flow field are energetic and of long period. Thus, the time-varying component of the flow may well be capable of transporting material across the Delta front.

Nutrient Inputs

The discharge waters of the Mississippi River system are a major source of dissolved nutrients for the shelf of the northern Gulf of Mexico. The annual discharge west of the delta is equivalent to approximately 10% of the volume of water on the entire Louisiana/Texas shelf out to 90 m (Dinnel and Wiseman, 1986), and nutrient concentrations in river waters are high; nitrate concentrations often exceed 100 μmol l^{-1} (Dagg and Whitledge, 1991). Indeed, if no other factor were limiting and riverine input of nitrate was uniformly distributed over the entire shelf, it would be sufficient to support "new" phytoplankton production of 30 g C m^{-2} y^{-1} (Table 4.1; Dagg and Whitledge, 1991). Patterns of nitrate use are of course not uniform. Rather, nitrate is stripped from the water within a short distance from the input sources (river discharge points), typically less than 100 km (Dagg and Whitledge, 1991; Ortner et al., submitted).

Because a significant fraction of the water in the coastal boundary layer originates from the river discharge plumes, these waters often contain high concentrations of dissolved nutrients. Exchange between the coastal boundary layer and the mid-shelf waters provides nutrients for mid-shelf production. An example of such an exchange was observed during a period of strong northerly and westerly wind during February 1984 when waters from the

Table 4.1. Potential stimulation of "new" production by the annual nitrate input to the Louisiana-Texas Shelf west of the Mississippi River delta. (From Dagg and Whitledge, 1991.)

Freshwater input[a]	382 x 10^9 m^{-3} y^{-1}
Nitrate concentration[b]	100 mg-at m^{-3}
Shelf area to 90 m[a]	10686 km^2
Nitrate input	3.57 x 10^2 mg-at m^{-2} y-1
Nitrogen input	50 x 10^2 mgN m^{-2} y-1
C equivalent (assumes C:N=6)	30 g C m^{-2} y^{-1}

[a]Dinnel and Wiseman (1986).
[b]Walsh et al. (1981) and Turner et al. (1987) suggest 150 mg-at m^{-3}. Concentrations in 1935 of less than 15 mg-at m^{-3} were reported by Riley (1937).

coastal boundary layer were displaced offshore at the surface by nutrient-poor upwelled water (Dagg, 1988).

High concentrations of dissolved nutrients are observed in subsurface and bottom waters of the mid-shelf region, particularly in areas strongly affected by discharge plumes during periods of the year when shelf waters are stratified, suggesting significant benthic regeneration of dissolved nutrients (see below). Processes by which these nutrients are supplied to the euphotic zone are not well studied in this region but permanent salt driven stratification around the discharge plumes suggests significant horizontal transport along the shelf usually must occur prior to winter overturn.

Processes that result in the exchange of shelf and slope waters are not likely to be a significant source of dissolved nitrogen for shelf production processes. Waters of the open Gulf of Mexico are typically impoverished of nitrate (<1 μmol l^{-1}) to a depth of 75-100 m throughout the year (references in Walsh et al., in press; Murrell and Dagg, 1987; Toon and Dagg, 1989). Nitrate concentrations equal to those often observed in bottom waters of the mid-shelf off Louisiana are not observed in open Gulf waters until depths of several hundred m are reached. Shelf-slope exchanges of water must primarily be viewed as contributing to nutrient losses from the shelf in this region. Quantification of these exchanges is not possible at this time.

One type of event that may cause significant exchange of shelf and slope or open Gulf waters is the abuttment of rings that break off from the Loop Current onto the outer shelf. Remote imagery indicates that these rings entrain shelf waters (Figure 4.2). As indicated above however, the nutrient concentrations of replacement waters are likely to be low.

Upwelling favorable winds off the Texas coast during summer, winds that reverse the flow of the coastal boundary layer, suggest that offshore water containing some significant source of dissolved nutrients might stimulate phytoplankton production on the Texas shelf. The existence of enhanced phytoplankton biomass in a bottom nepheloid layer between the coast and out as far as 70 m during summer (Kamykowski and Bird, 1981) is consistent with this view, although enhanced nutrient concentrations within the nepheloid layer were not observed.

Phytoplankton

The northern Gulf of Mexico is characterized by strong frontal boundaries, at the interface between river discharge plumes and receiving shelf waters, at the interface between the coastal boundary layer and mid-shelf waters, and at the interface between the outer-shelf waters and the open Gulf. Remotely sensed patterns of chlorophyll distribution in the northern Gulf of Mexico illustrate that surface phytoplankton stocks are highest within the discharge plumes of the Mississippi and Atchafalaya Rivers, and within the coastal boundary layer. Concentrations in mid- and outer-shelf waters are higher than those of the open Gulf. Concentrations in the shelf waters off Texas are generally lower than those of the Louisiana shelf. Regions of mixing and exchange between coastal waters and mid-shelf waters are indicated by enhanced offshore concentrations of surface phytoplankton. Similar frontal processes are observed at the shelf/slope boundary (Heron et al., 1989).

High levels of primary production in the northern Gulf of Mexico have been attributed to the input of new nutrients from the Mississippi River (Riley, 1937; Thomas and Simmons, 1960). However, the dynamic and heterogeneous nature of the Mississippi River plume (Lohrenz et al., 1990; Ortner et al., submitted) has complicated attempts to relate changes in levels of riverine nutrient inputs to corresponding changes in regional primary production. Nutrients (e.g., Riley, 1937; Boynton et al., 1982) and light (e.g., Pennock and Sharp, 1986; Cloern, 1987) are thought to be the principal factors regulating phytoplankton dynamics. Observations of initial limitation of production by light and subsequent limitation by nutrient supply along decreasing turbidity gradients in estuaries might be expected to apply to river plumes (e.g., Xiuren et al., 1988). Indeed, the spatial pattern of high production and biomass at intermediate salinities in the northern Gulf of Mexico (Figure 4.4) encourages such speculation. Further evidence comes from observations that water column light attenuation in the Mississippi River

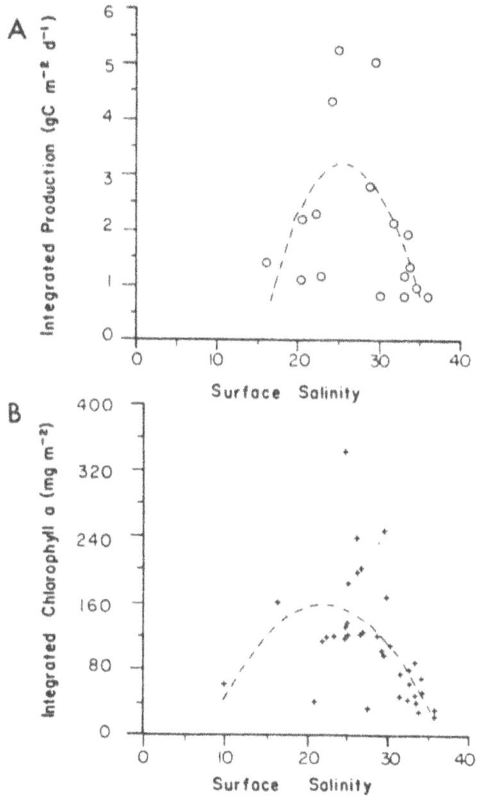

Figure 4.4. Phytoplankton productivity vs. salinity, near the Mississippi River delta during April 1988: ^{14}C primary production (A) and Chlorophyll *a* (B) integrated over the entire water column in relationship to surface salinity. (From Lohrenz et al., 1990).

plume was inversely related to nutrient concentrations along the salinity gradient (Lohrenz et al., 1990).

The hypothesis that primary production is light-limited in turbid areas of the Mississippi River plume was tested by Lohrenz et al. (1990) during an April 1988 study. The investigators found that a major proportion (>80%) of the spatial variation in production was accounted for on the basis of light and chlorophyll levels, supporting the view that light was an important controlling factor. However, a light-limitation model (Wofsy, 1983; Lohrenz et al., 1990) indicated light conditions in the surface mixed layer of the plume were adequate to support phytoplankton concentrations greater than were observed, even at the most turbid stations (Figure 4.5). These researchers suggested that factors in addition to light (e.g., sinking,

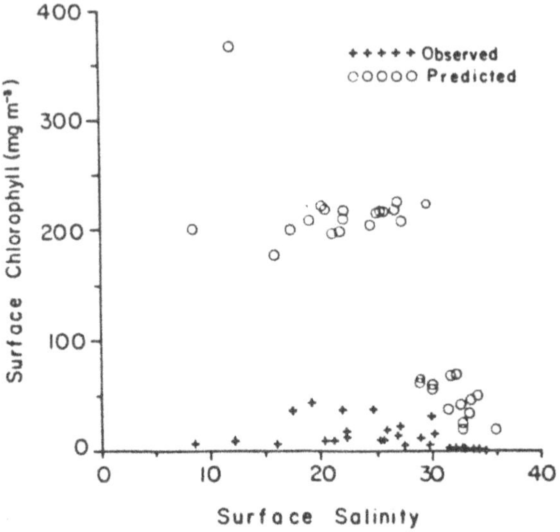

Figure 4.5. Comparison of observed surface Chlorophyll *a* concentrations with those predicted from Wofsy (1982) model as described in Lohrenz et al. (1990) plotted versus surface salinity.

grazing, physical dilution) contributed to regulation of phytoplankton concentrations, and hence primary production. Ortner et al. (submitted) found no evidence for shade adaptation in phytoplankton populations sampled from the base of the euphotic zone in areas of high turbidity during September 1982, supporting the view that phytoplankton growth may not be light-limited in the plume. Moreover, they estimated growth rates of 3-5 doublings d^{-1}, close to the maximum for phytoplankton at 28°C (Eppley, 1972).

In addition to uncertainties regarding the role of light in influencing production in turbid eutrophic areas, there are questions as to mechanisms of nutrient limitation in adjacent shelf waters. Concentrations of dissolved silicate and nitrate have been reported near detection levels (e.g., Shiller and Boyle, 1987; Lohrenz et al., 1990) and both nutrients have been suggested as limiting to phytoplankton along the plume/oceanic gradient (Sklar and Turner, 1981; Thomas and Simmons, 1960). It may be that different nutrients limit different taxonomic groups of phytoplankton (i.e., silica for diatoms and nitrogen for others). There are at least two distinct phytoplankton communities associated with the Mississippi River and Gulf of Mexico that mix to varying degrees in the plume (Thomas and Simmons, 1960). Phytoplankton from the Mississippi River are probably not limited by the same factors as phytoplankton from the

Gulf. In fact, even within the same community, different species may be limited by different factors.

It is clear that more work is needed to define relationships between regional primary production and Mississippi River nutrient inputs, and to determine the factor(s) which limit phytoplankton growth and production in northern Gulf of Mexico waters. Riley (1937) acknowledged the possibility that factors other than light and nutrients may contribute in regulating phytoplankton distributions in the northern Gulf of Mexico. The importance of other controlling factors has been subsequently noted in other coastal ecosystems (Pennock, 1985; Fransz, 1986).

Factors besides light and conventional nutrients which may play an important role in influencing primary production in the northern Gulf of Mexico include the high metal concentrations associated with the Mississippi River outflow (Shiller and Boyle, 1987). The toxicity of copper is well-known (e.g., Sunda and Guillard, 1976; Fitzwater et al., 1982), although the possible complexation of trace elements by organic matter in the plume would diminish effective toxicity. An additional factor which could inhibit phytoplankton growth is the steep gradient in salinity. Variations in salinity may significantly influence photosynthesis and growth of phytoplankton species (e.g., Smayda, 1969; Miller and Kamykowski, 1986a). Such effects may occur on relatively short time scales (Miller and Kamykowski, 1986b). These observations lead to the view that in areas of dynamic estuarine circulation, growth and production may be suppressed due to salinity variations.

Another potential impact of large changes in salinities would be loss of some species from the community (Filardo and Dunstan, 1985; Jackson et al., 1987). Such transitions may result in a lag as marine species colonize the newly mixed waters, and would significantly affect sinking characteristics of the phytoplankton community. Riley (1937) suspected an inhibitory effect of low salinity on phytoplankton in the Mississippi River plume, which could be explained by such a process. The combination of phytoplankton population and environmental changes can also affect grazer community characteristics (Dagg et al., 1987; Ortner et al., 1989).

Spatial and Temporal Patterns

Preliminary estimates of mixed layer integrated production suggest consumption of nutrients by phytoplankton production was comparable to riverine inputs within spatial scales of <100 km (Lohrenz et al., 1990). From these estimates, it was concluded that the ecosystem of the plume environs is eutrophic, with an abundant supply of new nutrients and production limited by other factors. As distances from the outflow region increase, the role of heterotrophic nutrient regeneration will likely become more important. Turner et al. (1987) noted that primary production beyond the plume is primarily nitrogen limited, and hypothesized that increases in riverine nutrient inputs result in increased inputs of phytoplankton carbon to bottom waters in those areas.

Data adequate to resolve seasonal patterns in phytoplankton production are currently lacking for the northern Gulf. However, in view of the large changes in river flow and corresponding nutrient outputs, seasonal variation in the extent of the river-influenced region is likely to be substantial (e.g., Fucik, 1974; Sklar and Turner, 1981). In addition to variation in the magnitude of freshwater and nutrient inputs, changes in ratios of nutrients may alter responses of the phytoplankton community. Also, seasonal variations in river-borne materials, such as dissolved organics and suspended particulates, could alter optical conditions in the plume (e.g., Sathyendranath et al., 1989; Lohrenz et al., 1990). Seasonal changes in temperature are also likely to influence ecosystem dynamics. Lower temperatures during winter months may constrain rates of primary production, as well as heterotrophic consumption (Boynton et al., 1982).

Phytoplankton Fates

Zooplankton Grazing

As pointed out recently by Ortner et al. (1989), little work has been done on zooplankton of the northern Gulf of Mexico. It has been well documented in other regions however, that species composition changes across shelves. Hopkins et al. (1981) showed this is the case for copepods on a transect of the west Florida shelf into the open Gulf; epipelagic species of copepods show farthest landward distribution while deeper dwelling forms remain more oceanic. Based on examination of three regions, Ortner et al. (1989) concluded there was a gradient in copepod species composition in the northern Gulf of Mexico between the estuarine community and the open Gulf community. Species characteristic of the shallow low salinity waters were *Acartia tonsa*, *Temora turbinata*, *Labidocera nerii*, *Centropages furcatus*, *Paracalanus quasimodo*, *Farranula carinata*, and *Temora stylifera*. Species characteristic of deeper more saline waters of the open Gulf of Mexico included *Eucalanus attenuatus*, *Phaenna spinifera*, *Candacia pachydactyla*, *Calanus minor*, *Eucalanus elongatus*, and *Calanus tenuicornis*. Because discharge waters from the Mississippi River frequently override oceanic waters in the vicinity south of the Mississippi Delta, species of the deep ocean community are often found beneath the estuarine community associated with the river plume. In addition, there is a distinct community in transition waters (as represented by samples from Cape San Blas, Florida), which included *Oithona plumifera*, *Corycaeus clausii*, *Oncea mediterranea*, *Oncea venusta*, and *Candacia curta*. Up to 87% of the total macrozooplankton numbers in the nearshore regions were copepods. Waters of the coastal boundary layer are likely to primarily contain the species of the "shallow, low salinity water" community of Ortner et al. (1989). Several years of collecting non-quantitative zooplankton samples for experimental work in coastal waters south of LUMCON support this expectation; the low salinity water community is most common, with occasional contributions from the transition community (Dagg, 1988; Dagg, unpublished observations). *Acartia tonsa* is typically the most abundant macrozooplanktonic species in estuarine areas of the northern Gulf of Mexico (Lesley, 1977; McIlwain, 1968), and observations from low salinity shallow waters near the Atchafalaya River mouth and from the Mississippi River plume indicate the *Acartia tonsa* is the dominant copepod in these waters (Dagg and Walser, submitted; Dagg, unpublished observations).

Evidence suggests that zooplankton production is high in the northern Gulf of Mexico. The highest individual sample biomass and copepod abundances measured by Ortner et al. (1989) were in the low salinity surface waters of the Mississippi River outflow, and concentrations of copepod nauplii are typically higher in the region around the Mississippi River Delta than in waters of similar depth and distance from shore farther to the east and west of the Delta in Florida and Texas (Dagg et al., 1987). Concentrations of nauplii away from the plume, in high salinity waters, were lower in all seasons than concentrations in the intermediate salinity range of the plume waters.

Limited data also suggest a strong seasonal pattern in the stocks of zooplankton, at least in the low-salinity community. Ortner et al. (1989) noted that springtime zooplankton stocks in the Mississippi River region were much higher than wintertime stocks. Similarly, concentrations of copepod nauplii in plume waters of the Mississippi River were between 10-100 l^{-1} during wintertime cruises (Dagg et al., 1987), between 50 and 100 l^{-1} during an April cruise (Dagg and Whitledge, 1991), and up to 1600 l^{-1} during a summer cruise (Dagg and Whitledge, 1991).

These limited data suggest a strong seasonal cycling of macrozooplankton stocks in the discharge plumes and probably in coastal and shelf waters influenced by river discharge. Such a cycle suggests that grazing by the macrozooplankton community is also strongly seasonal.

In a series of papers (see Turner, 1987 and references therein), Turner describes by SEM the contents of fecal pellets collected from common copepod species feeding in natural water from locations near the Mississippi River Delta. Collections were made during wintertime.

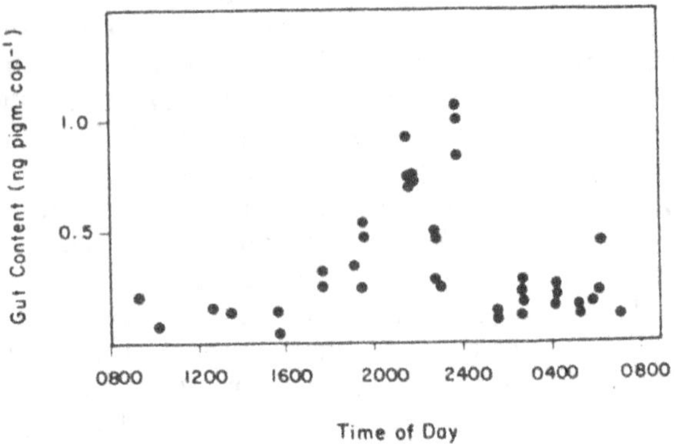

Figure 4.6. Level of phytoplankton pigments observed in the copepod *Acartia tonsa* from the Atchafalaya River discharge into Fourleague Bay during May 1989. Note the strong diel signal.

In most of the species he examined, including *Acartia tonsa*, *Paracalanus quasimodo*, *Temora turbinata*, *Temora stylifera*, and *Eucalanus pileatus*, pellet contents reflected the phytoplankton contents of the ambient water. In addition, copepods collected from highly turbid plume waters produced pellets containing large numbers of inorganic particles. Some species, including *Centropages velificatus* and *Labidocera aestiva*, contained crustacean remains in addition to phytoplankton remains and inorganic particles, indicating a more omnivorous diet. These studies are of interest but do not provide quantitative information on the feeding rates of these copepods.

Preliminary measurements of ingestion by *Acartia tonsa* adult females from Fourleague Bay (a bay receiving Atchafalaya River water at the head and Gulf of Mexico water at the mouth) were made during May 1989 (Figure 4.6). Ingestion of phytoplankton was 28.3 ng Chl d^{-1} during this 24 h sampling cycle, with most of the feeding occurring at night. Assuming a population density of 30 adults l^{-1}, commonly observed in Fourleague Bay (Dagg, unpublished), yields a total ingestion of 0.85 mg Chl m^{-3} d^{-1}. Chlorophyll stocks in surface waters at this time were 10-15 mg m^{-3}. Immature stages of *Acartia* greatly outnumbered adults at this time and would have contributed significantly to the total ingestion by this species of copepod. Therefore, the *Acartia tonsa* population probably ingested 20-30% of the phytoplankton stock daily during this period. Similar measurements for copepod populations in waters of the Mississippi River plume and Louisiana shelf have not been made but high copepod densities suggest that macrozooplankton grazing is an important source of phytoplankton mortality in these regions.

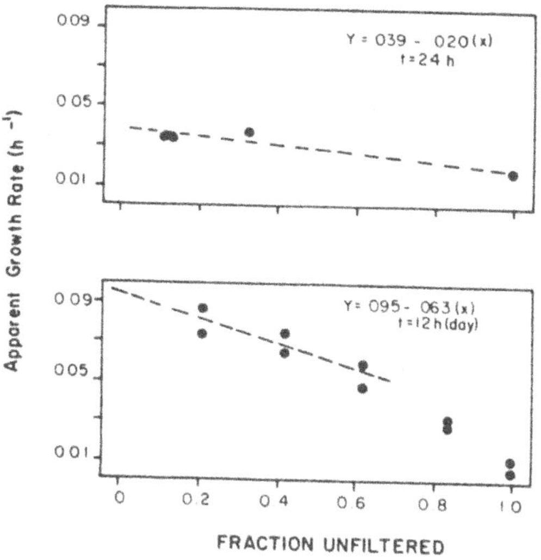

Figure 4.7. Grazing rates of the microzooplankton community from the Atchafalaya River discharge into Fourleague Bay during May 1989 (upper panel) and September 1989 (lower panel). Non-linearilty in the September experiment is attributed to nutrient limitation during the incubation.

Composition of the microzooplankton community in the northern Gulf of Mexico has not been documented. Some initial characterization was done by Gifford and Dagg (1988) in Terrebonne Bay, a coastal environment characterized by intermediate salinities, high phytoplankton concentrations and a macrozooplankton community dominated by *Acartia tonsa*. High concentrations of unicellular microzooplankton, including tintinnids, aloricate ciliates and zooflagellates, were observed during the summer and lower concentrations in the winter. During the summer, microzooplankton comprised a significant fraction of the diet of *Acartia tonsa* (Gifford and Dagg, 1988).

Preliminary measurements of ingestion by the microzooplankton community have been made in both Atchafalaya and Mississippi plume waters (Dagg, unpublished). During May and September 1989, the microzooplankton community ingested 38% and 53% (Figure 4.7) of the phytoplankton stock d^{-1}, respectively, in the region of the Atchafalaya River discharge. Similar, high, rates of ingestion by the microzooplankton community were observed in the Mississippi River plume during September 1989 (Dagg, unpublished). These experiments although preliminary, indicate the microzooplankton community is very active especially during the summer in the northern Gulf of Mexico.

In summary, macro- and microzooplankton grazing crop significant portions of the phytoplankton community in coastal and shelf waters of Louisiana. Losses are probably greater during the spring and summer than during the winter. Direct comparisons of phytoplankton mortality due to the total grazing by both communities with phytoplankton growth or productivity measurements are required.

Grazing on phytoplankton by the macrozooplankton community has additional consequences for the shelf ecosystem. A copepod defecates about 30% of the phytoplankton it ingests; the remainder is recycled (as CO_2 for carbon and primarily as NH_4 for nitrogen) or synthesized into new tissue as growth or reproductive products. Strong seasonality in grazing by macrozooplankton would significantly affect the flux of material to the bottom. If, as limited data suggest, winter and early spring are periods of comparatively low grazing mortality for phytoplankton, then direct sinking of phytoplankton cells to the benthos could dominate the vertical flux. In comparison, during the summer and fall when macrozooplankton stocks are high and grazing by the macrozooplankton community is a large source of mortality for the phytoplankton, little direct sedimentation of phytoplankton cells would occur and the dominant flux component would be copepod fecal pellets.

Fecal pellets are a major component of sinking particles beneath the plume of the Mississippi River during spring and fall (Nelson and Trefry, 1986). Summer fluxes of pellets would likely be even greater. Fecal pellet concentrations of $>1.5 \times 10^6$ m^{-2} have been observed in bottom samples on the Louisiana shelf (Dortch, pers. commun.). A population of only 10 adult *Acartia* spp. l^{-1} in river plume waters could produce 5×10^6 pellets m^{-2} d^{-1}.

Phytoplankton losses to microzooplankton grazing are more complete. Fecal remains from ingestion processes of microzooplankton organisms are small and effectively do not sink (Welschmeyer and Lohrenzen, 1985). Microzooplankton rapidly recycle a large fraction of the particulate carbon and nitrogen they ingest (Goldman et al., 1985). Ingested material partitioned to growth is thought to be quickly consumed by other microzooplankton components of the microbial food web and thereby mostly recycled (Azam et al., 1983), or ingested by macrozooplankton (Gifford and Dagg, 1988).

In summary, phytoplankton ingested by the microzooplankton community is primarily recycled within the water column, and the microzooplankton community contributes little to the flux of particulate material to the benthos. In contrast, approximately one third of the phytoplankton carbon and nitrogen ingested by the macrozooplankton community may reach the bottom as fecal pellets. This may have important implications for fueling near-bottom hypoxia, commonly observed during the summer in coastal waters of Louisiana (Rabalais et al., in press) and for the long-term fate of particulate carbon.

The high concentrations of nitrate entering the Gulf at the Mississippi River mouth are physically diluted and biologically stripped from the water within a comparatively small distance, <100 km (Dagg and Whitledge, 1991). Within this zone the phytoplankton community is dominated by large diatoms (Ortner et al., 1989; Turner, 1987 and references therein), and it is probable that a large fraction of the phytoplankton production is supported by "new" nitrogen (primarily nitrate but also some ammonium) input from the river. Macrozooplankton grazing is probably highest within this region. Vertical flux of particulate material, either live phytoplankton cells or copepod fecal pellets, is likely to be large. Outside of this zone of direct utilization of riverborne nitrogen, phytoplankton production decreases (Lohrenz et al., 1990) and is likely increasingly dependent on recycled nitrogen. The phytoplankton community is typically dominated by small cells, and microzooplankton grazing is likely predominant. Vertical flux of particulate material is small because particles, both phytoplankton cells and microzooplankton fecal pellets, are small. By implication, the same patterns would be observed near the Atchafalaya River mouth.

Sinking

Phytoplankton sinking is expected to be especially important in circumstances where the rapid response of phytoplankton to sudden introduction of new nutrients exceeds that of the zooplankton grazers. An example is the sedimentation of the spring diatom bloom in coastal areas (Platt and Subba Rao, 1970; Skjoldal and Lannergren, 1978). High phytoplankton sinking rates are also associated with nutrient depletion (Titman and Kilham, 1976; Fahnenstiel and Scavia, 1987) and with cells of high density (i.e., diatoms). Because diatoms appear to be the dominant phytoplankton assemblage in the Mississippi River plume and adjacent shelf regions (Thomas and Simmons, 1960; Fucik, 1974; Q. Dortch, pers. commun.) and nutrient availability varies tremendously across this gradient, losses through phytoplankton sinking may be significant. Evidence that phytoplankton sinking does occur comes from direct observations of intact cells on the sediment surface in the area (Q. Dortch, pers. commun.). High sediment pigment values, which have been associated with periodic low oxygen events to the west of the delta (Turner and Allen, 1982; Rabalais et al., in press), provide additional evidence for high rates of sedimentation of photosynthetically produced organic matter.

Deposition and Burial

Coastal and shelf sediments can be significant reservoirs of organic materials and nutrients that are deposited but not regenerated to the water column. The distinction in time scales between temporary versus permanent storage of particulate material in bottom sediments is important in determining the ultimate fate of materials in shelf ecosystems. The terms used to describe sedimentation on various time scales can be distinguished quantitatively using radiochemical techniques for establishing geochronologies within bedded sediments, and for examining rates of sedimentary processes. Deposition is the temporary emplacement of particulate material on a sediment surface. Burial (accumulation) is the sum of deposition and removal over a longer time scale (McKee et al., 1983). Within an environment, the magnitude of benthic nutrient processes (sedimentation and regeneration) depends on the time scale of interest.

The relative magnitude of burial is controlled by the quantity and quality of particulate material supplied to bottom sediments (Zeitzschel, 1980; Klump and Martens, 1983). Direct empirical relationships have been established between high sedimentation rates and the burial of organic nutrients (Toth and Lerman, 1977; Suess, 1980). A substantial proportion of the particulate material delivered to coastal ecosystems is supplied by a few large rivers such as the Mississippi/Atchafalaya system. In river-dominated shelf environments, areas with highest particle deposition rates are usually dominated by relatively refractory terrigenous materials (Aller et al., 1985). Using a transport-reaction model, Aller and Mackin (1984) hypothesized that the maximum loss of a given nutrient by burial occurs when $D_sK = W^2$, where D_s is the sediment mixing coefficient (dominated by bioturbation), K is the first-order decomposition rate constant for the organic fraction, and W is the deposition rate. On any time scale, it is the relationship between deposition and decomposition that determines the magnitude of particulate nutrient storage in bottom sediments.

The input of freshwater to land-margin ecosystems leads to the formation of density fronts that enhance sedimentation and lead to high inputs of nitrogen, phosphorus and organic matter to bottom sediments. A common observation in river-dominated environments is that most of the sediment discharge is initially deposited near the river mouth and sedimentation rates decrease with increasing distance from the source (DeMaster et al., 1985; Nittrouer et al., 1987). Therefore, a decrease in deposition rates from proximal to distal portions of dispersal systems is likely, especially during high flow periods when riverine discharge dominates. From high flow to low

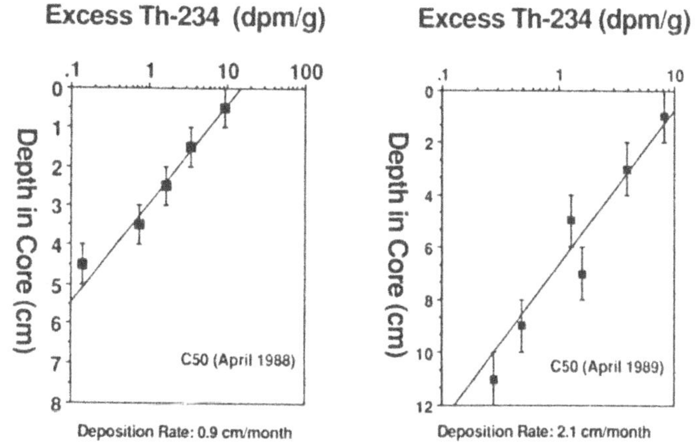

Figure 4.8. Particulate deposition rates at a shelf site (50 m water depth) proximate to Southwest Pass of the Mississippi River as determined by ^{234}Th geochronology. The deposition rate during April (high river-discharge) of the "dry" year, 1988, was a factor of two less than the deposition rate at the same site for April of a "normal" year, 1989.

flow periods, Mississippi River discharge decreases by about 85%, resulting in reduced riverine input to the shelf and leading to a more uniform dispersal of particulates throughout the dispersal system.

In a temporally variable environment such as this, a seasonal measure of particulate deposition may be much more important to understanding patterns of benthic nutrient regeneration and loss of nutrients via sedimentation than a measure of long-term accumulation rates that integrate over several decades. Net nutrient storage in bottom sediments is a balance between deposition and regeneration. The combination of high deposition rates and low sediment reactivity should result in a proportionally higher temporary storage of particulate nutrients, especially in the proximal portion of the dispersal system. Deposition in the more distal portions is dominated by reactive organic matter from phytoplankton and thus decomposition and regeneration in these regions may be higher.

Sediment cores collected from the shelf adjacent to the Mississippi River were analyzed for naturally occurring ^{234}Th and ^{210}Pb, to quantify sedimentary processes on time scales ranging from months to years (Figures 4.8 and 4.9, McKee et al., 1990). The absence of macrobenthos near the river mouth results in the preservation of primary sedimentary structure. Therefore, the distribution of ^{234}Th and ^{210}Pb in the seabed is controlled by sedimentation and these naturally occurring radionuclides can be used

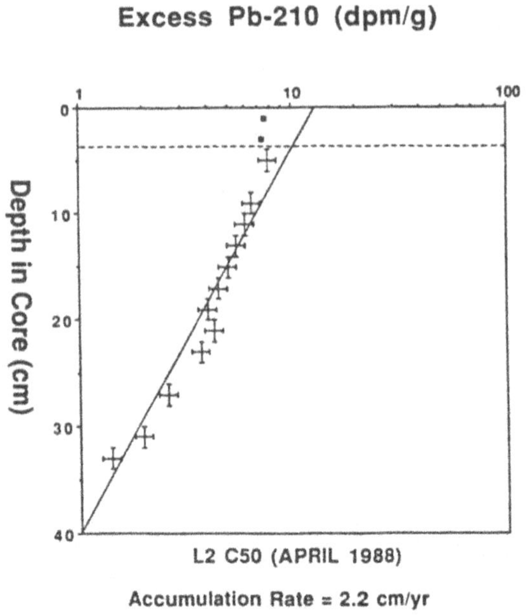

Figure 4.9. The rate of particulate accumulation (deposition minus removal) at the 50-m shelf site as determined by ^{210}Pb geochronology. This rate integrates over a period of approximately 100 years, and is a factor of 6 to 12 times less than deposition rates measured at this site.

to quantify sedimentation rates. Using ^{234}Th, sediment deposition rates (100-d time scale) were measured during the high river-discharge stage of two successive years. In 1988 (an unusually dry year in the Mississippi River basin), sediment deposition during the high river-discharge period was 0.90 cm/mo (Figure 4.8 upper panel). In 1989 (a "normal" year), the deposition rate was 2.10 cm/month, a factor of two higher (Figure 4.8 lower panel). Deposition rates (100-d time scale) at this site are 5-10 times greater than sediment accumulation rates (100-yr time scale) determined using ^{210}Pb (Figure 4.9). Therefore, the relative rates of deposition, redistribution and burial in the study area may vary on time scales of days to decades.

In addition to measurements of the quantities of organic matter that are deposited in bottom sediments, some estimate of the biological reactivity (quality) of the organic matter is needed to determine its potential for decomposition/remineralization. In other regions, lignin and carbohydrate analyses of particulate organic matter collected in sediment traps and in bottom sediments of a coastal marine bay have been used to determine the relative fluxes and reactivities of terrigenous and marine-derived organic matter (Hedges et al., 1988). These investigators concluded that about 67% of the organic carbon flux was marine derived, but that only about 33% of the organic matter in the bottom sediments was of marine origin. Thus, marine-derived organic matter was several-fold more reactive than terrigenous materials and contributed most significantly to benthic and water-column nutrient regeneration. Even though terrigenous organic matter is not as biologically reactive as its marine counterpart, the recycling of a small portion of the annual discharge of the Mississippi River (4 X 10^{12} g C; Malcolm and Durum, 1976) would have a major impact on coastal ocean productivity.

Benthic Remineralization

The contribution of nutrient remineralization from sediments to water column production of phytoplankton in shelf ecosystems is significant. In the North Sea, 75% of the nitrogen requirements are met by benthic regeneration (Billen, 1978), and in the Kiel Bight this flux may provide 100% of the phytoplankton demand for nutrients (Zeitzschel, 1980). Walsh et al. (1978) estimated that sediments contributed 38% of the total nitrogen regeneration in the Mid Atlantic Bight, equivalent to the regeneration by zooplankton in the water column. Rowe et al. (1975) suggested that sediments in shallow shelf ecosystems such as in the New York Bight, may provide "more than the total required nitrogen." In a later analysis, regenerated nitrogen in the Mid Atlantic Bight during the summer of 1980 was found to provide 50-80% of nitrogen productivity although the contribution by sediments was estimated to be only 7% of the total nitrogen regeneration (Harrison et al., 1983). Direct measurements of benthic ammonium flux in Christiansen Basin and stations in the New Jersey and Long Island shelf regions average 71 and 105 μmol m^{-2} h^{-1}, respectively (Rowe, 1978). Ammonium flux from sediments in the shelf off the mouth of the Changjiang River, East China Sea, ranged from -108 to 142 μmol m^{-2} h^{-1}, with an average of 29 μmol m^{-2} h^{-1} (Aller et al., 1985). Wide variation is observed in rates of benthic nitrogen regeneration in shelf ecosystems but it appears that benthic regeneration of nitrogen in most shelf ecosystems is about half that observed in shallow land margin systems such as estuaries (e.g., Boynton and Kemp, 1985; Nixon, 1981).

Fluxes of nutrients and dissolved oxygen across the sediment-water interface of the Louisiana shelf ecosystem are within the upper range of rates for most shelf ecosystems (Figure 4.10). These remineralization rates are based on measurements at 9 stations in the shelf region near the vicinity of the Mississippi River plume, during August 1987, April 1988, and April 1989. Rates of ammonium

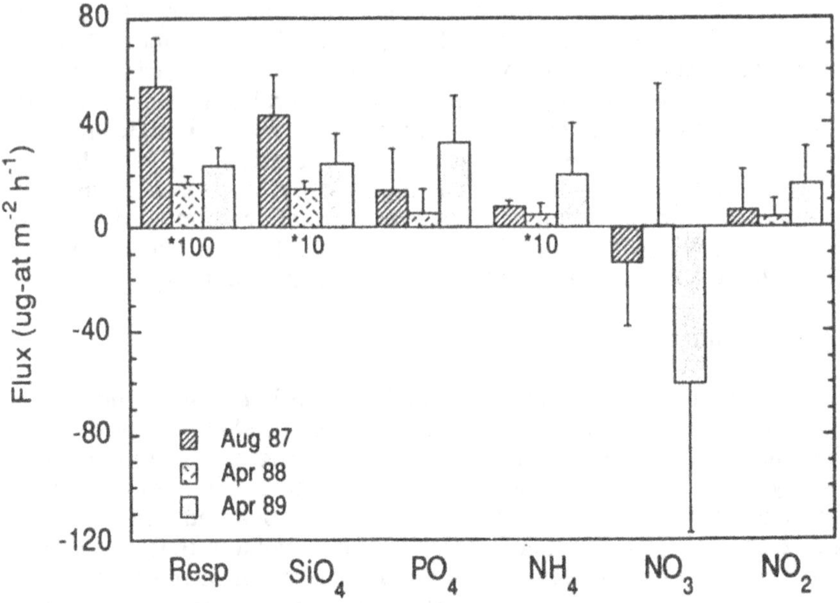

Figure 4.10. Rates of exchange of dissolved oxygen, silicate, phosphate, ammonium, nitrate, and nitrite across the sediment-water interface in a Louisiana shelf ecosystem.

regeneration ranged from 75-200 μmol m^{-2} h^{-1}. Lohrenz et al. (1990) have estimated phytoplankton demand for nitrogen (2.0 x 10^7 g N d^{-1}) based on models of primary productivity in a 1700 km^2 region of the Louisiana shelf in April 1988. Using a mean flux of 200 μmol m^{-2} h^{-1} of ammonium from sediments during this period, we calculate that benthic remineralization could contribute about 40% (0.8 x 10^7 g N d^{-1}) of the phytoplankton demand for nitrogen in this system. The contribution of sediments may be even higher in summer when river nutrient input is lower and benthic remineralization rates are higher. Thus in the Louisiana Shelf Ecosystem, sediments may be significant in sustaining high rates of primary productivity during the summer when allochthonous input is low.

Benthic nutrient remineralization is proportional to the quantity and quality of organic nutrients deposited to the benthos, and is influenced by in situ production rates (e.g., Hargrave, 1973; Klump and Martens, 1983) and by the density of animal burrows in sediments. A close association between riverine nutrient input and sediment regeneration has been observed in the Louisiana shelf. Large differences in remineralization rates, attributed to differences in river discharge, were observed between April 1988 and 1989

(Figure 4.10). River discharge was lower in 1988, resulting in less input of materials to the seabed. Regeneration rates of silicate, phosphate, and ammonium were all lower in 1988 than 1989, a near normal river discharge year (Figure 4.10). Based on models by Toth and Lerman (1977) and sedimentation rates ranging from 2-5 cm yr^{-1} on the Louisiana shelf (McKee, pers. commun.), ammonium regeneration rates should range from 51-128 $\mu mol\ m^{-2}\ h^{-1}$, similar to the range observed from direct measurements of sediment processes.

In river-dominated coastal environments, sedimentation of terrigenous materials and associated water column turbidity dilute the labile or reactive organic matter reaching bottom sediments and inhibit in situ production of particulate material in the overlying water column (DeMaster and Nittrouer, 1983; Aller et al., 1985). The reactivity (relative to decomposition) of bottom sediments in these environments is inversely proportional to sedimentation rate and rates of benthic nutrient remineralization may differ spatially due to the reactivity of deposited material (Blackburn, 1980; Aller et al., 1985). Thus maximum rates of ammonium and silicate benthic regeneration occurred off the shelf from the Changjiang River in areas of lower deposition. Rates of benthic regeneration of ammonium and silicate in the Louisiana shelf region are generally higher near the mouth of the Mississippi River, yet maximum flux is sometimes observed downstream from the river plume. Both ammonium and silicate show a significant correlation with pigment content in sediments, indicating the importance of the quality of material to rates of benthic regeneration.

Denitrification may also represent a loss of nutrients from marine ecosystems and reduce the potential of nitrogen regeneration. The theoretical ratio of particulate material in the water column according to Redfield (1958) is 106C:16N:1P. Stoichiometric signals of elements buried in the sediment that vary from this theoretical ratio can be used to model processes that selectively operate on nitrogen relative to phosphorus in pore waters (Grundmanis and Murray, 1977). Stoichiometric models of estuaries have concluded that benthic remineralization of organic matter yields inorganic nitrogen and phosphorus fluxes back to the water column which are low in nitrogen relative to phosphorus, and low in nitrogen relative to depositing organic matter. For example, material deposited to sediments in Narrangasett Bay had N:P ratios of 13.3, yet the ratio of sediment remineralization was 3.8 to 7.5 (Nixon, 1981). This shift in the N:P ratio of regenerated nutrients is caused by denitrification and significantly impacts the metabolism of marine ecosystems (Seitzinger et al., 1980). Yet the influence of denitrification and burial have not been interpreted relative to patterns of nitrogen regeneration from sediments.

Denitrification has not been directly measured in sediments of the Louisiana shelf. However, using stoichiometric models, the fluxes of dissolved oxygen, ammonium and phosphate across the sediment-water interface can be used to estimate this process. Based on the measured release of phosphate from sediment and a N:P ratio of 16, the expected flux of ammonium is 277 μmol m^{-2} h^{-1}. However, the measured rate is only 108 μmol m^{-2} h^{-1}, suggesting that nitrification is occurring at 169 μmol m^{-2} h^{-1}. Assuming that all except for the nitrite released from sediment is denitrified, and adding the sediment nitrate uptake rate, total denitrification is 185 μmol m^{-2} h^{-1} (86% of which is from coupled nitrification/ denitrification). The proximal portions of shelf systems in Louisiana are characterized by high concentrations of nitrate and here maximum rates of sediment uptake and denitrification should occur (Twilley and Kemp, 1986). From a mass balance perspective, this loss of nitrogen is nearly equivalent to that which is regenerated and represents a significant flux of nitrogen in this shelf ecosystem.

Fishes of the Northern Gulf of Mexico

Fishes of the northern Gulf of Mexico are generally considered by zoogeographers to be warm temperate (Carolinian), with close affinities to the ichthyofauna along the U. S. east coast from Cape Hatteras to about Cape Canaveral. There is considerable faunal change with depth and season, with tropical species being more common on the deeper portions of the continental shelf and juveniles of tropical species appearing throughout the northern Gulf during the warmer season (Briggs, 1974).

Some species are resident, relying upon resources within the northern gulf to successfully complete their life cycles, while other species are migratory and occur only seasonally in the northern gulf. Migratory species are able to take advantage of ephemeral resources that are seasonally available for adult feeding, or food and shelter of young stages.

Any of several ecological criteria (e.g., trophic position, habitat requirements, reproductive strategy, etc.) may be used to construct categories of estuarine dependent, coastal, reef and oceanic species.

Estuarine Dependent Species

The estuarine dependent group includes the Gulf menhaden, southern flounder, red drum, spot, Atlantic croaker, spotted sea trout and striped mullet. The life history pattern for this group typically involves spawning in neritic waters followed by transport of larvae and young juveniles to estuaries. Larval survival and recruitment to estuaries is dependent to a large extent upon oceanographic and meterological conditions (e.g., Nelson et al., 1977; Checkly et al., 1987). Many of these species feed at relatively low trophic levels in the estuaries, and tend to be more abundant than coastal species which frequently prey heavily upon them. However, important sources of variability in recruitment to the adult stock may also be present during the estuarine phase of the life history, when physical factors, food abundance and other habitat characteristics cause considerable variation in growth and mortality.

Perhaps the best studied species in this group is the Gulf menhaden, *Brevoortia patronus*. It has a relatively short food chain consisting on one or two trophic levels. Spawning takes place on the continental shelf during fall and winter (Shaw et al. 1985) and the pelagic larvae are transported into estuarine nursery areas in 6-10 wk (Deegan and Thompson, 1987). Juveniles and adults are filter feeders. The food repertoire changes with age from a predominantly herbivorous to a more omnivorous diet, with a wide size overlap (Friedland et al., 1984). Young larval Gulf menhaden feed on phytoplankton and zooplankton, especially dinoflagellates, tintinnids and copepod eggs, while older larvae eat larger zooplankton, including juvenile pelecypods, pteropods and copepod nauplii, copepodites and small adult copepods (Govoni et al., 1983). Juveniles may consume primarily detritus of marine macroplytes in estuaries (Lewis and Peters, 1984), and adults apparently feed on both large phytoplankton and zooplankton (Durbin, 1979). The extraordinary productivity of this species (it supports the largest fishery by weight in the U. S.) is partially accounted for by its short food chain, and partially by the highly productive environments, offshore and estuarine, in the northern Gulf of Mexico.

Coastal Species

The coastal group is comprised of a large number of ecologically and economically important species, including king mackerel, Spanish mackerel, bluefish, coastal herrings, little tuny, crevalle jack, blue runner, butterfish and hakes. The most important shared characteristic of this group is that members complete their life cycles in the pelagic or benthic realm associated with the continental shelf. Most are high trophic level predators, usually piscivores, as adults and juveniles. Most spawn pelagic eggs in marine waters, and although some species opportunistically inhabit or feed in estuaries, particularly as juveniles (e.g., bluefish and Spanish mackerel), none are strictly estuarine dependent. Principal spawning areas and factors that may determine reproductive success and ultimate recruitment are little known. Because spawning is in continental shelf waters, major oceanographic phenomena such as geostrophic current boundaries or eddies and riverine fronts may offer key habitats for successful reproduction and recruitment (e.g., Kirobe, 1986; Govoni et al., 1989).

The king mackerel is a good example of a coastal species in the northern Gulf of Mexico. Fish prey occurred in stomachs of over 95% of adult king mackeral collected off northwest Florida, Louisiana, and Texas and made up over 85% of the total volume of prey consumed (Saloman and Naughton, 1983). Generally, the most common prey were members of the estuarine dependent group. Among invertebrates, which accounted for up to 15% of the volume of prey consumed, the estuarine dependent penaeid shrimps were important. Juvenile king mackerel (100-400 mm FL) also consumed mostly fish prey (92% frequency of occurrence and 89% volume) that were mostly estuarine dependent species (Finucane et al., in press).

Early life stages of mackerel are rather exceptional in that they are almost entirely piscivorous. Larvae and small juveniles (3-23 mm SL) consumed mostly carangids (jacks), sciaenids (croakers), engraulids (anchovies), and clupeids (herrings). Similar to adults, the prey families are mostly estuarine dependent groups although a coastal group (Carangidae) is the most frequently consumed (Finucane et al., in press).

Locations where prey are concentrated could be important microhabitats offering conditions favoring growth, survival and recruitment of larvae of coastal species. For example, ichthyoplankton and macrozooplankton are concentrated in the frontal waters of the Mississippi River discharge plume (Govoni et al., 1989), and larval and small juvenile king mackerel appear to grow faster in the vicinity of the plume (DeVries and Grimes, 1989). Faster growth can lead to greater survival and recruitment provided survival gains due to faster growth are not offset by increased mortality from predation.

Reef Species

Reef fish occupy hard substrates scattered across the continental shelf from northwest Florida to Texas. Reef fauna constitute the most speciose and ecologically complex of marine communities. For example, continental reef fish communities off North Carolina and South Carolina include 200 species (Grimes et al., 1982).

Reef communities occur in clear shelf waters on outcropped sedimentary rocks richly overgrown by sponges, sea fans, soft and hard corals, and hermatypic corals. Most complete their life cycle on the shelf, with juveniles frequently found on the shallower portion of the shelf (Grimes et al., 1982). Some evidence suggests that a few species (e.g., gag and grey snapper) may be estuarine dependent (Keener et al., 1988).

Adults of the red snapper, *Lutjanidae campechanus* are high trophic level carnivores. Snapper larvae are rarely encountered in plankton surveys and little is known about the ecology of the young. Spawning is thought to occur over reefs and rock outcroppings (Grimes, 1987) of the mid- to outer-continental shelf (Powles, 1977; Houde et al., 1979). It is not clear if larvae are retained near their natal reefs by eddy-like circulation or are transported into and out of the area by major ocean currents (Powles, 1977).

Diet and feeding ecology of red snapper larvae are unkown although Richards and Saksena (1980) successfully reared larvae of *L. griseus* in the laboratory on a diet consisting of 35-73 μm and 73-110 μm wild zooplankton.

Once juvenile red snapper settle they begin to prefer benthic foods associated with the reef or nearby flats. Juveniles began to eat larger demersal prey by 100 mm FL (Bradley and Bryan, 1976). Shrimp occur in the diet of juveniles year round (Camber, 1955). Squids and octopus, as well as a wide variety of other mollusks, crustaceans and fishes are reported in the diet of juvenile snappers. Adults become exclusively benthic carnivores, feeding upon shrimp, small reef fish, crabs and gastropods, and occasionally tunicates (e.g., Bradley and Bryan, 1976; Futch and Bruger, 1976).

Oceanic Species

The oceanic group of fishes consists of large pelagic species that are top carnivores of oceanic ecosystems. Long range migrations, some even transoceanic, are common to all members. Fast growth, especially during early life, is typical. Many species school as juveniles and young adults but become solitary at older ages. Many oceanic species probably congregate for spawning along prominent oceanographic features (e.g., oceanic fronts) where conditions promote higher productivity, providing food for both adults and larvae alike. The principal fishes in this group are the tunas including bluefin, yellowfin, albacore, bigeye, and skipjack tuna, billfish, swordfish including blue and white marlin and sailfish, and sharks.

Oceanic species could be excluded from this discussion because their life cycles are thought to be largely restricted to oceanic ecosystems. While this is accurate for most members of this group, the yellowfin tuna, *Thunnus albacares*, may be an exception. There is little or no evidence from adults that spawning occurs in the Gulf of Mexico (Erdman, 1968; Goldberg and Herring-Dyal, 1981) and until recently few young stages (<50) had been collected in the Gulf of Mexico (e.g., Klawe and Schimada, 1959; Kelley et al., 1985). However, in 1987, 978 and 131 larvae provisionally identified as yellowfin tuna (Grimes et al., unpublished data; Shaw, unpublished data) were collected during research cruises in the discharge plume of the Mississippi River. Preliminary analyses of these data show that the most and largest larvae were associated with frontal waters of the

Mississippi discharge plume and that larvae were consuming mainly zooplankton prey (Grimes et al., unpublished data). These results suggest that at least this one oceanic species has an important connection to coastal processes and trophic pathways.

Adult yellowfin tuna are high trophic level carnivores on both vertebrate and invertebrate prey; invertebrates (85%) and fishes (77%) occur in the diet in about equal frequencies (Manooch and Mason, 1983). Invertebrate prey were mainly cephalopods and crustaceans; the latter being most often represented by large larvae such as megalopae. Overall diets suggest they are fast aggressive predators that also use their gill apparatus to strain small near-surface organisms from the sea.

Summary

Biological productivity in the northern Gulf of Mexico is dominated by discharge from the Mississippi River system. Annual discharge is approximately 10% of the volume of water on the entire Louisiana/Texas shelf, and much of this discharge is transported along shore in a westerly direction. High nutrient concentrations, for example >100 μmol l^{-1} NO_3, in river water in conjunction with the apparent small impact of nutrients from slope/shelf exchanges, further increase the biological impact of the river.

Waters of the northern Gulf of Mexico are characterized by strong fronts: at the boundaries between river plumes and receiving waters; at the boundary between the Coastal Boundary Layer and the mid-shelf water; and at the boundary between the waters of the outer shelf and the open Gulf. Cross frontal exchanges are not well understood.

Phytoplankton production and stocks within the region of the river plumes are high, often >5 mg C m^{-2} d^{-1} and >300 mg Chl m^{-2}, but models incorporating only nutrients and light indicate stocks are further limited by some combination of grazing, sinking, and advection.

Preliminary summertime measurements of microzooplankton grazing indicate it is a significant source of mortality for phytoplankton; between 35% and 60% of phytoplankton stock d^{-1}. Furthermore, measurements of grazing by the macrozooplankton community indicate an additional 20-30% of the stock is taken by these organisms on some occasions. Fall and winter grazing rates appear to be lower.

There is little available information on sinking fluxes but deposition rates near the Mississippi River Delta are high, between 1 and 2 cm mo^{-1}. Long-term burial rates are about 10% of deposition rates indicating significant resuspension, transport or remineralization.

Measured nutrient fluxes from the bottom to the water column near the Mississippi River Delta are high.

Fisheries production in the northern Gulf of Mexico is high. For example, the region supports the largest volume fishery in the U. S., the gulf menhanden *Brevoortia patronus*. Recruitment is apparently enhanced by the Mississippi River plume fronts and by the abundance of estuarine habitat in this region. Coastal, reef, and even some oceanic fish species are also abundant in the northern Gulf of Mexico.

Acknowledgments

The following are gratefully acknowledged for their support of portions of the work presented in this paper: Grant No. 86-LUM(1)-083-13 from the Louisiana Board of Regents, the Louisiana Universities Marine Consortium, the University of Texas, the Coastal Science Program of the Office of Naval Research in Arlington, Virginia, and ONR Contract No. N0014-88-K-0155. This document is contribution No. USM.CMS 119.

References

Aller, R. C., and J. E. Mackin. 1984. Preservation of reactive organic matter in marine sediments. Earth Planetary Sci. Letters 70:260-266.

Aller, R. C., J. E. Mackin, W. J. Ullman, C. H. Wang, S. M. Tsai, J. C. Jin, Y. N. Sui, and J. Z. Hong. 1985. Early chemical diagenesis, sediment-water solute exchange, and storage of reactive organic matter near the mouth of the Chang Jiang, East China Sea. Cont. Shelf Res. 4:227-251.

Azam, F., T. Fenchel, J. G. Field, J. S. Gray, L. A. Meyer-Reil, and F. Thingstad. 1983. The ecological role of water column microbes in the sea. Mar. Ecol. Prog. Ser. 10:257-263.

Billen, G. 1978. A budget of nitrogen recycling in North Sea sediments off the Belgian Coast. Estuar. Coast. Mar. Sci. 7:127-146.

Blackburn, T. H. 1980. Seasonal variations in the rate of organic-N mineralization in anoxic sediments. Pp. 173-183. *In* Biogeochimie de la matiere organique a l'interface eau-sediment marin, CNRS, Paris.

Boynton, W. R., and W. M. Kemp. 1985. Nutrient regeneration and oxygen consumption by sediments along an estuarine salinity gradient. Mar. Ecol. Prog. Ser. 23:45-55.

Boynton, W. R., W. M. Kemp, and C. G. Osborne. 1982. Nutrient fluxes across the sediment-water interface in the turbid zone of a coastal plain estuary. Pp. 93-109. *In* V. Kennedy (Ed.) Estuarine perspectives. Academic Press, New York.

Bradley, E., and C. E. Bryan. 1975. Life history and fishery of the red snapper (*Lutjanus campechanus*) in the northwestern Gulf of Mexico: 1970-1974. Proc. Gulf Caribb. Fish. Inst. 27:77-106.

Briggs, J. C. 1974. Marine zoogeography. McGraw-Hill Book Co., New York. 475 Pp.

Camber, C. I. 1955. A survey of the red snapper fishery of the Gulf of Mexico, with special reference to the Campeche Banks. Fla. Board Conserv. Mar. Res. Lab. Tech. Ser. 12. 64 Pp.

Checkley, D. M., Jr., G. A. Maillet, and K. M. Mason. 1987. Winter weather, the Gulf Stream, and the Atlantic menhaden. Annual Meeting Am. Fish. Soc., Winston Salem, NC. Pp.45 (abstract only).

Chuang, W.-S., and W. J. Wiseman, Jr. 1983. Coastal sea level response to frontal passages on the Louisiana-Texas shelf. J. Geophys. Res. 88:2615-2620.

Chuang, W.-S., W. W. Schroeder, and W. J. Wiseman, Jr. 1982. Summer current observations off the Alabama coast. Contr. Mar. Sci. 25:121-131.

Cloern, J. E. 1987. Turbidity as a control on phytoplankton biomass and productivity in estuaries. Cont. Shelf Res. 7:1367-1381.

Cochrane, J. D., and F. J. Kelly. 1986. Low-frequency circulation on the Texas-Louisiana continental shelf. J. Geophys. Res. 91:10645-10659.

Dagg, M. J. 1988. Physical and biological responses to the passage of a winter storm in the coastal and inner shelf waters of the northern Gulf of Mexico. Cont. Shelf Res. 8:167-178.

Dagg, M. J., and W. E. Walser. Patterns of egg production in the copepod *Acartia tonsa* Dana in a shallow estuary. Mar. Biol. (submitted).

Dagg, M. J., and T. E. Whitledge. 1991. Concentrations of copepod nauplii associated with the nutrient-rich plume of the Mississippi River. Cont. Shelf Res. (in press).

Dagg, M. J., P. B. Ortner and F. Al-Yamani. 1987. Winter-time distribution and abundance of copepod nauplii in the northern Gulf of Mexico. Fish. Bull., U.S. 86: 319-330.

Deegan, L. A., and B. A. Thompson. 1987. Growth rate and life history events of young-of-the-year Gulf menhaden as determined from otoliths. Trans. Am. Fish. Soc. 116: 663-667.

DeMaster D. J., B. A. McKee, C. A. Nittrouer, J. Qian, and G. Cheng. 1985. Rates of sediment accumulation and particle reworking based on radiochemical measurements from continental shelf deposits in the East China Sea. Cont. Shelf Res. 4:143-155.

DeMaster, K. J., and C. A. Nittrouer. 1983. Uptake, dissolution, and accumulation of silica near the mouth of the Changjiang River.

Pp. 235-240. *In* Proceedings of the International Symposium in Sedimentation on the Continental Shelf, with special reference to the east China Sea. China Ocean Press, Beijing.

DeVries, D. A., and C. V. Grimes. 1989. Aging of king mackerel, *Scomberomorus cavalla*, from the U.S. South Atlantic and Gulf of Mexico, 1988-1989. Unpublished MS, NOAA, Nat. Mar. Fish. Serv., Panama City, FL, 22 Pp.

Dinnel, S. P. 1988. Circulation and sediment dispersal on the Louisiana-Misissippi-Alabama continental shelf. Ph.D. dissertation, Louisiana State Univ., Baton Rouge. 173 Pp.

Dinnel, S. P., and W. J. Wiseman, Jr. 1986. Freshwater on the Louisiana and Texas shelf. Cont. Shelf Res. 6:765-784.

Drennan, K. L. 1968. Hydrographic studies in the northeast Gulf of Mexico. Gulf South Res. Inst. 110 Pp.

Durbin, A. G. 1979. Food selection by plankton feeding fishes. Pp. 203-218. *In* H. Clepper (Ed.) Predator-prey systems in fisheries management. Sport Fishing Inst., Washington, DC.

Erdman, D. S. 1968. Spawning seasons of some game fishes around Puerto Rico. Int. Oceanogr. Found., Twelfth Annual Int. Gamefish Conf., Nov. 17-18, 1967, Pp. 11-19.

El Sayed S. Z. 1972. Primary productivity and standing crop of phytoplankton. Pp. 8-13. *In* V. C. Bushnell (Ed.) Chemistry, primary productivity, and benthic algae of the Gulf of Mexico. Am. Geogr. Soc., New York.

Eppley, R. W. 1972. Temperature and phytoplankton growth in the sea. Fish. Bull., U.S. 70:1063-1085.

Fahnenstiel, G. L., and D. Scavia. 1987. Dynamics of Lake Michigan phytoplankton: Primary production and growth. Can. J. Fish. Aquat. Sci. 44:499-508.

Filardo, M. J., and W. N. Dunstan. 1985. Hydrodynamic control of phytoplankton in low salinity waters of the James River Estuary, Virginia, U.S.A. East Coast. Shelf Sci. 21:653-667.

Finucane, J. H., C. G. Grimes, and S. P. Naughton. In press. Diet of larval and juvenile king mackerel, *Scomberomorus cavalla*, and Spanish mackerel, *S. maculatus*, from the U.S. South Atlantic Bight and Gulf of Mexico. N.E. Gulf Sci.

Fitzwater S. E., G. A. Knauer, and J. H. Martin. 1982. Metal contamination and its effect on primary production measurements. Limnol. Oceanogr. 27:544-551.

Fransz, H. G. 1986. Effects of fresh water inflow on the distribution, composition and production of plankton in the Dutch coastal waters of the North Sea. Pp. 241-249. *In* S. Skreslet (Ed.) The role of freshwater outflow in coastal marine ecosystems. Springer-Verlag, Heidelberg.

Friedland, K. D., L. W. Haas, and J. W. Merriner. 1984. Filtering rates of the juvenile Atlantic menhaden, *Brevoortia tyrannus*

(Pisces: Clupeidae), with consideration of the effects of detritus and swimming speed. Mar. Biol. 84:109-117.

Fucik, K. W. 1974. The effect of petroleum operations on the phytoplankton ecology of the Louisiana coastal waters. Masters Thesis. Texas A&M Univ., College Station.

Futch, R. B., and G. E. Bruger. 1976. Age, growth and reproduction of red snapper in Florida waters. Pp. 165-185. *In* H. R. Bullis, Jr. and A. C. Jones (Eds.) Proceedings, Colloquium on snapper-grouper fishery resources of the western central Atlantic Ocean. Florida Sea Grant Coll., Rep. 17.

Gifford, D. J., and M. J. Dagg. 1988. Feeding of the estuarine copepod *Acartia tonsa* Dana: carnivory vs. herbivory in natural microplankton assemblages. Bull. Mar. Sci. 43:458-468.

Goldberg, S. R., and H. Herring-Dyal. 1981. Histological gonad analyses of late summer-early winter collections of bigeye tuna, *Thunnus obesus*, and yellowfin tuna, *T. albacares*, from the northwest Atlantic and the Gulf of Mexico. U.S. Dep. Commer., NOAA Tech. Memo. NMFS-SWFC 14, 9 Pp.

Goldman, J. C., D. A. Caron, O. K. Andersen, and M. R. Dennett. 1985. Nutrient cycling in a microflagellate food chain. I. Nitrogen dynamics. Mar. Ecol. Prog. Ser. 24:243-254.

Govoni, J. J., D. E. Hoss, and A. J. Chester. 1983. Comparative feeding of three species of larval fishes in the northern Gulf of Mexico: *Brevoortia patronis*, *Leistomous xanthurus*, and *Micropogonias undulatus*. Mar. Ecol. Prog. Ser. 13:189-199.

Govoni, J. J., D. E. Hoss, and D. R. Colby. 1989. The spatial distribution of larval fishes about the Mississippi River plume. Limnol. Oceanogr. 34:178-187.

Grimes, C. B. 1987. Reproductive biology of the Lutjanidae: a review. Pp. 239-294. *In* J. J. Polovina and S. Ralston (Eds.) Tropical snappers and groupers: biology and fisheries management. Westview Press, Inc., Boulder.

Grimes, C. B., C. S. Manooch, and G. R. Huntsman. 1982. Reef and rock outcropping fishes of the outer-continental shelf of North Carolina and South Carolina, and ecological notes on the red porgy and vermillion snapper. Bull. Mar. Sci. 32:277-289.

Grundmanis, V., and J. W. Murray. 1977. Nitrification and denitrification in marine sediments from Puget Sound. Limnol. Oceanogr. 2:804-813.

Hargrave, B. R. 1973. Coupling carbon flow through some pelagic and benthic communities. J. Fish. Res. Board Can. 30:1317-1326.

Harrison, W. G., D. Douglas, P. Falkowski, G. Rowe, and J. Vidal. 1983. Summer nutrient dynamics of the mid-Atlantic Bight: Nitrogen uptake and regeneration. J. Plankton Res. 5:539-556.

Hart, W. E., and S. P. Murray. 1978. Energy balance and wind effects in a shallow sound. J. Geophys. Res. 83:4097-4206.

Hedges J. I., W. A. Clark, and G. L. Cowie. 1988. Fluxes and reactivities of organic matter in a coastal marine bay. Limnology and Oceanography 33:1137-1152.

Heron, R. C., T. D. Leming, and J. Li. 1989. Satellite-detected fronts and butterfish aggregations in the northern Gulf of Mexico. Cont. Shelf Res. 9:569-588.

Hopkins, T. L., D. M. Milliken, L. M. Bell, E. J. McMichael, J. J. Heffernan, and R. V. Cano. 1981. The landward distribution of oceanic plankton and micronekton over the west Florida continental shelf as related to their vertical distribution. J. Plankon. Res. 3:645-658.

Houde, E. D., J. C. Leak, C. E. Dowd, S. A. Berkeley, and W. J. Richards. 1979. Ichthyoplankton abundance and diversity in the eastern Gulf of Mexico. Rep. Bur. Land Manage. Contract AA350-CT7-28. NTIS-PB-299-839, 546 Pp.

Huh, O. K., W. J. Wiseman, Jr., and L. J. Rouse, Jr. 1981. Intrusion of Loop Current water onto the West Florida continental shelf. J. Geophys. Res. 86:4186-4192.

Jackson, R., P. J. le B. Williams, and I. R. Joint. 1987. Freshwater phytoplankton in the low salinity region of the river Tamor estuary. East Coast. Shelf Sci. 25:299-311.

Kamykowski, D., and J. L. Bird. 1981. Phytoplankton associations with the variable nepheloid layer on the Texas continental shelf. East Coast. and Shelf Sci. 13:317-326.

Keener, P., C. D. Johnson, D. W. Stender, E. D. Brothers, and H. R. Beatty. 1988. Ingress of post larval gag, *Mycteroperca microlepis*, (Pisces; Serranidae), through a South Carolina barrier island inlet. Bull. Mar. Sci. 42:376-396.

Kelley, S., T. Potthoff, W. J. Richards, L. Ejsymont, and J. W. Gartner. 1985. SEAMAP 1983--ichthyoplankton. U.S. Dep. of Commer., NOAA Tech. Memo. NMFS-SEFC-167, 78 Pp.

Kirobe, T. 1986. Distribution and abundance of herring larvae in relation to North Sea oceanographic fronts. Am. Fish. Soc., Early Life History Sec., Annual Meeting, Miami, FL. (abstract only).

Klawe, W. L., and B. M. Schimada. 1959. Young scombroid fishes from the Gulf of Mexico. Bull. Mar. Sci. Gulf and Carib. 9:100-115.

Klump, J. V., and C. S. Martens. 1983. Benthic nitrogen regeneration. Pp. 411-457. *In* E. J. Carpenter and D. G. Capone (Eds.), Nitrogen in the marine environment. Academic Press, New York.

Lesley, D. E. 1977. A field and laboratory investigation of the salinity and temperature tolerance of an estuarine calanoid copepod *Acartia tonsa* Dana. Ph.D. dissertation. Tulane Univ., New Orleans. 124 Pp.

Lewis, V. P., and D. S. Peters. 1984. Menhaden-a single step from vascular plant to fishery harvest. J. Exp. Mar. Biol. Ecol. 84: 95-100.

Lohrenz, S. E., M. J. Dagg, and T. E. Whitledge. 1990. Enhanced primary production at the plume/oceanic interface of the Mississippi River. Cont. Shelf Res. 10:639-664.

Malcolm, R. L., and W. H. Durum. 1976. Organic carbon and nitrogen concentrations and annual organic load of six selected rivers of the United States. USGS Water Supply Paper 1817-F.

Manooch, C. S., and D. L. Mason. 1983. Comparative food studies of yellowfin tuna, *Thunnus albacares*, and the southeastern and Gulf coast of the United States. Brimleyana 9:33-52.

McGrail, D. W., and M. Carnes. 1983. Shelfedge dynamics and the nepheloid layer in the northwest Gulf of Mexico. SEPM Special Pub. No. 33, Pp. 251-264.

McIlwain, T. D. 1968. Seasonal occurrence of the pelagic Copepoda in Mississippi Sound. Gulf. Res. Rep. 2:257-270.

McKee, B. A., C. A. Nittrouer, and D. J. DeMaster. 1983. Concepts of sediment deposition and accumulation applied to the continental shelf near the mouth of the Yangtze River. Geology 11:631-633.

McKee, B. A., J. G. Booth, and P. W. Swarzenski. 1990. The fate of particulates and particle-reactive constituents in the Mississippi River/ocean mixing zone. Trans. Am. Geophys. Union 71:71.

Miller, R. L., and D. L. Kamykowski. 1986a. Effects of temperature, salinity, irradiance and the diurnal periodicity on growth and photosynthesis in the diatom *Nitzchia americana*: Light-limited growth. J. Plankton Res. 8:215-228.

Miller, R. L., and D. L. Kamykowski. 1986b. Short-term photosynthetic responses in the diatom *Nitzchia americana* to a simulated salinity environment. J. Plankton Res. 8:305-315.

Murrell, M., and M. Dagg. 1987. LaSER Oceanography: Data Report No. 1: CTD and hydrographic data. Louisiana Universities Marine Consortium Data Rep. No. 4.

Nelson, W. R., M. C. Ingham, and W. E. Schaaf. 1977. Larval transport and year class strength of Atlantic menhaden, *Brevoortia tyrannus*. Fish. Bull., U.S. 75:23-41.

Nelson, T. A., and J. H. Trefry. 1986. Pollutant-particle relationships in the marine environment: A study of particulates and their fate in a major river-delta shelf system. Rapp. P.-v. Reun. Cons. int. Explor. Mer 186:115-127.

Nittrouer, C. A., D. J. DeMaster, S. A. Kuehl, and B. A. McKee. 1987. Association of sand with mud deposits accumulating on continental shelves. Pp. 17-27. *In* R. J. Knight and J. R. McLean (Eds.) Shelf sands and sandstones. Can. Soc. Petroleum Geologists, Calgary Canada.

Nixon, S. W. 1981. Remineralization and nutrient cycling in coastal marine ecosystems. Pp. 111-138. *In* B. J. Neilson and L. E. Cronin (Eds.) Estuaries and nutrients. Humana Press, Clifton, NJ.

Nowlin, W. D., Jr., and C. A. Parker. 1974. Effects of a cold-air outbreak on shelf waters of the Gulf of Mexico. J. Phys. Oceanogr. 4:467-486.

Ortner, P. B., L. C. Hill, and S. R. Cummings. 1989. Zooplankton community structure and copepod species composition in the northern Gulf of Mexico. Cont. Shelf Res. 9:387-402.

Ortner, P. B., G. L. Hitchcock, R. L. Cuhel, and T. A. Nelsen. Photoautotrophic production and particulate distribution near the Mississippi River Delta. Cont. Shelf Res. (submitted).

Pennock, J. R. 1985. Chlorophyll distributions in the Delaware estuary: Regulation by light-limitation. East Coast. Shelf Sci. 21:711-725.

Pennock, J. R., and J. H. Sharp. 1986. Phytoplankton production in the Delaware estuary: temporal and spatial variability. Mar. Ecol. Progr. Ser. 34:143-155.

Platt, T., and D. V. Subba Rao. 1970. Primary production measurements on a natural phytoplankton bloom. J. Fish. Res. Board Can. 27:887.

Powles, H. 1977. Larval distributions and recruitment hypotheses for snappers and groupers of the South Atlantic Bight. Proceedings, Annual Conference, Southeastern Assoc. Fish Wildl. Agencies. 31:362-371

Rabalais, N. N., R. E. Turner, W. J. Wiseman, Jr., and D. F. Boesch. In press. A Brief summary of hypoxia on the northern Gulf of Mexico continental shelf: 1985-1988. *In* R. V. Tyson (Ed.) Modern and ancient continental shelf hypoxia. Geol. Soc. Lon.

Redfield, A. C. 1958. The biological control of chemical factors in the environment. Am. Sci. 46:205-221.

Richards, W. J., and V. P. Saksena. 1980. Description of larvae and early juveniles of laboratory-reared gray snapper, *Lutjanus griseus* (Linnaeus) (Pisces: Lutjanaidae). Bull. Mar. Sci. 30:515-521.

Riley, G. A. 1937. The significance of the Mississippi River drainage for biological conditions in the northern Gulf of Mexico. J. Mar. Res. 1:60-74.

Rowe, G. T. 1978. Benthic nutrient regeneration and high rate of primary production in continental shelf waters: Rowe replies. Nature 274:189-190.

Rowe, G. T., C. H. Clifford, K. L. Smith, and P. L. Hamilton. 1975. Benthic nutrient regeneration and its coupling to primary productivity in coastal waters. Nature 255: 215-217.

Saloman, C. H., and S. P. Naughton. 1983. Food of king mackerel, *Scomberomorus cavalla*, from the southeastern United States including the Gulf of Mexico. U.S. Dep. of Commer. NOAA Tech. Memo. NMFS-SEFC-126, 25 Pp.

Sathyendranath, S., L. Prieur, and A. Morel. 1989. A three-component model of ocean colour and its application to remote sensing of phytoplankton pigments in coastal waters. Int. J. Remote Sens. 10:1373-1394.

Schroeder, W. W., and W. J. Wiseman, Jr. 1986. Low-frequency shelf-estuarine exchange processes in Mobile Bay and other estuarine systems on the northern Gulf of Mexico. Pp. 355-367. *In* D. Wolfe (Ed.) Estuarine variability, Academic Press, New York.

Schroeder, W. W., S. P. Dinnel, W. J. Wiseman, Jr., and W. J. Merrel, Jr. 1987. Circulation patterns inferred from the movement of detached buoys in the eastern Gulf of Mexico. Cont. Shelf Res. 7:883-894.

Schroeder, W. W., O. K. Huh, L. J. Rouse, Jr., and W. J. Wiseman, Jr. 1985. Satellite observations of the circulation east of the Mississippi Delta: Cold-air outbreak conditions. Remote Sens. Environ. 18:49-58.

Seitzinger, S., S. C. Nixon, M.E.Q. Pilson, and S. Burke. 1980. Denitrification and N_2O production in near-shore marine sediments. Geochim. Cosmochim. Acta. 44:1853-1860.

Shaw, R. F., W. F. Wiseman, Jr., L. E. Turner, L. R. Rouse, and R. E. Condrey. 1985. Transport of larval Gulf menhaden, *Brevoortia patronus*, in continental shelf waters of western Louisiana: A hypothesis. Trans. Am. Fish. Soc. 114:452-460.

Shiller, A. M., and E. A. Boyle. 1987. Variability of dissolved trace metals in the Mississippi River. Geochim. Cosmochim. Acta 51:3273-3277.

Skjoldal, H. R., and C. Lannergren. 1978. The spring phytoplankton bloom in Lindaspollene, a land-locked Norwegian Fiord. II. Biomass and activity of net- and nanoplankton. Mar. Biol. 47: 313-323.

Sklar, F. H., and R. E. Turner. 1981. Characteristics of phytoplankton production off Barataria Bay in an area influenced by the Mississippi River. Contr. Mar. Sci. 24:93-106.

Smayda, T. J. 1969. Experimental observations on the influence of temperature, light and salinity on cell division of the marine diatom *Detonula confervacea* (Cleve) Gran. J. Phycol. 5:150-157.

Suess, E. 1980. Particulate organic carbon flux in the oceans-surface productivity and oxygen utilization. Nature 288:260-263.

Sunda, W., and R.R.L. Guillard. 1976. The relationship between cupric ion activity and the toxicity of copper to phytoplankton. J. Mar. Res. 24:511-529.

Thomas, W. H., and E. G. Simmons. 1960. Phytoplankton production in the Mississippi Delta. Pp. 103-116. *In* F. Shepard (Ed.) Recent sediments, Northwest Gulf of Mexico, American Association of Petrologists, Tulsa, OK.

Titman, D., and P. Kilham. 1976. Sinking in freshwater phytoplankton: Some ecological implications of cell nutrient status and physical mixing processes. Limnol. Oceanogr. 21:409-417.

Toon, R., and M. Dagg. 1989. LaSER Oceanography: Data Report No. 3: CTD and hydrographic data. Louisiana Universities Marine Consortium Data Rep. No. 10.

Toth, D. J., and A. Lerman. 1977. Organic matter reactivity and sedimentation rates in the ocean. Am. J. Sci. 277:465-485.

Turner, J. T. 1987. Zooplankton feeding ecology: contents of fecal pellets of the copepod Centropages velificatus from waters near the mouth of the Mississippi River. Biol. Bull. 173:377-386.

Turner, R. E., and R. L. Allen. 1982. Plankton respiration rates in the bottom waters of the Mississippi River delta bight. Contrib. Mar. Sci. 25:173-179.

Turner, R. E., R. Kaswadji, N. N. Rabalais, and D. F. Boesch. 1987. Long-term changes in the Mississippi River water quality and its relationship to hypoxic continental shelf waters. *In* Estuarine and Coastal Management--Tools of the Trade. Proceedings, 10th National Conference of the Coastal Society, 12-15 October, 1986, New Orleans.

Twilley, R., and W. M. Kemp. 1986. The relation of denitrification potentials to selected physical and chemical factors in sediments of Chesapeake Bay. Pp. 277-294. *In* D. A. Wolfe (Ed.), Estuarine variability. Academic Press, New York.

Walsh, J. J., D. A. Dieterle, M. B. Meyers, and F. Muller-Karger. In press. Nitrogen exchange at the continental margin: A numerical study of the Gulf of Mexico.

Walsh, J. J., T. E. Whitledge, F. W. Barvenick, C. D. Wirick, S. O. Howe, W. E. Esaias, and J. T. Scott. 1978. Wind events and food chain dynamics within the New York Bight. Limnol. Oceanogr. 23:659-683.

Warlen, S. M. 1988. Age and growth of larval gulf menhaden, *Brevoortia patronus*, in the northern Gulf of Mexico. Fish. Bull., U.S. 86:77-90.

Welschmeyer N. A., and C. J. Lorenzen. 1985. Chlorophyll budgets: zooplankton grazing and phytoplankton growth in a temperate fjord and the Central Pacific Gyres. Limnol. Oceanogr. 30:1-21.

Wiseman, W. J., Jr., and S. P. Dinnel. 1988. Shelf currents near the mouth of the Mississippi River. J. Phys. Oceanogr. 18:1287-1291.

Wiseman, W. J., Jr., J. M. Bane, S. P. Murray, and M. W. Tubman. 1976a. Small-scale temperature and salinity structure over the inner shelf west of the Mississippi River Delta. *In* J. C. J. Nihoul (Ed.) Memoires de la Societe Royale de Sciences de Liege, Belgium X:277-285.

Wiseman, W. J., Jr., J. M. Bane, S. P. Murray, and M. W. Tubman. 1982. Physical environment of the Louisiana Bight. Contr. Mar. Sci. 25:109-120.

Wiseman, W. J., Jr., L. D. Wright, L. J. Rouse, Jr., and J. M. Coleman. 1976b. Periodic phenomena at the mouth of the Mississippi River. Contr. Mar. Sci. 20:11-32.

Wofsy, S. C. 1983. A simple model to predict extinction coefficients and phytoplankton biomass in eutrophic waters. Limnol. Oceanogr. 28:1144-1155.

Xiuren, N. D. Vaulot, L. Zhensheng, and L. Zilin. 1988. Standing stock and production of phytoplankton in the estuary of the Changjiang (Yangtse) River and the adjacent East China Sea. Mar. Ecol. Prog. Ser. 49:141-150.

Zeitzschel, B. 1980. Sediment water interactions in nutrient dynamics. Pp. 195-218. *In* K. R. Tenore and B. C. Coull (Eds.), Marine benthic dynamics. Univ. South Carolina Press, Columbia.

M. P. Sissenwine and E. B. Cohen

5. Resource Productivity and Fisheries Management of the Northeast Shelf Ecosystem

Introduction

This symposium volume is about the relationship between food chains and management of large marine ecosystems, including fisheries. The conveners of the Symposium pose the question whether significant advances in the understanding of food chain dynamics have or should influence fisheries management. This paper examines the situation for the Northeast Shelf Ecosystem, with emphasis on Georges Bank. It provides a brief description of fishery resource productivity (i.e., net rate of change in biomass resulting from growth, reproduction, and natural mortality) as well as energy flow within the fish community. It also describes the biological basis for management of the important multispecies demersal fishery resource and how the fishery has responded to management. Lastly, it identifies potential future courses of action (based on both ecological and economic considerations) that should be considered to improve the resource situation and the fishery.

The paper is brief because most of the important information considered here is published elsewhere. Our intent in identifying potential future courses of action is to be thought-provoking, not comprehensive.

Productivity

The Northeast Shelf ecosystem has unusually high primary productivity that supports high productivity at all trophic levels (Sissenwine et al., 1984; Cohen and Grosslein, 1987). O'Reilly et al. (1987) reported levels of primary productivity in excess of 500 g C m^2 yr^{-1} along a narrow coastal band of the Mid-Atlantic region, which is influenced by nutrient-rich estuarine input. The well-mixed portion of Georges Bank has the highest offshore level of primary

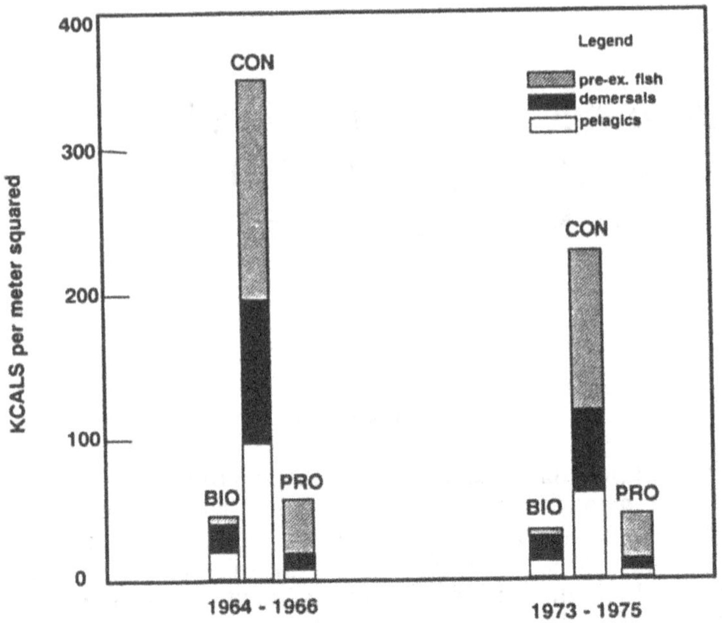

Figure 5.1. Biomass, production, and consumption of finfish on Georges Bank (after Sissenwine, 1986)

productivity, 455 g C m^{-2} yr^{-1}. The lowest levels of primary productivity--e.g., Gulf of Maine--are about 265 g C m^{-2} yr^{-1}. Overall primary productivity on Georges Bank is about 350 g C m^{-2} yr^{-1}, or about 4000 K cal/m^{-2} yr^{-1} (using 11.5 K cal/g C).

Energy flow through the Georges Bank fish community was reported (Sissenwine et al., 1984; Sissenwine, 1986; Figure 5.1) for two periods, mid-1960s and mid-1970s. The former was a period of high abundance. Biomass was reduced in the latter period following a decade of intense fishing, primarily by foreign vessels.

Figure 5.1 indicates the importance of pre-recruit fish (generally less than 1 yr old). Although they account only for 10% of the biomass, they are responsible for about half the consumption by fish and more than half the production. Fish production is about 1.7% (average of the two time periods) of primary production, which is surprisingly high considering that fish feed at the third to sixth trophic level (i.e., if they feed at the fourth trophic level, on average, that would imply an ecological efficiency of 36%). Cohen and Grosslein (1987) compared the multilevel trophic efficiency (phytoplankton to fish) of Georges Bank to other systems. They found that it was higher than several other ecosystems (e.g., 0.8% for

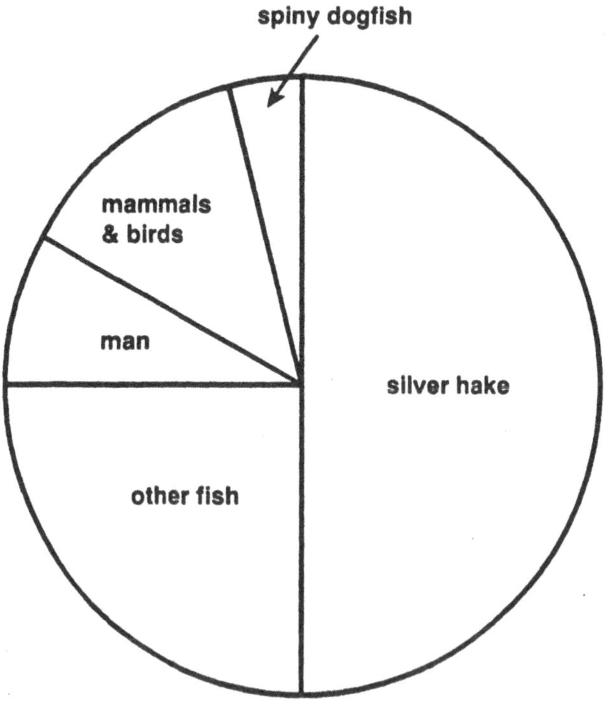

Figure 5.2. Consumption of finfish production on Georges Bank.

both North Sea and Scotian Shelf), but lower than the East Bering Sea (2.7%). Given the uncertainty in these gross calculations, it is dangerous to draw conclusions about the mechanisms that make trophic efficiency of one ecosystem higher than another.

One important aspect of understanding productivity of the fish community is the high degree of internal consumption (i.e., fish eat a large quantity of fish). Several species are predominantly piscivorous (e.g., silver hake, spiny dogfish, and goosefish). Silver hake alone consumed about half the total fish production during the period 1972-1975 (Figure 5.2, Cohen and Grosslein, 1987). Spiny dogfish was the second most important fish predator and several other species in aggregate consumed a large portion of production. Mankind, marine mammals, and sea birds consumed the rest. As will be discussed later, the relative importance of the fish predators is variable. Dogfish abundance has increased and it now appears that dogfish have replaced silver hake as the primary predator.

The importance of fish as consumers of fish production is not unique to Georges Bank. Bax (in press) compared the distribution of fish production to fish consumers for six ecosystems, including

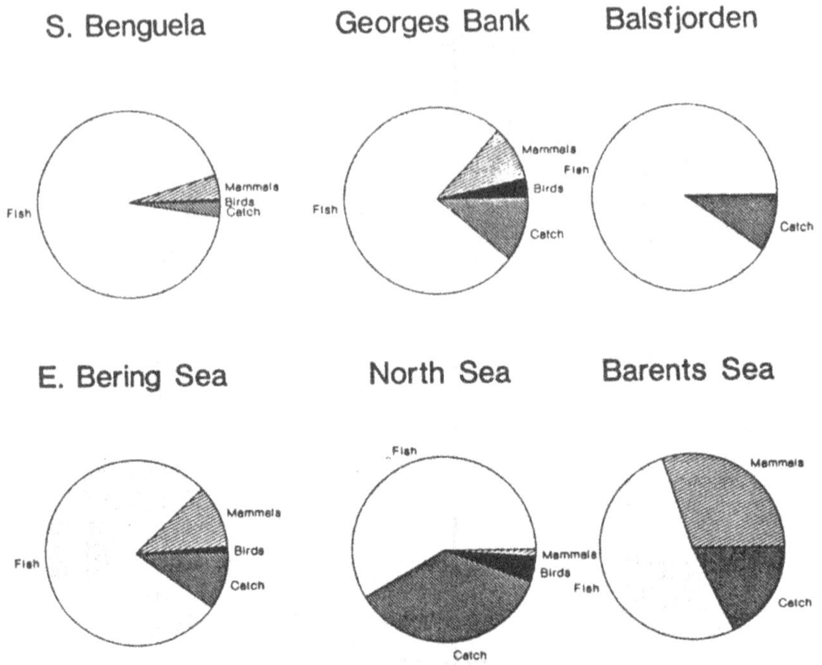

Figure 5.3. Consumption of finfish production by six ecosystems with "pies" proportional to production (after Bax, in press).

Georges Bank (Figure 5.3). Again, the comparisons should not be over-interpreted, since there are many uncertainties in the calculations, but it is clear that in general fish consume most of their own production. Data from the Peruvian upwelling ecosystem (Jarre, in press) and for the Gulf of Mexico (Browder, in press) support this conclusion.

Fish predation on fish is significant for several reasons:

(1) It is in part responsible for high pre-recruit mortality, particularly on post-larvae and juveniles (Sissenwine et al., 1984).

(2) There is negative feedback within the fish community (i.e., a large biomass results in a high predation mortality, which reduces population growth) which should act to stabilize the total biomass, although it may contribute to the variability of individual species.

(3) Fisheries management is implicitly multispecies (i.e., management of one species affects others indirectly), though most management decisions are based on single-species considerations (Sissenwine and Daan, in press).

The preceding description of typical levels of energy flow may give the impression that the fish populations are relatively constant; this is not the case. Most fluctuation in fish populations results from variability in recruitment. The causes of variability in recruitment have been the focus of much research. Cohen et al. (1988) have summarized research for Georges Bank.

Management Strategy

Recruitment variability complicates fisheries management (see Sissenwine et al. [1988] for an overview of the problem). The most important problem is that it obscures the relationship between spawning stock size and recruitment. In general, it is assumed that there is an ecosystem-carrying capacity that limits the number of recruits at high levels of spawning stock size. Recruitment may even decrease as spawning stock size increases above some critical level. Regardless of the specific form of the relationship, it is usually assumed to be compensatory (i.e., per capita recruitment is an inverse function of spawning stock size).

A logical fisheries management strategy when there is compensation is (1) to limit or curtail fishing when spawning stock size has been reduced to a specified minimum level which results in a significant reduction in recruitment (e.g., below the carrying capacity or critical level of spawning stock size); and (2) above the minimum level, to set the fishing intensity (in terms of an age-specific mortality vector) to a "target" level based on yield-per-recruit (YPR) considerations, regardless of the number of recruits (see Ricker [1975] for a review of YPR analysis). The target could be the fishing intensity that will maximize YPR (F_{max}), or a lower level that produces a somewhat lower YPR with a lower fishing effort and a higher catch-per-unit effort (e.g., $F_{0.1}$, Gulland and Boerema, 1973). The strategy would prevent both recruitment overfishing (i.e., fishing-down the stock to the point where recruitment fails [Cushing, 1976]) and growth overfishing (fishing-out recruits so rapidly that the potential growth of a year-class is wasted [Ricker, 1975]).

The problem with the strategy is that recruitment variability usually makes it difficult to specify the minimum spawning stock size. This can only be done objectively if there is a statistically significant fit to a compensatory spawner-recruit relationship. Quite often, if not usually, this is not the case. The lack of fit is probably in part due to measurement error in recruitment and spawning stock size (Walters and Ludwig, 1981), but, environmental factors undoubtedly play a major role. As a result, it is common for fisheries to be managed by applying a target fishing mortality based on YPR analysis without a minimum spawning stock size constraint. This approach implicitly assumes a compensatory spawner-recruit relationship (i.e., it assumes that average recruitment is unaffected by spawning stock biomass). Furthermore, it does not use spawner-recruit data in the formulation of the management strategy.

An alternative strategy was described by Sissenwine and Shepherd (1987). In theory (we will elaborate later), it was adopted for management of most Northeast Shelf groundfish (e.g., cod, haddock, redfish, pollock, and several species of flounder).

The strategy is based on the axiom that conservation requires that year-classes replace the spawning biomass of their parents on average. The strategy does not assume compensation. This does not mean there is no compensation, but it is more conservative not to assume it without a statistical basis for its quantification. One reason there might not be quantitative evidence of compensation is that all available data may have been collected during a period when a population had already been fished-down below the carrying capacity.

The approach is based on Shepherd's 1982 paper on stock recruitment functions, a report of the ICES Working Group on Stock Assessment Methods (Anonymous, 1983), and a Panel Report from the Dalhem Conference on "Exploitation of Marine Populations" (Beverton et al., 1984). Sissenwine and Shepherd (1987) used the approach introduced in these previous studies to demonstrate that YPR analysis and spawner-recruit data can be combined to estimate the fishing intensity that fulfills the conservative criteria stated above (i.e., year-classes replace the spawning biomass of their parents on average). The ratio of recruits to spawning biomass is a measure of survival for each year-class(es). Yield-per-recruit analysis can be used to calculate spawning stock biomass-per-recruit (SSBPR) as a function of fishing intensity. The fishing intensity that has an inverse value of SSBPR equal to the mean survival ratio ($SSBPR^{-1} = s$) satisfies the conservation criteria.

The method is illustrated for Georges Bank haddock in Figure 5.4. Lines with slope equal to the inverse of SSBPR for a range of annual exploitation ratios (i.e., fraction of a year-class that is harvested, u) are superimposed on spawner-recruit data. The SSBPR for each line (as a fraction of SSBPR with no fishing) is also given. The lines are known as replacement lines, since they are the loci of spawner-recruit points for which year-classes over their life span exactly replace the biomass of their parents (i.e., the spawning biomass during the year the year-class was produced), for the exploitation ratio that corresponds to the line. If the median is used to estimate s (other estimators, such as the mean, could be used), it follows that the line that bisects the points, fulfills the conservation criteria. The fishing mortality rate that corresponds to the median replacement line was designated as F_{med} by ICES (Anonymous, 1983). Sissenwine and Shepherd referred to it as F_{rep} to: (1) draw attention

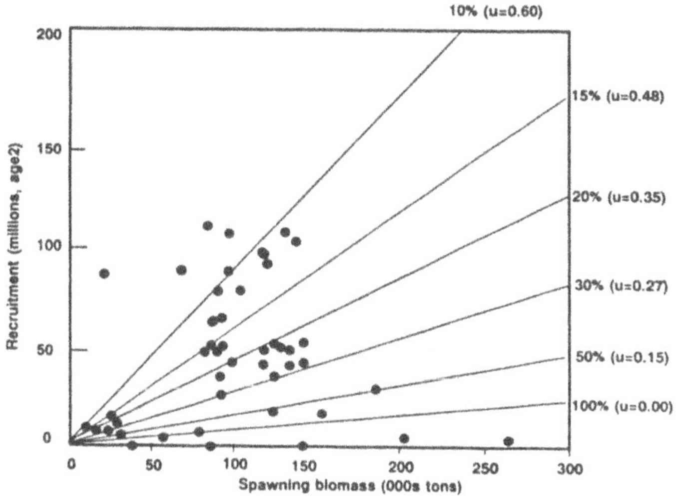

Figure 5.4. Spawner-recruit data for Georges Bank haddock.

to its significance with respect to replacement of successive generations in the spawning stock; and (2) distinguish the specific estimator that is used (i.e., the median) from the concept. There are several limitations to F_{rep} (or F_{med}) as a fisheries management strategy, including the following:

(1) It is not a target fishing mortality, since it ignores yield considerations. It is only a conservation constraint.

(2) It does not allow for rebuilding of depleted stocks. To do so will require a lower fishing mortality rate.

(3) It is a conservative constraint if there is compensation, in the sense that the population may be able to sustain more intense fishing.

(4) It assumes a stationary pattern of survival ratios.

(5) It ignores species interactions that are almost certainly important with respect to (4), above.

With respect to the actual application of F_{rep} to groundfish of the Northeast Shelf ecosystem, there are two additional problems. F_{rep} has only been estimated from spawner-recruit data for a few stocks. In most cases, it was set equal to a value corresponding to 20% SSBPR by analogy or in consideration of historic performance of the fisheries (i.e., they were relatively stable and productive at that level of SSBPR).

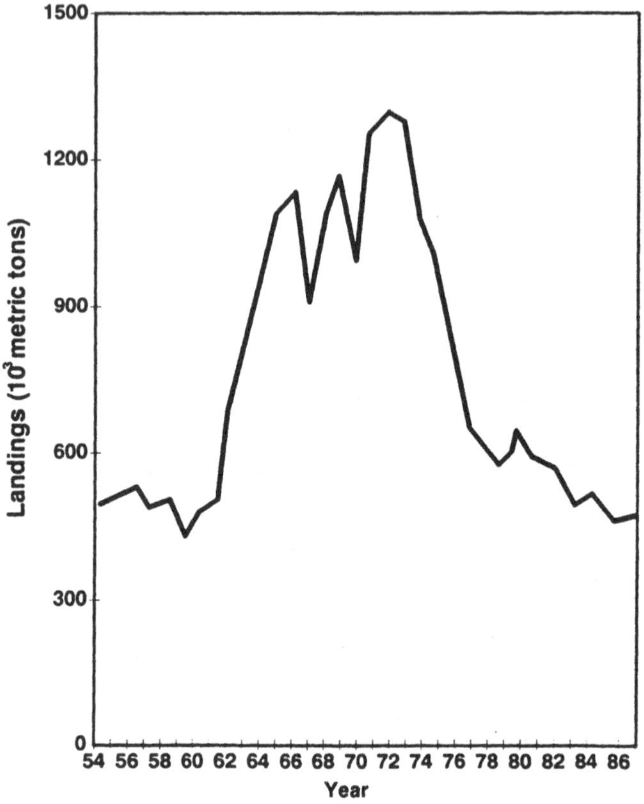

Figure 5.5. Total nominal catch of all species from NAFO Areas 5+6 (except menhaden, billfishes and sharks).

A more serious problem has been that fisheries management regulations have not been adequate to adhere to the conservation criteria since 1985, when it was adopted. This is reflected in the current condition of the fishery.

It should be recognized that there are several other fisheries (herring, mackerel, several invertebrates) in the region that are not managed with F_{rep} as a conservation constraint. It is not practical to describe the management strategies for these fisheries in this paper.

Fisheries

The fisheries off the northeastern U.S. date back prior to colonial times. Figure 5.5 gives landings since the mid-1950s. Before 1960, landings were relatively stable at 300,000 tons per year. During the

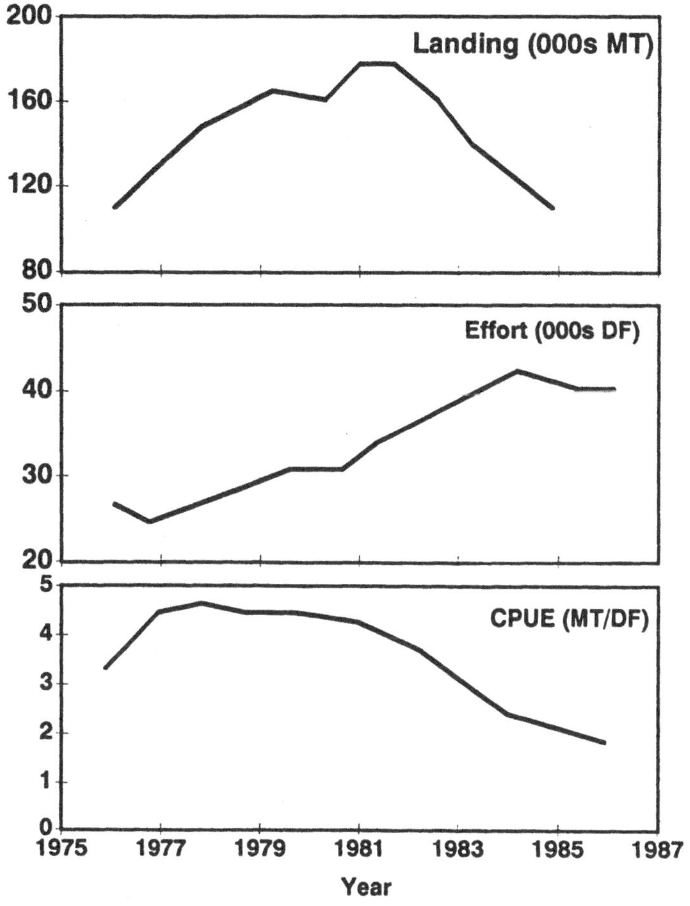

Figure 5.6. Total landings, effort and catch per unit effort, CPUE. (After NEFC, 1989).

1960s and early 1970s, distant-water foreign fishing vessels increased the yield rapidly to over 1,000,000. In 1977, the Magnuson Fishery Conservation and Management Act (MFCMA), which extended U.S. jurisdiction to 200 miles, eliminated most foreign fishing. Since MFCMA, landings have returned to the level of the traditional domestic fishery that preceded foreign fishing, but there has been a declining trend, particularly for groundfish.

Anthony (1990) recently reviewed the conditions of the fisheries. We rely heavily on his review. Figure 5.6 gives the landing, standardized fishing effort, and catch-per-unit effort (CPUE) for the bottom-trawl fishery. The trend in CPUE indicates the condition of the exploited groundfish resource. Stock assessments for several

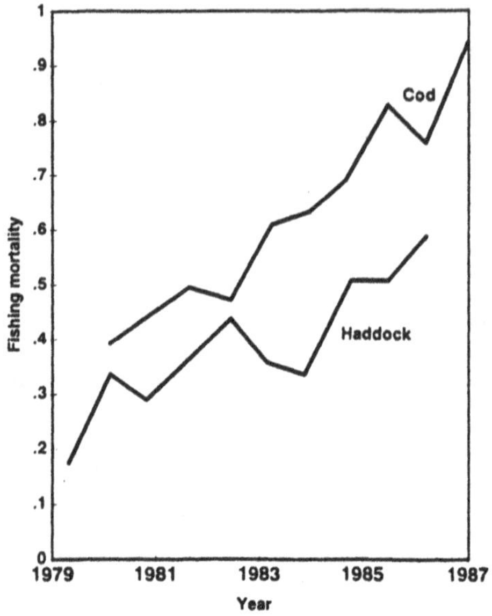

Figure 5.7. Estimate of fishing mortality rate for Georges Bank cod and haddock (After Anthony, 1990).

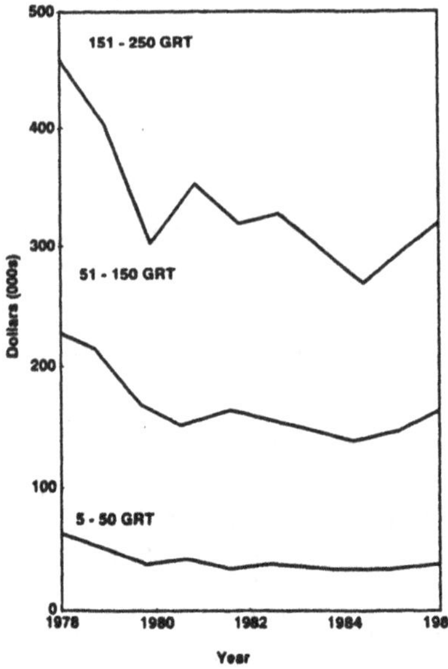

Figure 5.8. Deflated (adjusted) gross revenue per vessel by tonnage class for those vessels using otter trawl gear and landing in a New England port at least once during the year. All revenue, regardless of whatever gear was used or wherever catches were sold, is included. Revenue adjusted by the Consumer Price Index (CPI) with 1978 as the base year. (After NEFC, 1989).

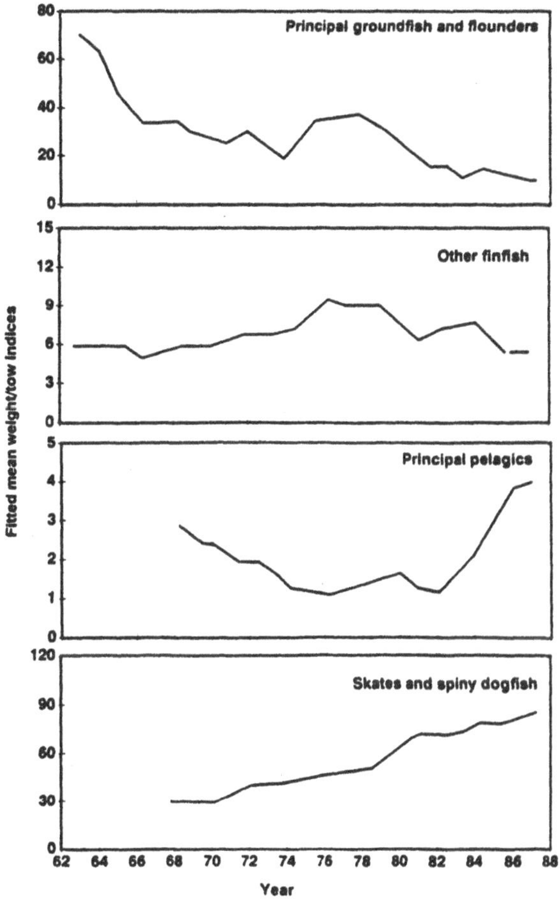

Figure 5.9. Trends in indexes of aggregate abundance (catch in weight per survey trawl haul) for four species groups, reflecting the major changes in fishery resources in recent decades. (After NEFC, 1989).

individual species (e.g., haddock, cod) show historically high levels of fishing mortality rate (Figure 5.7). Figure 5.8 indicates that the increases in fishing effort and decline in resource abundance has adversely affected revenues. The decrease in revenues would have been much sharper if not for increases in ex-vessel prices.

Research vessel bottom-trawl surveys confirm the decline in abundance of the principal groundfish and flounders that are exploited by the trawl fishery (Figure 5.9). But they also reveal that other finfish (e.g., scup, ocean pout, cusk, wolffish) have been relatively stable and skates and spiny dogfish have steadily increased

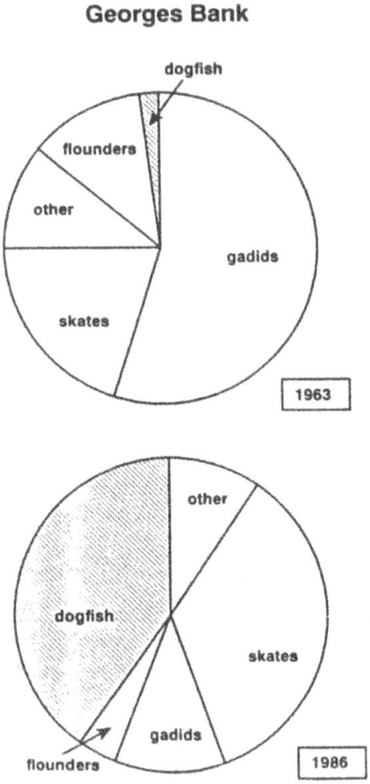

Figure 5.10. Species composition on Georges Bank.

to offset the decline in principal groundfish and flounders. Figure 5.10 indicates the change in species composition based on trawl surveys. In 1963, dogfish were a minor component of the trawl survey catch. In 1986, they were the most abundant species. One of the gadoids that has declined drastically is silver hake. Silver hake had consumed about 50% of the fish production (Figure 5.2) during the mid-1970s. Dogfish now fulfill the role as the dominant fish predator.

The demise of the bottom-trawl fishery began with excessive fishing pressure from a combination of foreign and domestic vessels prior to MFCMA. Catch quota regulations, imposed by the International Commission for the Northwest Atlantic Fisheries (ICNAF) and under the authority of the MFCMA, restricted fishing during most of the 1970s. There was evidence of some improvement in the condition of some species (e.g., haddock) under catch quota management. Since 1982, the fishery has been regulated by a combination of mesh regulations, closed seasons and areas, and minimum fish size limits. These regulations have not prevented the fishing intensity from exceeding the level that allows year-classes to replace the spawning biomass of their parents on average (i.e., criteria described in the previous section). Rebuilding depleted stocks will require more than replacement.

It would be incorrect to assume all the fisheries resources of the Northeast Continental Shelf ecosystem are in as poor condition as the principal groundfish and flounders. The pelagic resource (herring and mackerel) recovered from overfishing that occurred prior to MFCMA. During the period when herring and mackerel abundance was low, they were partially replaced by sand lance, which is unexploited (Sherman et al., 1981; Fogarty et al., in press).

Some invertebrate resources are in better condition than the principal groundfish and flounders (e.g., surf clams), but fishing intensity for some of these invertebrates may not be sustainable (e.g., sea scallops).

Perspectives on the Future

Two issues that merit detailed scientific evaluation are (1) the economic and social implications of reducing fishing mortality on groundfish and flounders, and (2) the effect of reducing dogfish biomass. We do not intend to be comprehensive in our discussion of these issues, rather our intent is to be thought-provoking.

There is widespread (although not unanimous) recognition that the fishing mortality on groundfish and flounders is too high. But there is not open recognition that severe economic dislocation will probably be necessary to solve the problem. Presumably, the fishery is operating at an open-access equilibrium, or beyond. This means that the average vessel cannot afford to reduce its catch to reduce fishing mortality. Regulations that are effective in reducing fishing mortality will force many vessels to leave the fishery for groundfish and flounders. In the long term, the fleet should be reduced by 50% or more, if potential economic benefits are to be realized. Clearly, there will be significant short-term losses associated with reducing fishing mortality. Someone must pay. It could be fishermen who leave the fishery, or banks that foreclose on vessels, or government loan guarantee funds.

Another option is to design an "economic rationalization" that extracts some future profits from the vessels that remain in the fishery to offset the short-term losses of the vessels that leave. In effect, this has been done in the rationalization of the New Zealand fisheries, as part of Individual Transfer Quota management

(Sissenwine and Mace, in press). At present, there are no plans to implement an economic rationalization nor is there legal authority to do so.

The impetus to consider reducing dogfish biomass is obvious. They have little present economic value, but they consume a large quantity of fish. But it should be recognized that: (1) it is not a trivial task to remove a significant portion of the dogfish population; (2) the ecosystem response cannot be predicted with a significant degree of certainty; and (3) it may be expensive to monitor the response, depending upon which ecosystem variables need to be measured.

The biomass of dogfish has not been estimated precisely, but it may approach one million tons. It would take a large fishing capacity to rapidly reduce dogfish biomass. A rapid reduction is desirable to maximize the probability of observing the ecosystem response. There is only a small market for dogfish accessible to domestic fishermen. It is possible that foreign vessels would want to harvest large quantities of dogfish.

Any plan to heavily exploit dogfish should consider the amount of bycatch of valuable species. This amount should be weighed against the amount the dogfish will consume if they are not harvested. Dogfish predation on fish is well documented. The amount of fish in the diet of dogfish ranged from 9-15% (by weight) during the period 1963-1983 (Bowman and Eppi, 1984). The percentage of fish varies between years and seasons. Commercially valuable species in the diet include mackerel, gadoids and flounders.

The response of the ecosystem to a reduction in dogfish biomass cannot be predicted without a much better knowledge of dogfish diet composition, and of feeding responses (i.e., Overholtz et al., in press) and prey preference. It seems reasonable to expect that some of the prey of dogfish will benefit, but unexpected responses are also possible. For example, dogfish predation may be holding down the abundance of another low-valued species that may turn out to be a significant predator of larvae and post-larvae of a valuable species.

A reduction in dogfish biomass could be viewed as an adaptive management experiment (Walters, 1986), but the value of the experiment will be limited by the lack of spatial replicates. Recruitment variability also will make it difficult to interpret results. One strategy is to design a research program to test specific hypotheses. Specific hypotheses are: (1) the mortality rate of the most significant prey species will be reduced; (2) dogfish growth and reproductive rate will increase as prey abundance increases; (3) dogfish will be replaced in their role as the dominant fish predator (e.g., by silver hake); and (4) dogfish growth and reproductive rate will revert back to current levels once they have been replaced.

It is unlikely that fisheries managers will implement an economic rationalization of the groundfish and flounder fishery along the lines discussed above, or reduce dogfish biomass in the near future. There are many legitimate reasons (beyond the scope of this paper) that preclude them from doing so. But, scientists should objectively evaluate the options. Otherwise, the options are not actually available to managers.

References

Anonymous. 1983. Report of the Working Group on Methods of Fish Stock Assessment. Int. Coun. Explor. Sea Coop. Res. Rep. 133. 135 Pp.

Anthony, V. C. 1990. The New England groundfish fishery after ten years of MFCMA. North Am. J. Fish. Mgt. 10:175-184.

Bax, N. J. In press. A comparison of the fish biomass flow to fish, fisheries and mammals for six marine ecosystems. *In* M. P. Sissenwine and N. Daan (Eds). Multispecies models relevant to management of living resources. Rapp. P.-v. Reun. Cons. int. Explor. Mer.

Beverton, et. al. (with 10 other authors). 1984. Dynamics of single species. Pp. 13-58. *In* R. M. May (Ed.), Exploitation of marine communities, Dahlem Konferenzen. Springer-Verlag, New York.

Bowman, R., and R. Eppi. 1984. The predatory impact of spiny dogfish in the Northwest Atlantic. ICES C.M. 1984/G:27. 20 Pp.

Browder, J., B. Brown, W. Nelson, and A. Bane. In press. Multispecies Fisheries in the Gulf of Mexico. *In* M. P. Sissenwine and N. Daan (Eds). Multispecies models relevant to management of living resources. Rapp P.-v. Reun. Cons. int. Explor. Mer.

Cohen, E. B., and M. D. Grosslein. 1987. Production on Georges Bank compared with other shelf ecosystems. Pp. 383-391. *In* R. H. Backus, R. L. Price and D. W. Bourne (Eds.) Georges Bank, MIT Press, Cambridge, MA.

Cohen, E. B., M. P. Sissenwine, and G. C. Laurence. 1988. The "recruitment problem" for marine fish populations with emphasis on Georges Bank. Pp. 373-392. *In* B. J. Rothschild (Ed.) Toward a theory of biological-physical interactions in the sea world ocean, NATO ASI Series C: Mathematical and Physical Sciences, Vol. 239. Kluwer Academic Publishers, Dordrecht. 650 Pp.

Cushing, D. H. 1976. In praise of Peterson. J. Cons. Int. Explor. Mer. 36(3):277-281.

Fogarty, M. J., E. B. Cohen, W. L. Michaels, and W. W. Morse. In press. Interactions among herring, mackerel and sand lance in the Northwest Atlantic. *In* M. P. Sissenwine and N. Daan (Eds)

Multispecies models relevant to management of living resources. Rapp P.-v. Reun. Cons. int. Explor. Mer.

Gulland, J. A., and L. K. Boerema. 1973. Scientific advice on catch levels. Fish. Bull., U.S. 71:325-335.

Jarre, A., P. Muck, and D. Pauly. In press. Two approaches for modeling fish stock interactions in the Peruvian upwelling ecosystem. *In* M. P. Sissenwine and N. Daan (Eds) Multispecies models relevant to management of living resources. Rapp. P.-v. Reun. Cons. int. Explor. Mer.

Northeast Fisheries Center [NEFC]. 1989. Status of the fishery resources off the Northeastern United States for 1989. NOAA Tech. Mem. NMFS-F/NEC-72, 110 Pp.

O'Reilly, J. E., C. Evans-Zetlin, and D. A. Busch. 1987. Primary production. Pp. 220-233. *In* R. H. Backus, R. L. Price, and D. W. Bourne, (Eds.) Georges Bank, MIT Press, Cambridge, MA.

Overholtz, W. J., S. A. Murawski, and K. L. Foster. In press. Impact of predatory fish, marine mammals, and seabirds on the pelagic fish ecosystem of the Northeastern U.S.A. *In* M. P. Sissenwine and N. Daan (Eds). Multispecies models relevant to management of living resources. Rapp. P.-v Reun. Cons. int. Explor. Mer.

Ricker, W. E. 1975. Computation and interpretation of biological statistics of fish populations. Bull. Fish. Res. Board. Can. 191: 382 Pp.

Shepherd, J. G. 1982. A versatile new stock-recruitment relationship of fisheries and construction of sustainable yield curves. Cons. perm. int. Explor. Mer 40(1):67-75.

Sherman, K., C. Jones, L. Sullivan, W. Smith, P. Berrien, and L. Ejsymont. 1981. Congruent shifts in sand eel abundance in western and eastern North Atlantic ecosystems. Nature 291:486-489.

Sissenwine, M. P. 1986. Perturbation of predator controlled continental shelf ecosystem. Pp. 55-85. *In* K. Sherman and L. M. Alexander (Eds.) Variability and management of large marine ecosystems. AAAS Selected Symposium 99. Westview Press, Boulder.

Sissenwine, M. P., and N. Daan (Editors). In press. Multispecies Models Relevant to Management of Living Resources. Rapp. P.-v. Reun. Cons. int. Explor. Mer.

Sissenwine, M. P., and P. M. Mace. In press. ITQs in New Zealand: The first three years. Mar. Res. Econ. J.

Sissenwine, M. P., and J. Shepherd. 1987. An alternative perspective on recruitment overfishing and biological reference points. Can. J. Fish. Aquat. Sci. 44:913-918.

Sissenwine, M. P., E. B. Cohen, and M. D. Grosslein. 1984. Structure of the Georges Bank ecosystem. Rapp. P.-v. Reun. Cons. int. Explor. Mer 183:243-254.

Sissenwine, M. P., J. J. Fogarty, and W. J. Overholtz. 1988. Some fisheries management implications of recruitment variability. Pp. 129-152. *In* J. Gulland (Ed.) Fish population dynamics. John Wiley & Sons Limited, Sussex, England.

Walters, C. J. 1986. Adaptive Management of Renewable Resources, Macmillan, New York. 374 Pp.

Walters, C. J., and D. Ludwig. 1981. Effects of measurement errors on the assessment of stock-recruitment relationships, Can. J. Fish. Aquat. Sci. 38:704-10.

Bradford E. Brown, Joan A. Browder, Joseph Powers, and Carole D. Goodyear

6. Biomass, Yield Models, and Management Strategies for the Gulf of Mexico Ecosystem[1]

Abstract

The Gulf of Mexico has been one of the last frontiers for fisheries in United States waters. But traditional Gulf fisheries, such as those for penaeid shrimp[2] and menhaden, have reached their harvesting limits, as have many newer fisheries, such as those for reef fish, coastal migratory pelagic fish, and large oceanic pelagics. Now fisheries are developing for the smaller, lesser known species, which appear to be very abundant but may be the food source sustaining many currently valuable fishery species, such as king and Spanish mackerel. Ecological approaches are needed to determine how the growing harvests of prey species will affect the populations and, consequently, the harvests of their predators.

In this report we quantify the consumption requirements of upper-trophic-level species in the Gulf of Mexico fisheries of both the United States and Mexico. We compare this to first approximations of the biomass of the middle-trophic-level species that are their prey. We estimate that the total consumption requirements are approximately 6.6 million metric tons (mmt) annually, whereas the biomass of potential prey species that we have quantified is approximately 8.4 metric tons (mt). This suggests that approximately 78% of the estimated standing biomass of prey species may be

[1]Contribution No. MIA-89/90-04.

[2]Scientific names for all common names mentioned in text or tables are given in the Appendix, Table 6.13. Common and scientific names of fish follow the usage of the American Fisheries Society (Robins et al., 1980).

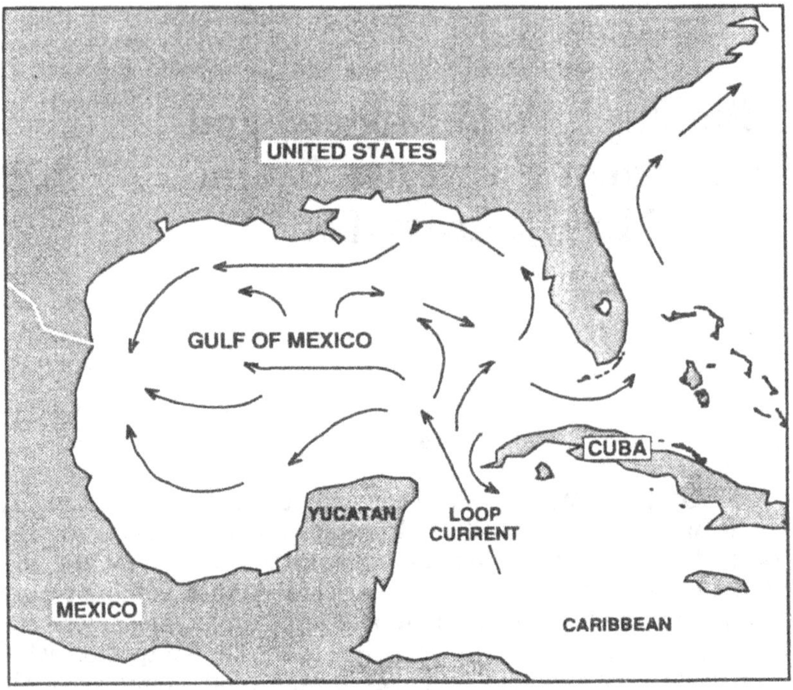

Figure 6.1. General surface current patterns in the Gulf of Mexico.

consumed annually by fishery species. Fisheries currently catch approximately 15% of estimated prey biomass. This means that reproduction and growth rates of prey species (the production/ biomass (P/B) ratio) must be at least 0.93 to maintain biomass stability.

The estimated total catch of prey species in both U.S. and Mexican waters of the Gulf of Mexico is 1.3 mmt, including discards. Menhaden, with an annual catch of 0.6 mmt in 1988, accounts for approximately one-half the catch of prey species. The by-catch of the shrimp fishery, most of which is discarded, is estimated at 0.5 mmt.

Introduction

The Gulf of Mexico as a large marine ecosystem supporting fisheries has been described by Richards and McGowan (1989). The Gulf is a Mediterranean Sea of approximately 1.6 million square kilometers, bounded by the United States, Mexico, and Cuba (Figure 6.1). Rivers entering the Gulf drain two-thirds of the United States and one-half of Mexico. The largest is the Mississippi-Atchafalaya drainage, which accounts for two-thirds of the 30,000 m^3/sec that flow into the Gulf (NOAA, 1985).

The Gulf of Mexico exchanges water with the Caribbean Sea and the Atlantic Ocean by means of the Loop Current, which is a part of the Gulf Stream system (Figure 6.1). Water enters the Gulf by the Yucatan Channel and exits through the Straits of Florida (Vukovich and Maul, 1985). Both channels are about 160 km wide, but the Yucatan Channel has a sill depth of 1,750 m, whereas the Straits of Florida have a sill depth of only 800 m. Because of the difference in sill depth, water exits the Gulf of Mexico not only through the Straits of Florida but also by means of countercurrents in the Yucatan Channel (Maul et al., 1985).

The Loop Current is the primary oceanographic feature of the eastern and central Gulf of Mexico. Eddies that separate from the Loop Current and drift westward are important to the circulation of the western basin. The Mississippi-Atchafalaya drainage also influences the circulation of the western basin.

The Gulf of Mexico has valuable fishery resources, which were described by Richards and McGowan (1989). The Gulf traditionally has supported extensive fisheries for shrimp and menhaden. The shrimp fisheries have an extensive by-catch of species such as Atlantic croaker. Recently, intense efforts have been directed toward other fisheries, particularly for predatory species such as king mackerel, snappers, groupers, yellowfin tuna, swordfish, and sharks. Many of these fisheries are now under restrictive regulations because of declining availability and increasing fishing pressure, both commercial and recreational. User conflicts have resulted in allocations that are shifting the harvests toward recreational fishing.

The potential for increasing the commercial harvest now rests with small pelagic prey species that have been estimated as having a biomass of 5 million metric tons (mmt) or more (Sanders et al., 1990). However, the harvest of these species raises questions concerning their role as the food for predatory species that already support highly developed fisheries. This overview provides a first approximation of the biomass of the predatory and prey species in the Gulf of Mexico and the annual food consumption of the predatory species. These estimates provide a perspective with which to evaluate the extent to which the Gulf of Mexico system might be expected to support additional exploitation without affecting predator stocks. The available data are extremely limited, and thus it is expected that the primary use of these estimates will be to guide decisions on the initiation, scope, and direction of predator/prey research in the Gulf of Mexico.

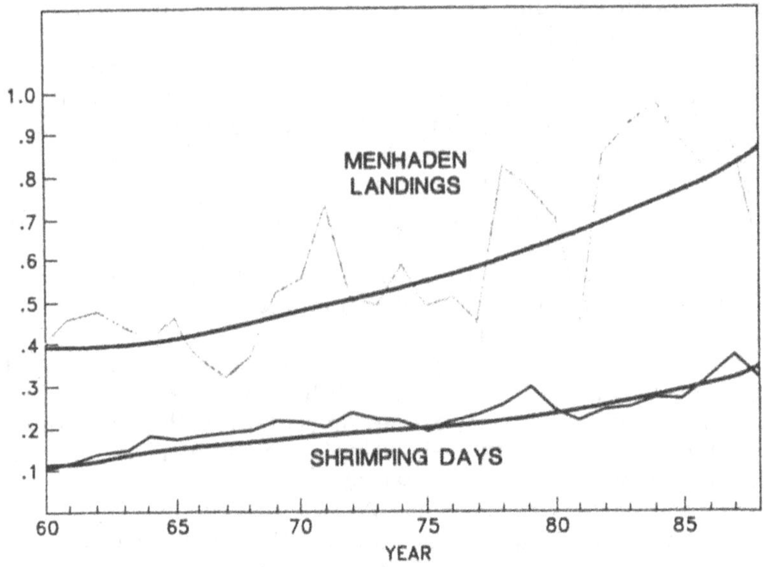

Figure 6.2. Trends in indices of fishing activity in the Gulf of Mexico's two traditional fisheries (from Browder et al., in press).

Present Status and Trends in Gulf of Mexico Fishery Resources

Browder et al. (in press) describe recent trends in Gulf of Mexico fisheries. The traditional fisheries have been menhaden and shrimp. The catch of menhaden has more than doubled in the past 20 years (Figure 6.2), reflecting the growth of activity in that fishery. Activity in the shrimp fishery also has intensified, with effort approximately doubling (Figure 6.2). Shrimp trawls catch many more tons of fish than shrimp, and most of this is discarded dead. Therefore, fishing mortality rates for the by-catch species have likely increased at least proportionally to the increase in effort directed at shrimp.

Historically, coastal regions of the Gulf of Mexico were not heavily populated. While commercial fishing for a number of fish and macroinvertebrate species was of interest to resource managers, there was little apprehension about overfishing. This has changed in the last quarter century, and now recreational and commercial harvesters eagerly compete for these resources, some of which are

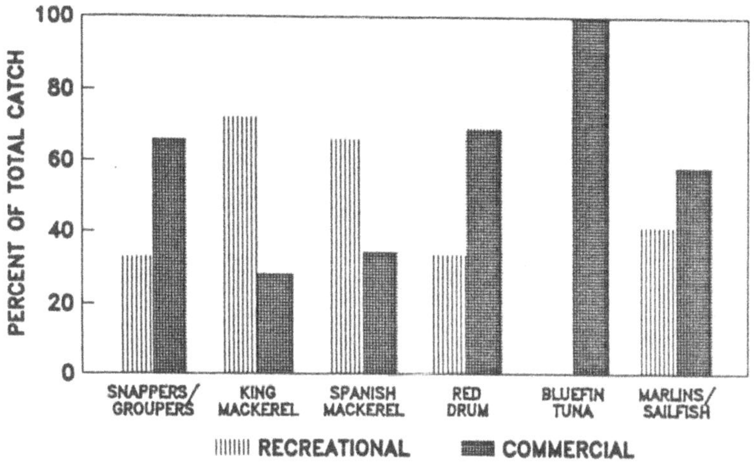

Figure 6.3. Commercial and recreational proportions of the harvest of selected key species in U. S. Gulf of Mexico fisheries in 1986. (Compiled from Miyake et al., 1987; Goodyear, 1988; Goodyear, 1989; and SEFC, 1990.)

declining. Figure 6.3 presents the commercial and recreational proportions of the annual harvest of selected key species in 1986. For some species, such as king mackerel, the recreational harvest greatly exceeds the commercial harvest in U.S. waters.

Figures 6.4 and 6.5 present recent trends in landings or stock-assessment indices for large oceanic pelagic species. Increased landings, as in the case of yellowfin tuna, sharks, and swordfish, are due to increased fishing effort, not increased stocks. Assessments have shown significant declines in swordfish (ICCAT SCRS, 1989a; SEFC, 1989a) and bluefin tuna (ICCAT SCRS, 1989b).

Growing concern about the health of oceanic pelagic resources throughout the western North Atlantic has led to action. Regulations are in place or under development for several large oceanic pelagic fish in the Gulf of Mexico. The National Marine Fisheries Service, at the request of the Regional Fishery Management Councils, is preparing an emergency plan to prevent overfishing of sharks. Restrictions on swordfish catches in the western North Atlantic, including the Gulf of Mexico and the Caribbean, also are under consideration by the Regional Councils. Bluefin can now be harvested in the Gulf of Mexico only as by-catch in long-line fisheries for yellowfin tuna and swordfish. The marlin, a highly prized recreational fish, no longer can be fished commercially. Concerns over possible declines in marlin stocks (Browder and Prince,

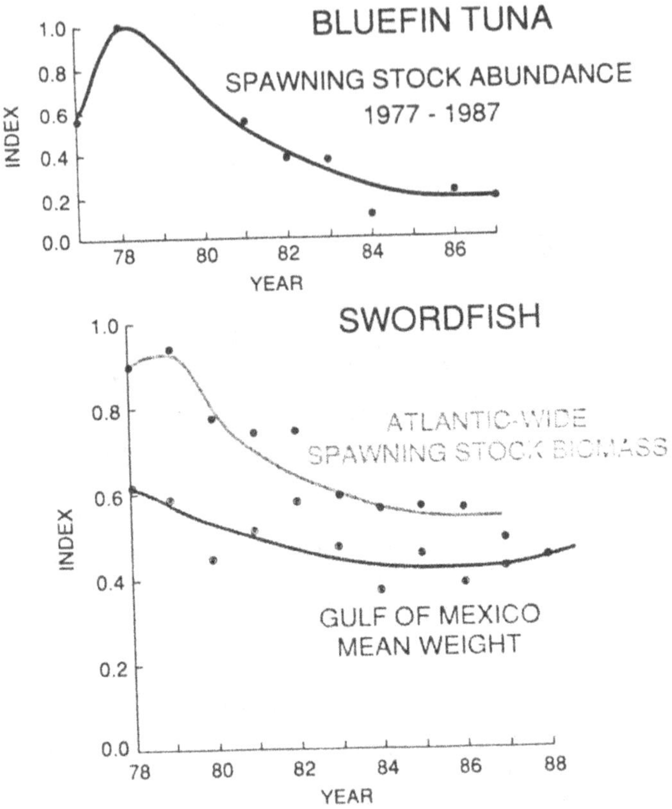

Figure 6.4. Trends in stock indices for bluefin tuna and swordfish. (From ICCAT, 1989a, 1989b).

1991) led the International Commission for Conservation of Atlantic Tunas (ICCAT) to initiate a special program of Enhanced Billfish Research to collect data in preparation for an assessment. Of the oceanic pelagics group in the Gulf of Mexico, yellowfin tuna is the only major fishery species that is not currently regulated or under immediate consideration for regulation.

The status of coastal pelagic species in the Gulf of Mexico also is of concern. Figure 6.5 shows the decline in king mackerel spawning stock, as documented in SEFC assessments (SEFC, 1990). Spanish mackerel have shown similar declines (SEFC, 1990). Red drum fishing has been greatly curtailed in the aftermath of studies stimulated by the rapid development of an offshore purse seine fishery. The studies determined that intensive fishing in inshore

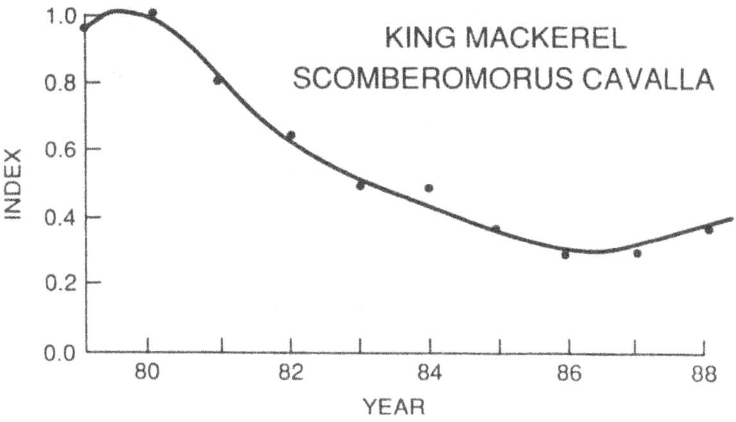

Figure 6.5. Trends in the index of spawning stock biomass for king mackerel in the U. S. Gulf of Mexico. (From SEFC, 1990.)

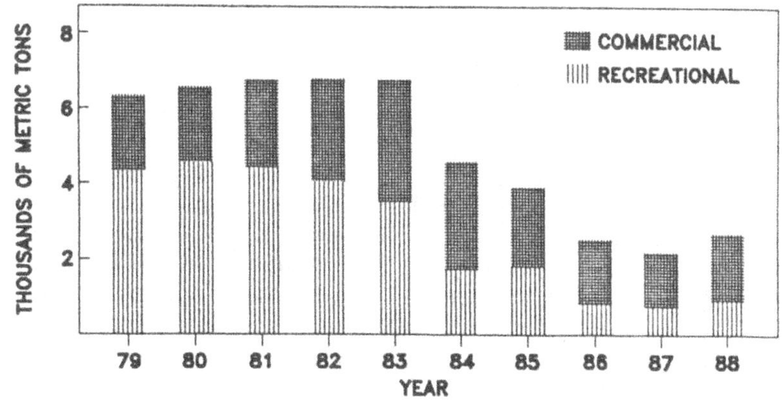

Figure 6.6. Annual catches of red snapper in the U. S. Gulf of Mexico, showing commercial and recreational portions, 1979 to 1988. (From Goodyear and Phares, 1990.)

waters limited escapement to offshore spawning stocks, resulting in major stock declines (Goodyear, 1989). Presently, industrial-scale fishing for red drum is prohibited, and inshore fishing is restricted by the states.

Declines in stocks of red snapper (Figure 6.6) also have resulted in stricter regulations. This decline has been accentuated by high mortality rates on juveniles in shrimp trawls (Goodyear and Phares, 1990). Declines in other grouper and snapper species also have been seen. While the data on these species are not as complete as for red snapper, the preliminary indications have induced implementation of regulations by the Gulf of Mexico Fishery Management Council.

Thus, with the full exploitation of traditional prey species and declining stocks of predator species, the future for increased commercial harvests would appear to rest with smaller, prey species not yet significantly exploited. These are of two types, demersal and pelagic.

Efforts to develop fisheries for stocks of small demersal species in the Gulf of Mexico have not been entirely successful. A pet food industry developed in the 1970s using small bottomfish species such as Atlantic croaker, but most of the petfood plants closed due to inability to catch sufficient quantities of desirable species. This resource also was the target for the potential development of the surimi industry, and an experimental plant was funded for this purpose in the early 1980s. However, species such as croaker that are heavily fished as by-catch in the shrimp fishery have shown large declines in recent years (Herron, 1988), paralleling the increase in shrimping effort (Figure 6.7).

Small pelagic prey species, unharvested and potentially exploitable, provide another alternative for fishery development in the Gulf of Mexico. Traditionally, small fisheries for some of these species have existed where they could be found near shore, such as along the northwest coast of Florida. The inability to find high concentrations on a regular basis has hampered full development of these fisheries.

Improved gear and new fish-locating technology may make it commercially feasible to fish farther offshore and more intensively in the future. For example, a butterfish fishery has developed recently, harvesting several hundred metric tons annually in 1986 and 1987 (Herron, 1988). The development of techniques for shore-to-vessel transmission of sea-surface temperature charts derived from satellite data provide fishermen with a new tool for locating harvestable fish concentrations (Herron et al., 1989). The butterfish fishery is one of several new operations that could turn to other small pelagic species of the Gulf of Mexico and rapidly increase harvests. Whether large harvests of both these species and the predatory species they support are sustainable remains to be determined. The following overview is a first step at addressing that question.

Estimation Methods and Results

In the preparation of this overview, a variety of methods were used to calculate the biomass of predator species, annual consumption

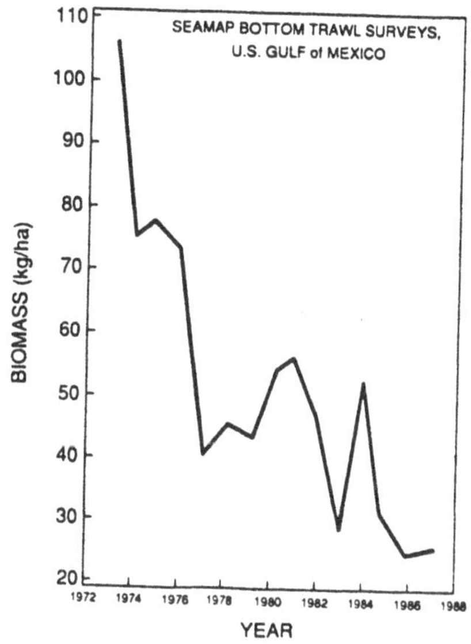

Figure 6.7. Trend in relative biomass of fish caught in bottom trawls during fall resource survey cruises in the North-central Gulf of Mexico, 1972-1987. (From Herron et al., 1988.)

by predator species, and the biomass of prey species. Details of the calculations, input data, and sources are given within each of the following sections.

Predator Biomass

The principal fishery species in the Gulf of Mexico that are predators on small forage fish and macroinvertebrates are listed in Tables 6.1 and 6.2 with landings statistics and estimated biomasses. In most cases, we estimated biomass from fishery landings by the relationship:

$$P = C/F, \tag{1}$$

where P is population biomass, C is catch, and F is instantaneous fishing mortality. A separate calculation was made for each species or species group.

Catch statistics obtained for the five Gulf states in 1988 (Table 6.1) and Mexico in 1987 (Table 6.2) provided the bases for most of the biomass estimates. A fishing mortality specific to each taxon was obtained from published or unpublished reports or was approximated

Table 6.1. 1988 U.S. catch, estimated fishing mortality (F), and estimated biomass of major predator species in the Gulf of Mexico. (Except where indicated, biomass was estimated by dividing catch by estimated fishing mortality.)

	Catch mt	F	Biomass mt
OCEANIC PELAGICS			
Yellowfin tuna	[a]7,284	.35	20,811
Bluefin tuna	[a]143	.30	477
Bigeye tuna	[a]15	.10	150
Albacore tuna	[a]2	.40	5
Skipjack	[a]1	.20	5
Blackfin tuna	[a]34	.30	113
Other tuna	[a]27	.30	90
Swordfish	[a]1,022	.80	1,278
Blue marlin	[a]36	.20	180
White marlin	[a]10	.20	50
Sailfish	[a]3	.20	15
Sharks	[b]5,363	.40	13,408
COASTAL PELAGICS			
King mackerel	[c]1,720		[c]15,830
Spanish mackerel	[c]1,982		[c]20,301
Bluefish	[d]568	.35	1,623
Dolphinfish	[d]764	.40	1,910
Little tunny	[a]106	.30	353
Wahoo and bonito	[e]222	.30	740
Cobia	[d]648	.30	2,160
Misc. carangids	[f]2,175	.30	7,250
Ladyfish	[d]1,877	.30	4,693
Red drum			[g]55,800
Black drum	[e]4,749	.40	11,873
REEF-FISH			
Red Snapper	[h]3,283		[i]8,950
Other Snapper	[j]2,641	.25	10,564
Grouper	[k]7,938	.20	39,690
Amberjack	[l]2,778	.40	6,945
CETACEANS			
Bottlenose dolphin			[m]6,055

Table footnotes follow on next page.

Table 6.1 footnotes:

[a]1988 commercial and recreational catches, compiled by NOAA, National Marine Fisheries Service, Southeast Fisheries Center, Miami Laboratory, Miami, Florida, for 1989 ICCAT report.

[b]Preliminary data on 1988 commercial and recreational catch from NMFS (1989). The recreational catch estimate, 1,406 mt, is based on "type A" fish (fish actually seen by interviewers) only.

[c]1988 estimates, computed from unpublished background calculations for king and Spanish mackerel stock assessment workshop, April 4-7, 1989. Biomass is calculated as VPA stock number at age, multiplied by weight at age (in kg.).

[d]Commercial and recreational catches, 1988. Commercial landings from unpublished data of Guy Davenport, NOAA/NMFS/SEFC, Miami, Fla. Recreational catches from National Recreational Survey, corrected by estimating the weight of unweighed sets of fish. Dolphinfish includes *C. hippurus*, *C. equisetis*, and *Coryphaena* spp.

[e]Wahoo and bonito; blackdrum catch: commercial catch only. Unpublished statistics, NOAA/NMFS/SEFC, Miami, Fla.

[f]Miscellaneous carangids catch: Includes blue runner, crevalle jack, permit, and pompano. Commercial catch only. Unpublished statistics, NOAA/NMFS/SEFC, Miami, Fla.

[g]Red drum biomass: Estimate of offshore stock biomass (age 4 and older), Gulf of Mexico, fall, 1987 (from Goodyear, 1989, page 28, based on mark and recapture).

[h]Red snapper catch: 1988 commercial landings, 1,854 mt (unpublished data, NOAA/NMFS/SEFC, Miami, Fla). 1986 estimated recreational catches, 1,429 mt (Goodyear, 1988, based on National Recreational Survey) (does not include fish in groups where individual weights were not obtained).

[i]Red Snapper biomass: Calculated from age-specific population number estimates and average weights for 1988 from Goodyear (unpublished).

[j]Other snapper catch: 1988 commercial landings, 1,881 MT (unpublished data, NOAA/NMFS/SEFC, Miami, Fla.). 1986 estimated recreational catches, 760 mt (does not include fish in groups where individual weights were not obtained and is obviously an underestimate) (Goodyear, 1988, from Marine Recreational Fisheries Statistics Survey).

[k]Grouper catch: 1988 commercial landings, 4,618 mt (includes seabass and hogfish) (unpublished data, NOAA/NMFS/SEFC, Miami, Fla.). 1986 estimated recreational catches of grouper and seabasses, 3,320 mt (does not include fish in groups where individual weights were not obtained) (Goodyear, 1988, based on Marine Recreational Fisheries Statistics Survey).

[l]Amberjack catch: 1988 commercial landings, 1,165 mt (unpublished data, NOAA/NMFS/SEFC, Miami, Fla.). 1986 estimated recreational catches, 1,613 mt (does not include fish in groups where individual weights were not obtained) (Goodyear, 1988, based on the National Recreational Survey).

[m]Bottlenose dolphin biomass: Calculated from approximately 40,000 individuals in Gulf of Mexico (Scott et al.,1989), size frequency distribution in the Gulf of Mexico (Hansen, unpublished) and length-weight relationships for male and female from Odell and Asper (1982).

Table 6.2. 1987 Mexican catch, estimated fishing mortality (F), and estimated biomass of predator species in the Gulf of Mexico. (Except where indicated, biomass was estimated by dividing catch by estimated fishing mortality.)

	Catch mt	F	Biomass mt
COASTAL PREDATOR			
King mackerel	2,387	.30	7,957
Spanish mackerel	6,455	.40	16,138
Atlantic cutlassfish	1,416	.40	3,540
Jacks	2,946	.40	7,365
ESTUARINE-RELATED			
Red drum	321		[a]3,000
Seatrout	3,058		
Snook	3,227		
OCEANIC PREDATOR			
Tunas	33	.35	94
Sharks	11,018	.40	27,545
REEF-FISH			
Red snapper	4,622	.32	14,444
Lane snapper	548	.40	1,370
Yellowtail snapper	2,506	.16	15,663
Wenchman, etc. (by-catch)	462	1.00	462
Other snappers	1,553	.25	6,212
Grouper	12,627	.20	63,135
Barracuda	362	.20	1,810
Amberjack	423	.40	1,058

NOTE: 1987 landings statistics from Secretaria de Pesca de Mexico, 1988.

[a]Roughly estimated based on biomass in U.S. Gulf of Mexico (Goodyear, 1989).

from the best available information. Table 6.3 lists the values and sources for the fishing mortality estimates given in Tables 6.1 and 6.2. In the few cases indicated in footnotes to Tables 6.1 and 6.2, biomass estimates based on virtual population analyses (VPAs) were obtained directly from published or unpublished reports, rather than being calculated with Equation 1. Predator biomass estimated from U. S. and Mexican catch statistics and VPAs totalled about 400 thousand metric tons.

Predator Consumption

Tables 6.4 and 6.5 display our consumption estimates for predators caught in the fisheries of the United States and Mexican Gulf, respectively. Consumption was calculated from three variables that also are shown on the tables: estimated months in the Gulf of Mexico (MOS), consumption per day as a proportion of body weight (BWD), and biomass (BIO). The annual maintenance consumption (C_m) was calculated by:

$$C_m = MOS \times 30.4 \times BWD \times BIO \qquad (2)$$

where 30.4 is the average number of days in a month. Estimates of BWD that went into our calculations are documented in Table 6.6. We made two types of revisions in the original estimates of BWD: (1) where routine (Hoar et al., 1979) values were reported, we multiplied by 2 as an approximate conversion from routine to field, based on Winberg (1960); (2) where estimates were based on oxygen consumed, results were divided by 0.65 to account for losses due to egestion (E, 1 - assimilation), excretion (U), and specific dynamic action (SDA), after the manner of Kitchell et al. (1978). These loss rates are directly related to the amount of energy consumed, whether measured as oxygen or food. Consumption estimates based on food intake usually are corrected for these losses, whereas those based on oxygen intake are not. Kitchell et al. (1978) defined SDA as the energetic cost of converting food for catabolic and/or anabolic processes. Kitchell's estimates for the above loss rates were E = 0.15, U = 0.05, and SDA = 0.15. The total loss rate is 0.35. Therefore, the energy in only 65% of the food consumed can be used for maintenance and growth, and a quantity 1/0.65 in addition to that used for maintenance and growth will be ingested.

Under the condition of a stable population biomass, recruitment and growth balance natural mortality (M) and fishing mortality (F). Therefore, as a first approximation, we estimated the additional consumption required for growth of each predator (C_g) with the following equation:

Table 6.3. Estimated total (Z), fishing (F), and natural (N) mortalities of various predator fish taxa in the U. S. Gulf of Mexico.

	Z	M	F
OCEANIC PELAGICS			
Yellowfin tuna	.95	.6	.35
Bluefin tuna	.4	.1	.3
Bigeye tuna	.5	.4	.1
Albacore tuna	.7	.4	.3
Skipjack	.8	.6	.2
Blackfin tuna	.9	.6	.3
Other tuna	.7	.4	.3
Swordfish	1.0	.2	.8
Blue marlin	.3	.1	.2
White marlin	.4	.2	.2
Sailfish	.4	.2	.2
Sharks	.6	.2	.4
COASTAL PELAGICS			
King mackerel	.45	.15	.3
Spanish mackerel	.7	.3	.4
Bluefish	.7	.35	.35
Dolphinfish	1.0	.6	.4
Little tunny	.9	.6	.3
Wahoo and bonito	.9	.6	.3
Cobia	.7	.4	.3
REEF-FISH			
Red snapper	.52	.20	.32
Vermilion snapper	.73	.20	.53
Gray snapper	.42	.22	.40
Yellowtail snapper	.36	.20	.16
Weighted other snapper			.25
Red grouper	.28	.15	.13
Black grouper	.30	.10	.20
Gag grouper	.31	.10	.21
Yellowedge grouper	.14	.10	.04
Weighted grouper			.09
Assumed grouper			.20
Amberjack	.60	.20	.40

Table notes follow on bottom of next page.

$$C_g = BIO \times \{1 - \exp[-(M+F)]\} / 0.65 \quad (3)$$

F and M are from Table 6.3. Values for BIO are given on Tables 6.4 and 6.5. Division by the constant, 0.65, is explained above.

Results of the calculations are listed in Tables 6.4 and 6.5. Even at current fishing rates, estimated "growth" consumption represents a very small part of total consumption. For instance, "growth" consumption constituted only 3.4% of the total annual consumption by U.S. predators, whereas "maintenance" consumption accounted for the rest. (E, U, and SDA are components of both growth consumption and maintenance consumption, as defined here.)

Predator-Prey Relations

Food habits studies of major predator species give us a view of specific energy sources. The most extensive data are for the coastal pelagics, which have been the subjects of intensive investigation by the Southeast Fisheries Center of the National Marine Fisheries Service. In most cases, we have quantitative information for the same predator species across several regions within the Gulf of Mexico. Additionally, we have information for several predator species within each region. Therefore, we can examine the food habits of a given predator across several geographic areas, and we can determine roughly, for a given region, the principal prey species of a set of predators. These then can be compared to the dominant potential

---

Notes for Table 6.3:

Mackerel	From unpublished background calculations for king and Spanish mackerel stock assessment workshop, April 4-7, 1989.
Red snapper	Red snapper annual fishing mortality rate: A weighted average for red snapper, estimated from P. Goodyear's unpublished data on F for 1988 by age, total population number, distributed by age, and average weight.
Vermilion snapper . . .	Unpublished estimates of Z, M, and F by P. Goodyear, NOAA/NMFS/SEFC, Miami Laboratory.
Gray snapper	Same as vermilion.
Red grouper	Same as vermilion.
Black grouper	Same as vermilion.
Gag grouper	Same as vermilion.
Yellowedge snapper . .	Same as vermilion.
Weighted grouper . . .	Estimated by weighting above grouper F estimates by weight of the grouper species in the commercial catch for 1988.
Assumed grouper	Weighted grouper seems too low to be accurate. Assumed value seems more realistic, relative to snapper F.
Amberjack	Same as vermilion.
Swordfish	Estimates of Atlantic-wide Z, M, and F from Swordfish Stock Assessment (SEFC, 1989a).
Bluefin tuna	Estimates of western Atlantic Z, M, and F from Bluefin Tuna Stock Assessment (ICCAT, 1989b).
Bluefish	Mid-Atlantic Fishery Management Council, 1989.

Table 6.4. Predator fishery species of the U.S. Gulf of Mexico and their calculated maintenance and maintenance-plus-growth consumption, assuming stable biomass, with growth balancing natural and fishing mortality.

Species	MOS	BWD	BWY	BIO	Annual Maintenance consumption	Annual Maintenance plus growth consumption
Yellowfin tuna	6	0.067	12.22	20,811	254,327	273,962
Bluefin tuna	4	0.100	12.16	477	5,800	6,042
Bigeye tuna	6	0.073	13.3	150	1,997	2,088
Albacore tuna	6	0.073	13.32	5	67	70
Skipjack tuna	6	0.131	23.89	5	119	124
Blackfin tuna	12	0.073	26.63	113	3,009	3,112
Other tuna	6	0.073	13.32	90	1,198	1,268
Swordfish	6	0.016	2.92	1,278	3,730	4,973
Blue marlin	6	0.025	4.56	180	821	893
White marlin	6	0.035	6.38	50	319	345
Sailfish	6	0.040	7.30	15	109	117
Sharks	12	0.020	7.30	13,408	97,825	107,132
King mackerel	12	0.070	25.54	15,830	404,235	413,060
Spanish mackerel	12	0.070	25.54	20,301	518,406	534,129
Bluefish	12	0.070	25.54	1,623	41,445	42,702
Dolphinfish	8	0.070	17.02	1,910	32,516	34,373
Little tunny	12	0.070	25.54	353	9,014	9,336
Wahoo and bonito	12	0.070	25.54	740	18,897	19,572
Cobia	12	0.020	7.30	2,160	15,759	17,432
Misc. carangids	12	0.100	36.48	7,250	264,480	269,003
Ladyfish	12	0.100	36.48	4,693	171,201	174,835
Red drum	12	0.028	10.21	55,800	569,964	585,525
Black drum	12	0.028	10.21	11,873	121,276	129,517
Red snapper	12	0.046	16.78	8,950	150,188	155,771
Other snapper	12	0.046	16.78	10,564	177,272	83,162
Grouper	12	0.028	10.21	39,690	405,410	425,540
Amberjack	12	0.100	36.48	6,945	253,354	258,174
Bottlenose dolphin	12	0.112	40.86	6,055	247,393	248,279
TOTAL				231,319	3,770,131	3,800,536

NOTE: MOS = months, BWD = daily consumption as percent body weight, BWY = yearly consumption as percent body weight, and BIO = total weight (biomass) of the group.

Maintenance consumption = MOS x BWD x BIO

Growth consumption = {1 - exp[-(M+F)]} x BIO/0.65

Units for BIO and consumption are metric tons.

See Table 6.3 for M and F.

Table 6.5. Predator fishery species of Mexican Gulf of Mexico and their calculated maintenance and maintenance-plus-growth consumption, assuming a stable biomass, with growth balancing natural and fishing mortality.

Species	MOS	BWD	BWY	BIO	Annual Maintenance consumption	Annual Maintenance plus growth consumption
Tunas	6	0.067	12.22	94	1,149	1,237
Sharks	12	0.020	7.30	27,545	200,968	220,088
King mackerel	12	0.070	25.54	7,957	203,190	207,626
Spanish mackerel	12	0.070	25.54	16,138	412,100	424,599
Jacks	12	0.100	36.48	7,365	268,675	273,788
Red drum	12	0.028	10.21	3,000	30,643	31,839
Atl. cutlassfish	12	0.070	25.54	3,540	90,397	93,139
Red snapper	12	0.046	16.78	14,444	242,382	251,392
Lane snapper	12	0.046	16.78	1,370	22,990	23,844
Yellowtail snapper	12	0.046	16.78	15,663	262,838	270,123
Wenchman, etc.	12	0.046	16.78	462	7,753	8,073
Other snappers	12	0.046	16.78	6,212	104,242	108,554
Grouper	12	0.028	10.21	63,135	644,886	676,908
Barracuda	12	0.028	10.21	1,810	18,488	19,406
Amberjack	12	0.100	36.48	1,058	38,596	39,330
Bottlenose dolphin	12	0.112	40.86	2,000	81,715	82,008
TOTAL				171,793	2,631,012	2,731,954

NOTE: MOS = months, BWD = daily consumption as percent body weight, BWY = yearly consumption as percent body weight, and BIO = total weight (biomass) of the group.

Maintenance consumption = MOS x BWD x BIO
Growth consumption = {1 - exp[-(M+F)]} x BIO/0.65
Units for BIO and consumption are metric tons.
See Table 6.3 for M and F.

prey species in resource-survey data and in existing and developing fisheries. This information will become increasingly valuable as more predator species are added, by region, to our food habits data set.

Tables 6.7, 6.8, and 6.9 show the dominant prey of several coastal pelagic species from U.S. Gulf waters off Louisiana, Texas, and northwestern Florida, respectively. The prey taxa are shown as a percentage of stomach volume; this value is more closely related to the percentage of energy from each food source than the percentage of number or frequency. Only those taxa making up at least 1.0% of volume are listed.

Four main generalizations can be observed from Tables 6.7-6.9. First, the same prey species often is important in the diets of several

Table 6.6. Documentation of energy-requirement estimates (BWD) listed in Tables 6.4 and 6.5.

Species measured	Species applied to	Method	Condition	Size	Treatment of original data	Source
Yellowfin tuna	Yellowfin	Cesium	field	1-2 yr		Olson & Boggs (1986)
Skipjack tuna	Skipjack	O_2 consump.	routine	1 kg	x2/.65	Kitchell et al. (1978)[a]
Albacore	Albacore Bigeye Blackfin Other tuna	O_2 consump.	routine	10 kg	x2/.65	Graham & Laurs (1982)
Swordfish	Swordfish	Stomach cont.	field			Stillwell & Kohler (1985)
Lemon shark	Shark[b]	O_2 consump.	routine	small	x2/.65	Nixon & Gruber (1988)
Chub mackerel	K. mack. Sp. mack. Bluefish Dol. fish Lt. tunny Wahoo & bonito	Food consump.	feeding		x2	Hatanaka & Takahashi (1960)
Seatrout	Drum Grouper	O_2 consump.	routine		x2/.65	Wohschlag & Wakeman (1978)
Red snapper	Reef-fish	O_2 consump.	routine		x2/.65	Wakeman et al. (1979)
Amberfish	Amberjack Jacks	Food consump.	feeding		x2	Hatanaka & Murakawa (1958)
Tursiops	Dolphin	O_2 consump.	resting	170 kg	x2/.65	Irving et al. (1941)

[a] Kitchell et al. (1978) made these revisions in the original data they used.
[b] The value for BWD that we used in calculations of energy consumption by sharks (Tables 6.4 and 6.5) was lower than the rate of 0.042 that we calculated based on Nixon and Gruber's (1988) figure from respirometer studies (at 23°C). We used approximately one-half that amount--0.02--to generalize for both sluggish and active species. Values for spiny dogfish were approximately 0.02, but are not exactly comparable because the temperature was 10°C.
NOTE: BWD = daily energy requirement, expressed as proportion of body weight.

Table 6.7. Major prey species of selected predators in Louisiana, ranked in order of relative volume in stomachs.

Species	Rank (%)
KING MACKEREL PREY SPECIES	
Sand seatrout	18.2
Atlantic croaker	12.4
Gulf menhaden	9.5
Atlantic thread herring	7.8
Blue runner	5.0
Atlantic cutlassfish	3.1
Other (<1% each)	44.0
SPANISH MACKEREL PREY SPECIES	
Gulf menhaden	10.4
Anchovies	7.1
Atlantic croaker	3.6
Pinfish	3.1
Atlantic bumper	2.9
Scaled sardine	2.8
Spanish sardine	2.9
Other (<1% each)	67.2
BLUEFISH PREY SPECIES	
Round scad	16.2
Gulf menhaden	15.9
Atlantic croaker	8.1
Squid	3.1
Blue runner	1.3
Other (<1% each)	55.4
LITTLE TUNNY PREY SPECIES	
Butterfish	25.0
Penaeid shrimp	9.7
Kingfish	6.1
Seabass	4.9
Silver seatrout	4.1
Squilla	3.3
Longspine porgy	3.0
Other (<1% each)	43.9

NOTE: Compiled from Saloman and Naughton (1983), Saloman and Naughton (1985), Naughton and Saloman (1984), and Manooch et al. (1985).

Table 6.8. Major prey species of selected predators in Texas, ranked in order of relative volume in stomachs.

Species	Rank (%)
KING MACKEREL PREY SPECIES	
Atlantic cutlassfish	24.9
Atlantic bumper	6.9
Spanish sardine	3.9
Round scad	3.8
Atlantic thread herring	2.1
Spanish mackerel	1.8
Penaeid shrimp	1.8
Other (<1% each)	45.2
SPANISH MACKEREL PREY SPECIES	
Anchovies	21.5
Atlantic bumper	10.5
Atlantic cutlassfish	9.0
Atlantic croaker	6.5
Round scad	5.2
Squid	2.4
Spanish sardine	2.2
Gulf menhaden	2.1
Penaeid shrimp	2.0
Seatrout	1.0
Other (<1% each)	61.4
LITTLE TUNNY PREY SPECIES	
Atlantic cutlassfish	42.6
Pink shrimp	12.6
Squilla	3.6
Squid	1.9
Gulf butterfish	1.3
Other (<1% each)	62.0

From Browder et al., in prep.

Table 6.9. Major prey species of selected predators in Northwest Florida, ranked in order of relative volume in stomachs.

Species	Rank (%)
KING MACKEREL PREY	
Round scad	29.9
Squid	10.6
Spanish sardine	9.6
Atlantic bumper	1.5
Blue runner	1.2
Other (<1% each)	47.2
SPANISH MACKEREL PREY	
Round scad	10.6
Spanish sardine	6.3
Anchovies	6.0
Squid	2.3
Scaled sardine	1.6
Atlantic bumper	1.2
Other (<1% each)	72.0
BLUEFISH PREY	
Atlantic bumper	11.1
Atlantic croaker	7.2
Menhaden	6.8
Round scad	1.8
Spot	1.7
Squid	1.3
Other (<1% each)	70.1
LITTLE TUNNY PREY	
Spanish sardine	61.0
Round scad	1.8
Anchovies	8.8
Squid	2.8
Other (<1% each)	15.6

From Browder et al., in prep.

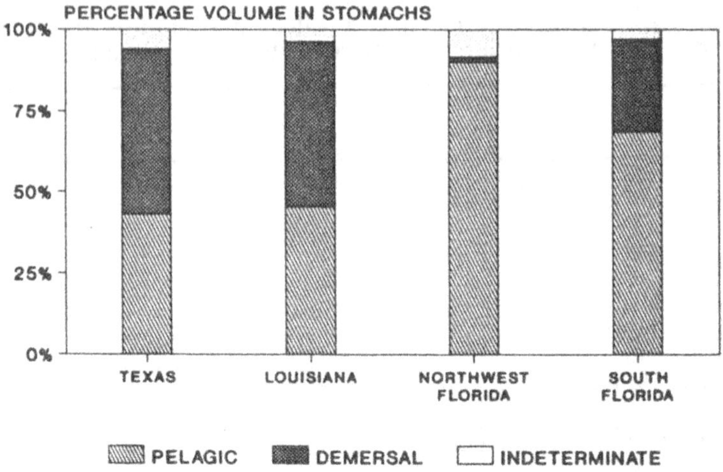

Figure 6.8. Relative volumes of demersal and pelagic prey in the diet of king mackerel in three regions of the Gulf of Mexico. (Redrawn from Browder et al., in prep.)

predator species within a region, possibly reflecting its local abundance. This is certainly true in the case of Gulf menhaden, which appears prominently in the diets of king mackerel, Spanish mackerel, and bluefish in Louisiana waters (Table 6.7). An analysis by Browder et al. (in prep.) indicated that demersal species are major components of the king mackerel's diet in Louisiana and Texas; whereas, off northwestern Florida, the major prey species are predominantly pelagic (Figure 6.8). Predators eating demersal species do not necessarily feed on the bottom; many demersal species move vertically in the water column and can be taken at any depth.

Second, there are some differences in the diets of various predator species operating in the same region. For instance, the diet of little tunny in Louisiana emphasizes butterfish and penaeid shrimp, whereas the other three predator species in Table 6.7 all eat significant quantities of both Gulf menhaden and Atlantic croaker.

Third, the diet of a given predator species differs somewhat among areas. Some of the same species are eaten in one area as in another, but in different proportions. A species that is important in the predator's diet in one area may be insignificant in another. Dietary differences undoubtedly reflect regional differences in availability. This is best reflected in the the Spanish mackerel diet, which emphasizes Atlantic croaker and Gulf menhaden in Louisiana and Texas, but not in northwestern Florida.

Fourth, usually half or more of the diet of each predator in the tables is composed of minor prey species, each of which comprise no more than 1% of stomach contents by volume. This component of each predator's diet is composed of hundreds of prey species, each present in a very small proportion. An exception in is little tunny off northwestern Florida, in which major prey species make up 84% of stomach contents volume. The opposite extreme is illustrated by Spanish mackerel off northwestern Florida, whose diet is only 28% major species.

Prey Biomass

Estimates of prey biomass in U.S. and Mexican waters are based primarily upon fishery independent data (Tables 6.10-6.12). With the exception of shrimp and menhaden, the prey species are not heavily fished; therefore reliable estimates of fishing mortality and VPAs for these species are lacking. Fishery-independent data (e.g., trawl surveys) were used to make specific estimates for each of four regions of the U.S. Gulf (Table 6.12). Prey biomass in both U.S. and Mexican waters totals approximately 8.4 mmt. This biomass is confined primarily to the coastal shelf, inside 200 m. The prey biomass estimates should be viewed as first approximations, which will be improved as our resource-survey program continues and expands, and survey design data analysis techniques become further refined. Demersal surveys have been conducted in the Gulf of Mexico by NOAA/NMFS/SEFC for many years, but surveys directed at pelagic species are relatively recent and still in the experimental stage.

Discussion and Conclusions

The information presented here, although limited, adds to our understanding of the Gulf of Mexico ecosystem. Because of the eclectic diet of many Gulf predators, it is useful, as a first-cut approach, to quantify their annual consumption in a generalized way. We have shown, however, that there are important differences in prey consumption between areas and between species feeding in the same area. Furthermore, feeding is to some extent species-, size-, and habitat-specific. Relative size affects who eats whom. Behavior, which differs by species, affects catchability. Habitat determines the spectrum of accessible prey. Food quality, which differs among species, influences assimilation and growth rates. While it seems legitimate to compare total prey biomass and total predator consumption, it is likely that comparisons taking species, size, and habitat into account will be more realistic.

The rate of consumption relative to the rate of production, rather than to total biomass, is the relevant issue. Production is a function

Table 6.10. 1988 catch and estimated biomass of major prey species in the U.S. Gulf of Mexico.

Taxon	Catch mt	Biomass mt
COASTAL PELAGICS		
Spanish sardines	[a]1,590	[b]420,000
Atl. thread herring	[a]723	[b]350,000
Round scad (cigarfish)	[a]845	[b]350,000
Unclassified bait	[a]1,729	
Scaled sardine		[b]515,000
Round herring		[b]490,000
Atlantic bumper		[b]270,000
Anchovies		[b]180,000
Rough scad		[b]70,000
Silver driftfish		[b]6,000
Ballyhoo	[a]221	[c]10,000
Squid	[a]71	[d]100,000
Butterfish	[a]2,833	[e]177,000
ESTUARINE-RELATED		
Gulf menhaden	[f]623,700	[g]2,500,000
Striped and white mullet	[h]11,410	[i]100,000
Shrimp	[j]60,360	[k]10,060
Flounder, spot, croaker, etc.	[a]874	
Small croakers etc.	[a]895	
Bottomfish discards from shrimp trawls	[b]511,000	[b]806,000
REEF-FISH		
Grunts and other small reef-fish	[l]4,009	[m]20,000
TOTAL	[n]709,260 [o]1,220,260	6,374,060

Table footnotes follow on next page.

Footnotes to Table 6.10.

[a]Small pelagics landings: Species such as scaled sardine, round herring, Atlantic bumper, etc., mainly caught in purse seines. 1988 data from unpublished statistics of NOAA/NMFS/SEFC, Miami, FL.

[b]Small pelagics biomass and bottomfish discards and bottomfish biomass: See Table 6.12.

[c]Ballyhoo biomass: Gross assumption.

[d]Squid biomass: Gross assumption.

[e]Butterfish biomass: Herron, R.C. (1988) Estimate is based on 1985 cruise data.

[f]Menhaden landings: For 1988 (SEFC, 1989b).

[g]Menhaden biomass: Value is for January 1, 1983, and was derived from a VPA analysis (Vaughan, 1987).

[h]Mullet landings: For 1988 (Unpublished data, NMFS/SEFC, Miami, FL).

[i]Mullet biomass: Gross assumption.

[j]Shrimp catch: Unpublished 1988 landings data of NOAA/NMFS/SEFC, Miami, FL, converted from tail weight to whole weight by multiplying by 1.6, as roughly estimated from data in Gulf of Mexico Fishery Management Council (1981).

[k]Shrimp biomass: Estimated by dividing landings by a fishing mortality (F) of 6, based on stock assessment (Nance and Nichols, 1988).

[l]Grunts and other small reef-fish catches: Commercial landings from unpublished landings statistics of NOAA/NMFS/SEFC, Miami, FL. Recreational landings from National Recreational Survey, compiled by P. Goodyear (1988). Catch in weight is underestimated because weight is not calculated for fish groups having no determined average weights.

[m]Grunt and small reef-fish biomass: Gross assumption.

[n]Total catch, excluding discards.

[o]Total catch, including discards.

Table 6.11. 1987 catches and estimated biomass of prey species in the Mexican Gulf of Mexico.

Taxon	Catch mt	Biomass mt
COASTAL PELAGICS		
Herring, sardines, and anchovies	[a]7,036	[b]350,000
Striped mullet	[a]6,803	[b]200,000
Small carangids		[c]150,000
ESTUARINE-RELATED		
Catfish	[a]4,397	[d]20,000
Mojarra	[a]7,219	[d]50,000
Sheepshead	[a]878	[d]10,000
Flounder	[a]128	[b]800,000
Atlantic croaker and spot	[a]1,558	[e]200,000
Shrimp	[a]22,272	[f]8,000
REEF-FISH		
Angelfish, etc.	[a]105	[g]10,000
Ray	[a]595	[g]40,000
Grunt	[a]1,040	[g]50,000
OTHER PREY SPECIES (probably shrimp by-catch)	[h]9,801	[g]10,000
Shrimp discards	[i]13,000	[g]15,000
Non-harvested groups		[g]100,000
TOTAL	[j]75,000	[k]2,013,000

Table footnotes follow on bottom of next page.

of biomass, but the ratio of production to biomass (i.e., the P/B ratio) is the critical determinant of whether a given biomass can support consumption without declining. Our first approximations suggest that the P/B ratio must be at least 1 in this system. In follow-up work, estimates of P/B ratios of major species should be obtained.

Prey biomass probably is an underestimate, but the biomass and resultant consumption estimates of the predators probably also were underestimated. The estimates based on fishery data, including those derived from VPAs, underestimate the age classes not fully recruited to the fishery. This shortcoming affects the estimates of predator biomass more than that of prey biomass, because the majority of the prey biomass estimates are based on resource survey data. The uncertainty of catchability is a weakness of the biomass estimates based on resource survey data. Catchability probably differs by species and is not well measured for any species. We used a catchability coefficient of 4, which is the best available estimate without species-specific trawl-efficiency studies.

Footnotes to Table 6.11.

[a]1987 landings statistics from Secretaria de Pesca de Mexico, 1988.

[b]Roughly estimated based on landings and ichthyoplankton surveys.

[c]Roughly estimated based on landings and ichthyoplankton surveys.

[d]Roughly estimated.

[e]Roughly estimated based on general ecology of area.

[f]Estimated by dividing landings by fishing mortality = 3, approximately one half the U.S. Gulf of Mexico level.

[g]Roughly estimated.

[h]According to R. Juhl (pers. comm.) this "unidentified" part of landings probably is the by-catch of the shrimp fleet and consists of small, primarily demersal, species.

[i]Estimated by assuming the fish/shrimp ratio is approximately 1:1. If shrimp = 22,272 and the rest of the catch = 9,801, then the discards = approximately 13,000.

[j]Rounded total of above.

[k]Rounded total of above.

Table 6.12. Estimates of prey biomass (metric tons) in the Gulf of Mexico, by area and taxon.

Taxon	Southeast Gulf	Northeast Gulf	N. Central Gulf	Northwest Gulf	Total U.S. Gulf
By-catch in shrimp trawls					511,000
Biomass of by-catch species	50,000	50,000	262,000	444,000	806,000
SMALL PELAGICS					
Thread herring	240,000	50,000 (a)	50,000	10,000 (a)	350,000
Round herring	380,000	10,000	100,000	0	490,000
Spanish sardine	250,000	100,000	20,000 (a)	50,000 (a)	420,000
Scaled sardine	185,000	20,000	50,000	260,000	515,000
Atlantic bumper	0	20,000 (a)	100,000	150,000 (a)	270,000
Round scad	100,000	100,000 (a)	100,000 (a)	50,000 (a)	350,000
Rough scad	0	20,000	50,000 (a)	0	70,000
Silver driftfish		6,000	0	0	6,000
Anchovies	10,000 (a)	20,000 (a)	50,000 (a)	100,000 (a)	180,000
Total small pelagics	1,115,000	346,000	520,000	620,000	2,651,000

NOTE: Some biomass estimates are roughly based on catch per unit area, multiplied by the total area of each subregion of the U.S. Gulf of Mexico. Inshore of 100 meters, the southeast, northeast, north-central, and northwest subregions encompass 34, 23, 13, and 30% of the total area, respectively (based on areal estimates by Patella [1974]). Inside approximately 200 meters, they cover 31, 24, 12, and 33% of the total area, respectively. The total area inside 100 m is roughly 26.4 million hectare. The total area inside 200 m is roughly 30 million hectare. Subregion boundaries are as follows: Eastern Gulf is area south of 29°N. Northeastern Gulf is area north of 29°N and east of 88°W. North-central Gulf is area west of 91°W.

Shrimp by-catch estimates are from Pellegrin (1982) and are based on observer data on fish/shrimp weight ratios from the 1970s. Although biomass may have declined in the north-central Gulf since that time, shrimp trawling effort has increased to such an extent that the magnitude of by-catch probably has remained about the same. This stability of by-catch magnitude was suggested in an analysis by Nichols et al., 1987) of by-catch and resource survey data for 13 by-catch species in the north-central and western Gulf.

Biomass of shrimp by-catch species was calculated for the North-central Gulf and for Western Louisiana and Texas using kilogram per hectare caught in research trawls, area covered by the research cruise, and an assumed catchability coefficient of 0.25. The area covered by North-central Gulf cruise data extended from 88°-91°30' W and from 5 to 50 fathoms. The western Louisiana and Texas cruise data extended from 91°30' to the Mexican border. From Patella (1975), we determined that the North-central Gulf cruise area covered 2,551,769 hectare. Biomass of shrimp by-catch species in the Eastern and Northeastern Gulf was roughly approximated based on knowledge that the fish/shrimp ratio in the by-catch was only about 1.5 in these regions, compared to about 3.5 in Texas and even greater in the North-central Gulf. In addition to fish, the by-catch biomass consists of a significant quantity of invertebrates, principally crabs and non-penaeid shrimp.

Coastal pelagic biomass estimates for the Eastern Gulf were rounded from Houde (1976), which were estimated from larval surveys.

Coastal pelagic biomass estimates for the Northeast Gulf and North-central Gulf were obtained from unpublished data compiled by C. Gledhill, NOAA/NMFS/SEFC, Pascagoula, MS. We assumed a trawl catchability coefficient of 0.25 and multiplied Gledhill's cruise data by 4 to use in our estimates. Items indicated with an 'a' have been adjusted upward from the original estimates. The upward adjustment was based on (1) landings statistics for the area, (2) stomach contents analyses of predators in the area, (3) comments of cruise-data experts.

Coastal pelagic biomass estimates for the Central and Western Gulf were very roughly approximated from Shaw et al. (1989) from relative number of larvae per unit area in the central and western areas to that in the eastern Gulf and also considering the relative areal extent and general environments of the three areas.

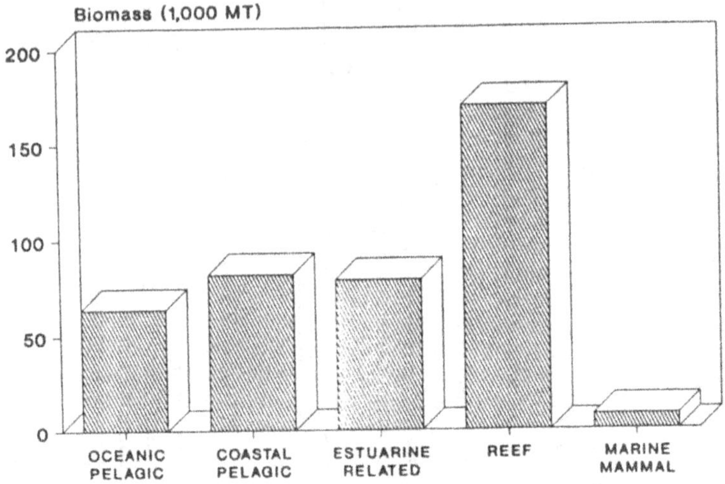

Figure 6.9. Distribution of the Gulf of Mexico biomass of predatory fishery species among ecological groups.

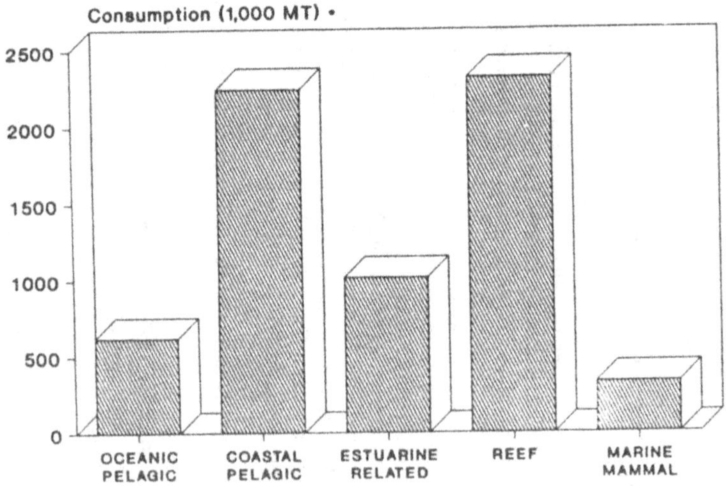

Figure 6.10. Distribution of consumption, or food demand, among predator ecological groups.

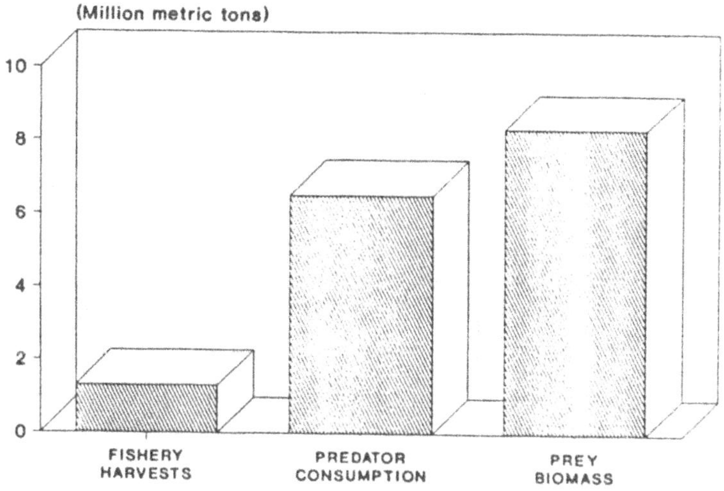

Figure 6.11. Consumption by predators, current fishery catches of prey species (including discards), and prey biomass.

We have given some special attention to the coastal pelagic species because they are the group most likely to be affected by presently developing fisheries on small pelagic prey species. Figures 6.9 and 6.10 allow the estimated biomass and estimated consumption of the coastal pelagic predators to be compared to that of other predator groups. The coastal pelagic predators seem to require a higher proportion of prey relative to their biomass than the other groups. If their support prey are reduced, it is unlikely they will find substitutions elsewhere in the ecosystem.

Estimated total consumption by Gulf predators is 6.6 mmt, and estimated prey biomass is 8.4 mmt (Figure 8.11). Most of the estimated consumption is by fish species supporting valuable, fully developed fisheries. Present harvests of prey species, which include discards from shrimp trawls, as well as shrimp and menhaden, are estimated at 1.3 mmt. In order to meet predator demand and maintain biomass stability over time, prey stocks would have to replace their biomass annually (i.e., P/B ~1) just to maintain biomass stability. This suggests a "tight" system in which prey production may already be fully utilized (see Sissenwine et al. [1984]) for a discussion of the Georges Bank ecosystem in this context. Of course, prey biomass is never fully utilized because feeding efficiency decreases as a function of prey density.

These first-approximation results indicate that expanded harvests of prey species should be accompanied by research and evaluation of impacts on predator species. The rate of increase in harvests should not be allowed to exceed the ability of research to provide the appropriate evaluation of expected results.

Increased fishing pressure on newly exploited stocks may lead to some increased Gulf of Mexico harvests overall; however, harvests of prey species may have long-term negative effects on production of currently harvested species. Some substitution of lower-trophic-level species for current target species might be possible in commercial markets, but much of the Gulf harvest of predator species is recreational (Figure 6.3), and substitution of lower-trophic-level species for currently sought-after recreational species is less feasible. The recreational fisheries of the Gulf of Mexico are important to local and state economies.

Fishery development should proceed cautiously and should be accompanied by research on important ecological relationships. Multispecies effects should be incorporated into fisheries research, building upon and expanding beyond the tradition of fisheries research aimed at individual species.

In the Southeast Fisheries Center, the overall objective of fisheries research is to provide scientific information to promote management decisions that will maintain the robustness of stocks without overly restricting harvests. Learning more about the roles of fishery species in the ecosystem is of more than theoretical interest. This information is critical in determining how the harvest of one species will affect another. The full-scale pursuit of a better quantitative understanding of this issue is urgent, given what we already know about the ecological roles of the existing and emerging fishery species. Needed is a combined approach of research, modelling, management, and monitoring with continual feedback loops throughout so that decisions are made on the basis of current scientific understanding, results are monitored, models are improved, and management measures adjusted.

Acknowledgments

This paper was supported by information provided by many persons. In particular, we would like to thank Christopher Gledhill, Rex Herron, and Perry Thompson at the Mississippi Laboratories of the Southeast Fisheries Center in Pascagoula and Bay St. Louis, Mississippi, for providing data from resource surveys. We also are particularly indebted to Phil Goodyear and Nancie Parrack at the Miami Laboratory of the Southeast Fisheries Center for access to both published and unpublished material from stock assessments.

References

Browder, J. A., B. E. Brown, W. Nelson, and N. Bane. In press. Multi-species fisheries in the Gulf of Mexico. *In* M. Sissenwine (Ed.). Multi-species fisheries symposium proceedings. 1989 Annual Meeting of International Council for the Exploration of the Sea.

Browder, J. A., and E. D. Prince. 1991. Standardized estimates of recreational fishing success for blue and white marlin in the western North Atlantic Ocean, 1972-1986. *In* R. H. Stroud (Ed.). Proceedings 1988 Billfish Symposium.

Browder, J. A., C. H. Saloman, S. P. Naughton, and C. S. Manooch, III. In prep. Trophic relations of king mackerel in the coastal shelf ecosystem.

Goodyear, C. P. 1988. The Gulf of Mexico fishery for reef fish species--a descriptive profile. NOAA, National Marine Fisheries Service, Southeast Fisheries Center, Miami Laboratory, Coastal Resources Division, Miami, FL. Contribution CRD 87/88-19. 6 Pp., 23 figures, and 102 tables.

Goodyear, C. P. 1989. Status of the red drum stocks of the Gulf of Mexico report for 1989. NOAA, National Marine Fisheries Service, Southeast Fisheries Center, Miami Laboratory, Coastal Resources Division, Miami, FL. Contribution CRD 88/89-14. 63 Pp.

Goodyear, C. P., and P. Phares. 1990. Status of the red snapper stocks of the Gulf of Mexico report for 1990. NOAA, National Marine Fisheries Service, Southeast Fisheries Center, Miami Laboratory, Miami, FL. Contribution CRD 89/90-05.

Graham, J. B., and R. M. Laurs. 1982. Metabolic rate of the albacore tuna *Thunnus alalunga*. Mar. Biol. 72:1-6.

Gulf of Mexico Fishery Management Council. 1981. Fishery Management Plan for the shrimp fishery of the Gulf of Mexico, United States waters. Gulf of Mexico Fishery Management Council. Tampa, FL. Separately numbered sections.

Hatanaka, M. A., and M. Takahashi. 1960. Studies on the amounts of the anchovy consumed by the mackerel. Tohoku J. Agric. Res. 11:83-100.

Hatanaka, M. and G. Murakawa. 1958. Growth and food consumption in young amberfish, *Seriola quinqueradiata* (T. et S.). Tohoku J. Agric. Res. 9:69-79.

Herron, R. C. 1988. Annual report on latent resources. NOAA, National Marine Fisheries Service, Southeast Fisheries Center, Mississippi Laboratories, Stennis Space Center, MS.

Herron, R. C., T. L. Leming, and J. Li. 1989. Satellite-detected fronts and butterfish aggregations in the northeastern Gulf of Mexico. Cont. Shelf Res. 9:569-588.

Hoar, W. S., D. J. Randall, and J. R. Brett. 1979. Fish Physiology, Vol. VIII: Bioenergetics and Growth. Academic Press, New York. 786 Pp.

Houde, E. D. 1976. Abundance and potential for fisheries development of some sardine-like fishes in the eastern Gulf of Mexico. Proc. Gulf Carib. Fish. Inst. 28:73-82.

ICCAT SCRS. 1989a. Swordfish stock assessment for the year 1989. Standing Committee Report. Int. Comm. Conserv. Atl. Tunas. Madrid, Spain.

ICCAT SCRS. 1989b. Bluefin tuna stock assessment for the year 1989. Standing Committee Report. Int. Comm. Conserv. Atl. Tunas. Madrid, Spain.

Irving, L., P. F. Scholander, and S. W. Grinnell. 1941. The respiration of the porpoise, *Tursiops truncatus*. J. Cell. Comp. Physiol. 17:145-168.

Kitchell, J. F., W. H. Neill, A. E. Dizon, and J. J. Magnuson. 1978. Bioenergetic spectra of skipjack and yellowfin tunas. Pp. 357-368 *In* G. D. Sharp and A. E. Dizon (Eds.). The physiological ecology of tunas. Academic Press, New York.

Manooch, C. S., III, D. L. Mason, and R. S. Nelson. 1985. Foods of little tunny, *Euthynnus alletteratus*, collected along the southeastern and gulf coasts of the United States. Bull. Jpn Soc. Sci. Fish. 51:1207-1218.

Maul, G. A., D. A. Mayer, and S. R. Baig. 1985. Comparisons between a continuous 3-year current-meter observation at the sill of the Yucatan Strait, satellite measurements of Gulf Loop Current area, and regional sea level. J. Geophys. Res. 90:9089-9096.

Mid-Atlantic Fishery Management Council. 1989. Fishery Management Plan for the Bluefish Fishery. Mid-Atlantic Fishery Management Council and Atlantic States Marine Fisheries Commission in cooperation with the National Marine Fisheries Service, the New England Fisheries Management Council, and the South Atlantic Fishery Management Council. Dover, Delaware. Separately numbered pages. 75 Pp.

Miyake, P. M., P. Kebe, D. da Rodda, S. Martin, and J. L. Gallego. 1986. Statistical Bulletin 17. Int. Comm. Conserv. Atl. Tunas. Madrid, Spain. 162 Pp.

Nance, J. M., and S. Nichols. 1988. Stock assessments for brown, white, and pink shrimp in the U.S. Gulf of Mexico, 1960-1986. NOAA Tech. Memo. NMFS-SEFC-203.

Nichols, S., A. Shaw, G. Pellegrin, Jr., and K. Mullin. 1987. Estimates of annual shrimp fleet by-catch for thirteen finfish species in the offshore waters of the Gulf of Mexico. 7 Pp.

Nixon, A. J., and S. H. Gruber. 1988. Diel metabolic and activity patterns of the lemon shark (*Negaprion brevirostris*). J. Exper. Zool. 248:1-6.

Odell, D. K. and E. O. Asper. 1982. Live capture, marking, and resighting of bottlenose dolphin, *Tursiops truncatus*. Final report, Contract No. NA80-GA-C-00063.

NMFS. 1989. Draft Secretarial Shark Fishery Management Plan for the Atlantic Ocean. NOAA, National Marine Fisheries Service, Washington, DC. 116 Pp.

Olson, R. J., and C. H. Boggs. 1986. Apex predation by yellowfin tuna (*Thunnus albacares*): Independent estimates from gastric evacuation and stomach contents, bioenergetics, and cesium concentrations. Can. J. Fish. Aquat. Sci. 43:1760-1775.

Patella, F. 1975. Water surface area within statistical subareas used in reporting Gulf coast shrimp data. Mar. Fish. Rev. 37(12):22-24.

Pellegrin, G., Jr. 1982. Fish by-catch - bonus from the sea: report of a technical consultation on shrimp by-catch utilization held in Georgetown, Guyana, 27-30 October, 1981. Ottawa, Ont., IDRC, 1982. 163 Pp.

Richards, W. J., and M. F. McGowan. 1989. Biological productivity in the Gulf of Mexico: identifying the causes of variability in fisheries. Pp. 287-325 *In* K. Sherman and L. M. Alexander (Eds.). Biomass yields and geography of large marine ecosystems. AAAS Selected Symposium 111. Westview Press, Boulder.

Robins, C. R., R. M. Bailey, C. E. Bond, J. R. Brooker, E. A. Lachner, R. N. Lea, and W. B. Scott. 1980. A List of Common and Scientific Names of Fishes from the United States and Canada. Am. Fish. Soc. Spec. Pub. No. 12. 174 Pp.

Saloman, C. H., and S. P. Naughton. 1983. Food of Spanish mackerel, *Scomberomorus maculatus*, from the Gulf of Mexico and southeastern seaboard of the United States. NOAA, National Marine Fisheries Service, Southest Fisheries Center, Panama City Laboratory, Panama City, FL. NOAA Tech. Memo. NMFS-SEFC-128. 22 Pp.

Saloman, C. H., and S. P. Naughton. 1985. Food of king mackerel, *Scomberomorus cavalla*, from the southern United States, including the Gulf of Mexico. NOAA, National Marine Fisheries Service, Southeast Fisheries Center, Panama City Laboratory, Panama City, FL. NOAA Tech. Memo. NMFS-SEFC-126. 25 Pp.

Sanders, N. J., Jr., T. Van Devender, and P. A. Thompson (Editors). 1990. SEAMAP Environmental and Biological Atlas of the Gulf of Mexico, 1986. Gulf States Marine Fisheries Commission Publication No. 20. Ocean Springs, MS. 327 Pp.

Scott, G. P., D. M. Burn, L. J. Hansen, and R. E. Owen. 1989. Estimates of bottlenose dolphin abundance in the Gulf of Mexico from regional aerial surveys. Contribution No. CRD-88/89-07. NOAA, National Marine Fisheries Service, Southeast Fisheries Center, Miami Laboratory, Miami, FL. 24 Pp.

Secretaria de Pesca de Mexico. 1988. Anuario estadistico de pesca 1987. Direccion Genereal de Informatica, Estadistica y Documentacion. Mexico, D.F. 351 Pp.

Southeast Fisheries Center [SEFC]. 1989a. Report of the NMFS swordfish stock assessment workshop. NOAA, National Marine Fisheries Service, Southeast Fisheries Center, Miami Laboratory, Miami, FL. 53 Pp.

SEFC. 1989b. Forecast of 1989 Gulf and Atlantic menhaden purse-seine landings. NOAA/National Marine Fisheries Service, Southeast Fisheries Center, Beaufort Laboratory, Beaufort, NC.

SEFC. 1990. 1990 report of the mackerel stock assessment panel. NOAA, National Marine Fisheries Service, Southeast Fisheries Center, Miami, FL. Miami Laboratory Contribution MIA-89/90-7. 30 Pp.

Shaw, R. F., J. G. Ditty, J. Lyczkowski-Shultz, and R. H. Blanchet. 1989. Fisheries-independent data on coastal herrings, carangids, and red drum from the northern Gulf of Mexico. Final Report. MARFIN Project No. NA 88-WC-H-MF198. 106 Pp.

Sissenwine, M. P., E. B. Cohen, and M. D. Grosslein. 1984. Structure of the Georges Bank ecosystem. Rapp. P.-v. Reun. Cons. int. Explor. Mer 183:243-254.

Stillwell, C. E., and N. E. Kohler. 1985. Food and feeding ecology of the swordfish *Xiphias gladius* in the western north Atlantic Ocean with estimates of daily ration. Mar. Ecol. Prog. Ser. 22:239-247.

Vaughan, D. S. 1987. Stock assessment of the gulf menhaden, *Brevoortia patronus*, fishery. U.S. Dep. Commer. NOAA Tech. Rep. NMFS 58. 18 Pp.

Vukovich, F. M., and G. A. Maul. 1985. Cyclonic eddies in the eastern Gulf of Mexico. J. Phys. Oceanogr. 15:105-117.

Wakeman, J. M., C. R. Arnold, D. E. Wohlschlag, and S. C. Rabalais. 1979. Oxygen consumption, energy expenditure, and growth of the red snapper (*Lutjanus campechanus*). Trans. Am. Fish. Soc. 108:288-292.

Winberg, G. G. 1960. Rate of metabolism and food requirements of fishes. Fish. Res. Bd. Can. Translation Series No. 194.

Wohlschlag, D. E., and J. M. Wakeman. 1978. Salinity stresses, metabolic responses, and distribution of the coastal spotted seatrout, *Cynoscion nebulosus*. Contrib. Mar. Sci. 21:171-185.

Appendix

Table 6.13. Scientific names for common names used in text or tables (fish names are based on Robins et al. [1980]).

Common name	Scientific name
Albacore	*Thunnus alalunga*
Amberjack	*Seriola* spp.
Anchovies	*Anchoa* spp.
Angelfish	Pomacanthidae
Atlantic bonito	*Sarda sarda*
Atlantic bumper	*Chloroscombrus chrysurus*
Atlantic croaker	*Micropogonias undulatus*
Atlantic cutlassfish	*Trichiurus lepturus*
Atlantic thread herring	*Opisthonema oglinum*
Ballyhoo	*Hemiramphus brasiliensis*
Barracuda	*Sphyraena* spp.
Bigeye tuna	*Thunnus obesus*
Black drum	*Pogonias cromis*
Black grouper	*Mycteroperca bonaci*
Blackfin tuna	*Thunnus atlanticus*
Bluefin tuna	*Thunnus thynnus*
Bluefish	*Pomatomus saltatrix*
Blue marlin	*Makaira nigricans*
Blue runner	*Caranx crysos*
Bottlenose dolphin	*Tursiops truncatus*
Catfish	Ariidae
Chub mackerel	*Scomber japonicus*
Cobia	*Rachycentron canadum*
Crevalle jack	*Caranx hippos*
Dolphin (fish)	*Coryphaena* spp.
Flounder	Bothidae
Gag	*Mycteroperca microlepis*
Groupers	Serranidae
Grunts	Haemulidae
Gulf menhaden	*Brevoortia patronus* or *gunteri*
Gulf butterfish	*Peprilus burti*
Jacks (includes crevalle jack, blue runner, permit, pompano, etc.	Carangidae
Kingfish	*Menticirrhus* spp.
King mackerel	*Scomberomorus cavalla*
Ladyfish	*Elops saurus*
Lane snapper	*Lutjanus synagris*

Table 6.13 continued.

Common name	Scientific name
Lemon shark	*Negaprion brevirostris*
Little tunny	*Euthynnus alletteratus*
Longspine porgy	*Stenotomus caprinus*
Mojarra	*Eucinostomus* spp.
Permit	*Trachinotus falcatus*
Pinfish	*Lagodon rhomboides*
Pink shrimp	*Penaeus duorarum*
Pompano	*Trachinotus* spp.
Red drum	*Sciaenops ocellatus*
Red grouper	*Epinephelus morio*
Red snapper	*Lutjanus campechanus*
Ray	Dasyatidae, Myliobatidae, Mobulidae
Rough scad	*Trachurus lathami*
Round scad	*Decapterus punctatus*
Round herring	*Etrumeus teres*
Sailfish	*Istiophorus platypterus*
Sand seatrout	*Cynoscion arenarius*
Scaled sardine	*Harengula jaguana*
Seabass	*Centropristis* spp. *Diplectrum* spp.
Seatrout	*Cynoscion* spp.
Sharks	Carcharhinidae
Sheepshead	*Archosargus probatocephalus*
Shrimp	*Penaeus spp.*
Silver driftfish	*Psenes maculatus*
Silver seatrout	*Cynoscion nothus*
Skipjack tuna	*Euthynnus pelamis*
Snappers	Lutjanidae
Snook	*Centropomus spp.*
Spot	*Leiostomus xanthurus*
Spanish mackerel	*Scomberomorus maculatus*
Spanish sardine	*Sardinella aurita*
Spiny dogfish	*Squalus acanthias*
Squilla	*Squilla* spp.
Squid	Loliginidae
Striped mullet	*Mugil cephalus*
Swordfish	*Xiphias gladius*
Vermillion snapper	*Rhomboplites aurorubens*
Wahoo	*Acanthocybium solanderi*

Table 6.13 continued.

Common name	Scientific name
Wenchman	*Pristipomoides aquilonaris*
White marlin	*Tetrapturus albidus*
White mullet	*Mugil curema*
Yellowedge grouper	*Epinephelus flavolimbatus*
Yellowfin tuna	*Thunnus albacares*
Yellowtail snapper	*Ocyurus chrysurus*

Michael M. Mullin

7. Spatial-Temporal Scales and Secondary Production Estimates in the California Current Ecosystem

Abstract

Most larval and juvenile fish depend on secondary production by zooplankton, rather than primary production, as a source of food. The relevant spatial/temporal scales range from seconds and centimeters for capture of individual prey to thousands of kilometers and interannual variability due to El Niño events, and there are good examples at both extremes in the California Current. The conceptual problem is to understand how these scales are related to variation in recruitment to fish populations.

Introduction

Primary production is usually defined in terms of a distinctive mode of nutrition (photoautotrophy), and is generally measured by an accepted, if not entirely understood, technique (uptake of $H^{14}CO_3^-$ by unconcentrated phytoplankton--actually, all seston--under specified illumination). The primary producing biomass is measurable, with little ambiguity, as chlorophyll retained on a fine filter.

For secondary production, which refers conceptually to heterotrophic growth based on phagotrophic ingestion of primary producers (i.e., herbivory), both traditional sampling and measurement are much more ambiguous. In fact, there are few exclusively herbivorous, free-living, pelagic animals, because some detrital organic carbon particles, protozoans, and young stages of metazoans are indistinguishable in size from phytoplanktonic cells, and undoubtedly contribute to the nutrition of most so-called herbivores (which are more properly called particle-grazers). Hence, secondary production as applied to the pelagic environment usually means growth in biomass (both somatic growth of individuals and gonadal growth resulting in gametes) of those types of zooplankton which do not appear to be exclusively macrophagous carnivores. Further, "zooplankton" is usually defined operationally, by the catch of a net of some particular mesh.

In addition to these definitional problems, the concept of secondary production does not encompass the subsequent fate of the biomass produced; biomass changes through time because of the difference between production and mortality (and, particularly in the pelagic, immigration minus emigration).

Beyond whatever intrinsic interest it may have as a step in the functioning of a food chain, the rate of secondary production is generally thought to affect the recruitment of commercially important species of finfish. Put simply, secondary production is viewed as the rate of supply of food for young fish, and therefore to be significant whenever recruitment is food-limited. The pragmatic goal in studying secondary production, from this perspective, is to understand rather large-scale processes, such as year-class success over the range of a population. Recruitment success may be statistically correlated with large-scale, average secondary production (though in fact unambiguous evidence on this point is quite sparse), but the causal mechanisms--how some fraction of secondary production actually becomes healthy young fish--act on much smaller scales. Further, knowledge of crucial, small-scale processes may be essential to understanding *variability* in the food web, and in recruitment, even if the steady-state situation seems comprehensible by homogenizing or averaging over these small scales.

In a sense, current societal concern over global climate change raises the converse problem. Given the likelihood that global climate models can predict large-scale changes in the atmosphere and ocean, can we deduce successfully from these any large-scale changes in secondary production and the recruitment for which it provides the fuel, or must we reason down the scale to the fundamental events of secondary production (the balance between an individual's assimilation and respiration of organic matter), understand how physics and chemistry on that scale will be altered, and how this will affect the fundamental events, and then reason back up the scale to whole populations and ecosystems?

Given these heady issues, I will first outline the general methods which have been used to estimate secondary production, showing by examples how they have been used to investigate different scales of secondary production in the California Current, and I will then apply the methods to data from the ongoing California Cooperative Oceanic Fisheries Investigations (CalCOFI). I must confess to a sense of *deja vu*, however, since the first major review I wrote 20 years ago concerned secondary production (Mullin, 1969), and, frankly, though many more estimates have now been published, relatively little has changed fundamentally in the intervening years.

General Methods and Examples of Applications in the California Current

Extrapolation from Primary Production

Since secondary production is defined as that heterotrophic production fueled by primary production, the former could be calculated from measurements of the latter if the efficiency of transfer is either relatively constant or predictably variable. Unfortunately, the same complexity of the pelagic food web which makes pure herbivory difficult to define also creates variation in the ratio of secondary to primary production (Steele, 1965; Isaacs, 1972).

The attractiveness of what can be called the "Lindeman" (1942) or "Ryther" (1969) approach is that primary production is measured by a relatively standard, simple, and direct incubation technique (though interpretive issues are not fully resolved, e.g., Peterson, 1980). Each measurement is made on a few hundred ml of seawater and its seston, but to calculate secondary production on this scale makes an unwarranted assumption of steady state. Given the differences in generation times, daily ambits, and spatial distributions between phytoplankton and zooplankton, this approach is intrinsically unsuitable to estimate secondary production on equally small scales (except possibly for protozoans). For herbivores or for zooplankton as a whole, one must integrate over many estimates of primary production, weighted for the actual spatial distributions of secondary producers, to achieve a single estimate for secondary production on rather large scale, without a direct estimate of its variance. This averages over the mismatches in spatial and temporal distributions on smaller scales.

There are also, however, indirect estimates of primary production over medium or large scales of space and time, such as those derived from mass balances of oxygen or nutrients (the former increased, the latter depleted by net primary production; e.g., Craig and Hayward, 1987) or from an algorithm which converts remotely sensed chlorophyll into a rate of primary production (e.g., Eppley et al., 1985). From satellite images particularly, a seductively rich pattern of secondary production could be calculated, but one whose precision and accuracy would depend on the validity of at least four algorithms--satellite-sensed signal to ocean color to chlorophyll to primary production to secondary production.

The issue of spatial/temporal scales is related to the variability of food chain efficiency in another way. Because primary production may change more rapidly than does the distribution of biomass of the secondary producers (which is a combination of behaviors and the balances of growth and death of many species), the conversion of primary into secondary production (especially secondary production utilizable by young fish) is likely to be less efficient when changes in primary production are rapid, highly localized, and ephemeral (as in sporadic, topographically restricted upwelling events) than when changes are slower and more widespread, so the herbivorous populations can "keep up."

Thus, we get at best a rather large-scale estimate, with unknown precision, by deriving secondary from primary production, and any pattern in its variability on small or meso-scales may be illusory. Such an estimate may be perfectly adequate for fisheries applications, such as the calculation of the likely rate of supply of food for young fish, and the sheer volume of available data, particularly concerning distributions of chlorophyll, may more than compensate for the uncertainties outlined above. Perhaps the most serious difficulty from the perspective of fisheries is that the form of the secondary production--whether it is as gelatinous salps, micro-copepods, or euphausiids--is unknown.

Metabolic Mass Balance

Because relatively straightforward techniques have been developed to measure rates of ingestion, assimilation, and respiration of zooplankters, indirect estimates of secondary production can be derived as assimilation minus respiration. This approach can be applied to any scale on which the metabolic rates can be estimated, and for ingestion at least, this can be as small as minutes and centimeters, using the gut fluorescence approach (e.g., Mackas and Bohrer, 1976; Huntley et al., 1987). Most such estimates have been of the secondary production of single species, or extrapolated to total zooplanktonic biomass from data on single species, since most of the metabolic measurements have been made on "representative" (= large, hardy, and easily identified) organisms. Indeed, this approach is probably better for estimating and understanding the small-scale variability in secondary production than for estimating the large-scale mean (e.g., Mullin and Brooks, 1976; Cox et al., 1983).

The Productivity/Biomass (P/B) Ratio

The basis of this approach is the quite reasonable assumption that, for zooplankton at least, the rate of growth of each unit of biomass is less variable than is the biomass itself. Hence, the pattern of production can be determined by multiplying a map of biomass by a P/B ratio. A major advantage of this approach is that past secondary production can be estimated from historical samples of zooplankton. This is particularly significant for the California Current, where the seasonal, interannual, and large-scale geographic variation in zooplanktonic biomass, and its correlation with large-scale physical transport, are particularly well known (Chelton et al., 1982) because of the California Cooperative Oceanic Fisheries Investigations (CalCOFI), which is now forty years old.

P/B can either be a simple or a temperature-dependent ratio (implying that secondary production is not food-limited), or a function dependent on temperature, food (usually determined as chlorophyll), and individual size of the organisms of concern (Huntley and Boyd, 1984). This last approach is based on generalizations derived from measurements similar to those used in the metabolic mass balance approach. Calibration measurements for enclosed populations not limited by food were performed by Huntley (1985) and McClatchie (1987).

The simple ratio can be used to determine secondary production on all scales at which biomass can be estimated, which can be as small as seconds and meters, using acoustic devices or towed particle counters (e.g., Dickey, 1988). Both these technologies can now provide crude information on individual size as well (e.g., Richter, 1985; Greene et al., 1989), and the towed devices also measure temperature and concentration of chlorophyll, so even the most complex and metabolically realistic calculation of P/B ratios can be made on very small scales and applied to the measured biomasses (see, e.g., Figure 6 of Huntley and Boyd, 1984; Boyd, 1985). To the degree that size is the primary determinant of whether a zooplankter is or is not acceptable food for young fish, this approach holds the most promise for fisheries biologists, but its initial cost in instrumentation is quite high. A "low-tech" application of metabolically realistic P/B ratios was used by Mullin and Brooks (1970) to calculate the production of *Calanus helgolandicus* (= *pacificus*) off La Jolla, California, from semiweekly measurements of biomasses and size distributions of the population and instantaneous P/B ratios which depended on individual size, temperature, and supply of food, and were derived from growth under controlled conditions in the laboratory.

Biochemical Measures

This is another indirect approach, in which biochemical properties related to secondary production--properties analogous to the fisheries biologists' "condition factor" in the broadest sense--are assayed in samples of zooplankton removed from the sea. The scale of resolution is set by the range of sizes of samples which can be collected and assayed, but the general approach has the possibility to include several spatial/temporal scales on a single sample, by measuring different properties whose response time to environmental change differs. Thus, for instance, ATP content or energy charge (= (2[ATP] + [ADP])/([ATP] + [ADP] + [AMP])) may integrate the organism's environment on the scale of minutes and meters, RNA/DNA ratio on hours and a few km, and lipid content on days and 10s-100s of km.

Hakanson (1984, 1987) has shown how the lipid contents of *Calanus pacificus* are related to its supply of food during growth, and to the concentration of phytoplankton on meso-scales in the California Current. Cox et al. (1982, 1983) used activity of the digestive enzyme, laminarinase, in zooplankton as an indicator of recent herbivorous feeding. Though neither of these examples provides a direct measure of the rate of secondary production, they illustrate the utility of biochemical techniques in understanding spatial patterns related to it.

The chief difficulty which has limited the application of this approach is the time and equipment required for the analyses once the samples are taken. Also, some early attempts fell discouragingly short of their promise (e.g., Dagg and Littlepage, 1972, for RNA/DNA).

Direct Measurements

The most direct measurement of secondary production is to determine the increase in size and change in abundance through time of individuals of a recognizable group, such as a cohort, and either to calculate an instantaneous P/B ratio from these measurements or to construct an "Allen curve" (see Mann, 1969). This approach is normally applied to particular species, rather than to total biomass, and can work very well for an annually breeding population with a distinct reproductive season, or one in which ages as well as sizes of individuals can be determined quickly and accurately, and which can be re-sampled through the year. Unfortunately, quasi-continuous breeding and advection through the sampling site make the method difficult to apply to most zooplanktonic secondary producers in the California Current, though Brinton's (1976) analysis of the population of *Euphausia pacifica* in the Southern California Bight suggests that such a calculation is possible for this important species.

The problem of re-sampling is obviated by enclosing the population under "near-natural" conditions. For example, Heinle (1966) enclosed young stages of the copepod, *Acartia tonsa*, in 4-l jugs to estimate growth in Patuxent River estuary water, and Kimmerer and McKinnon (1987) applied the method to *A. tranteri* in an Australian bay. Some quite large containers have been used (reviewed by Mullin, 1982). Interpretive difficulties are created by the degree to which the conditions used are only "near" rather than "natural."

The production of eggs by gravid female copepods can be measured through brief enclosure, or from the ratio of eggs to females in the water column and knowledge of the hatching time of eggs. This provides a measure of secondary production by females of a particular species, and Checkley (1980) used this general method to investigate the pattern of food-limitation of production by the copepod, *Paracalanus parvus*, in the Southern California Bight. Similarly, Peterson (1988) was able to correlate the fecundity of *Calanus marshallae* off Oregon with the concentration of chlorophyll, and reasoned from this how production might respond to onset and relaxation of coastal upwelling.

Berggrenn et al. (1988) and Peterson et al. (1991) argue that the rate of egg production responds to the supply of food as does somatic growth of juveniles, and so can be a general indicator of food-limitation of secondary production. However, Peterson et al. (1991) noted that the specific productivity represented by egg production--its P/B ratio--was generally less than that of juvenile growth.

Comparison of Methods in the Southern California Sector of the California Current

Several of the approaches to estimating secondary production can be compared on meso- to large-scales for cruises of the California Cooperative Fisheries Investigations (CalCOFI), which covered a fixed grid of stations in the southern California sector of the California Current, since chlorophyll, primary production, biomass of zooplankton in the upper 200 m, and rate of egg production by a particular species of zooplanktonic copepod, *Calanus pacificus*, were measured in October (cruise 8810), January (cruise 8901), April (cruise 8904), and July (cruise 8907). Methods of measurement are discussed in Mullin (1991), together with a discussion of potential ambiguities resulting from discrepancies between what was actually measured and what would be measured ideally. It suffices here to note that the reported egg production rates are probably underestimates of the *in-situ* rates.

Extrapolation from Chlorophyll (Primary Production)

The distributions of chlorophyll were characterized by high values around Point Conception, often extending southward along the outer edge of the Southern California Bight (roughly the Santa Rosa-Cortes Ridge) and along the Bight's nearshore zone (Figures 7.1-7.4, panels A). Biomasses less than 20 mg m^{-2} tended to occur in the

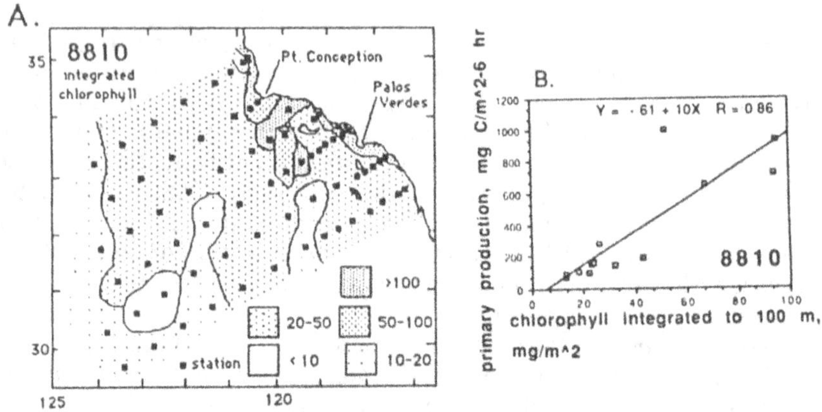

Figure 7.1. Chlorophyll in the California Current off southern California in October 1988 (Cruise 8810). (A) Chlorophyll integrated through the upper 100 m, mg m^{-2}. (B) Primary production, mg $C(m^2\ 6\ h)^{-1}$ versus integrated chlorophyll. Correlation is significant at $P<0.01$.

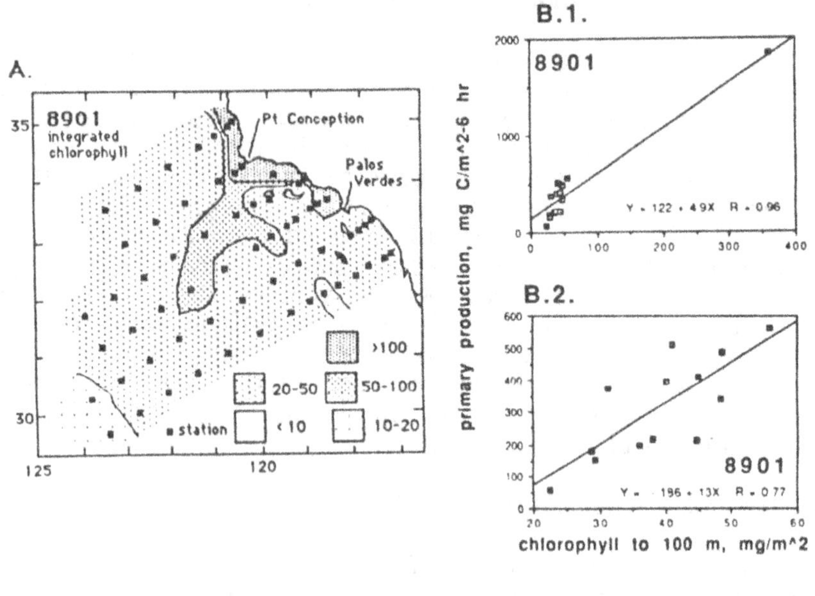

Figure 7.2. Chlorophyll in the California Current off southern California in January 1989 (Cruise 8901). (A) Chlorophyll integrated through the upper 100 m, mg m^{-2}. (B) Primary production, mg $C(m^2\ 6h)^{-1}$ versus integrated chlorophyll; (1) All data, (2) Single large value eliminated. Both correlations are significant at $P<0.01$.

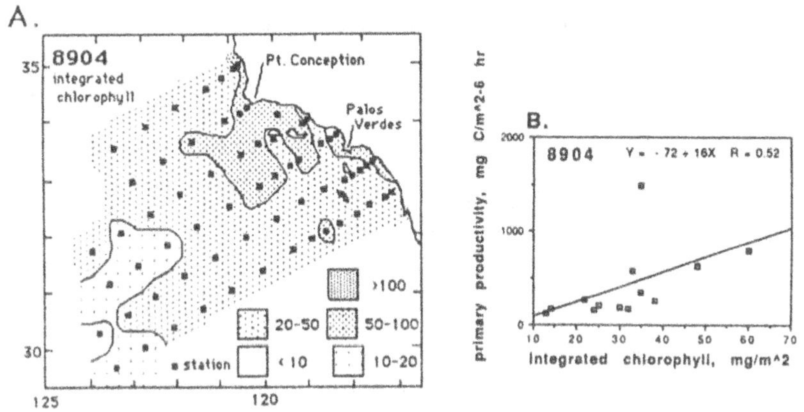

Figure 7.3. Chlorophyll in the California Current off southern California in April 1989 (Cruise 8904). (A) Chlorophyll integrated through the upper 100 m, mg m^2. (B) Primary production, mg $C(m^2\ 6\ h)^{-1}$ versus integrated chlorophyll. Correlation is significant at $P<0.05$ (1-tailed).

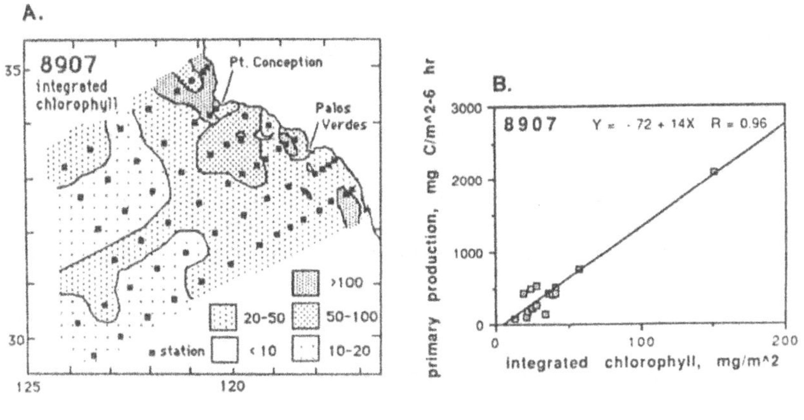

Figure 7.4. Chlorophyll in the California Current off southern California in July 1989 (Cruise 8907). (A) Chlorophyll integrated through the upper 100 m, mg m^{-2}. (B) Primary production, mg $C(m^2\ 6\ h)^{-1}$ versus integrated chlorophyll. Correlation is significant at $P<0.01$.

southwestern portion of the area surveyed, while biomasses in the northern offshore region, the central southern region, and the center of the Southern California Bight were typically 20-50 mg m^{-2}. Primary production (which was measured only at stations occupied near local noon) was significantly correlated on all cruises with the biomass of phytoplankton (as chlorophyll) at those stations (Figures 7.1-7.4, panels B), and it is therefore reasonable to assume that maps of primary production would resemble those of chlorophyll (Figures 7.1-7.4, panels A), and further to assume (from the "Lindeman" approach) that the spatial patterns of secondary production would also resemble these maps.

However, in addition to the variability in actual coupling between primary and secondary production, outlined above, only 27% of the variability in primary production was accounted for by the biomass of chlorophyll in the worst case (cruise 8904). This means that the pattern of primary production on this cruise particularly could differ from that of chlorophyll; the seriousness of this problem depends on where in the pattern the major departures from the overall relation occurred. Finally, it is necessary to remember that secondary production calculated in this way could actually be "packaged" in any of several quite different animal forms.

Extrapolation from Zooplanktonic Biomass

Zooplanktonic biomasses were determined as displacement volumes of catches of an 0.5-mm mesh net for the 0-200-m (or 0-bottom, if shallower) layer, and I multiplied values from stations taken at night by 0.75 to make them comparable to daytime values (based on comparisons summarized in Mullin, 1986). The resulting distributions (Figures 7.5-7.8, panels A) were broadly similar to those of phytoplankton in that biomasses were greatest in the northeast around Point Conception and least in the southwest. However, the correlations with chlorophyll were not strong (indeed, non-significant for 8810), and less than 50% of the variability in zooplanktonic biomass was explainable (Figures 7.5-7.8, panels B), so the spatial patterns of zooplankton and phytoplankton differed in detail on the scale of tens of km.

Conversion of zooplanktonic biomasses to secondary production by the simplest production/biomass ratio--a constant--would not alter the patterns. Maximal growth rates of 22-65% d^{-1} are reasonable (e.g., Figure 4 of Huntley and Boyd, 1984), so that daily rates of secondary production without limitation by food would be of the order of half the biomass. If the positive correlation between temperature and production were the only modifying factor, the patterns of secondary production would be less strong than those of biomass, since highest zooplanktonic biomasses tended to occur in the

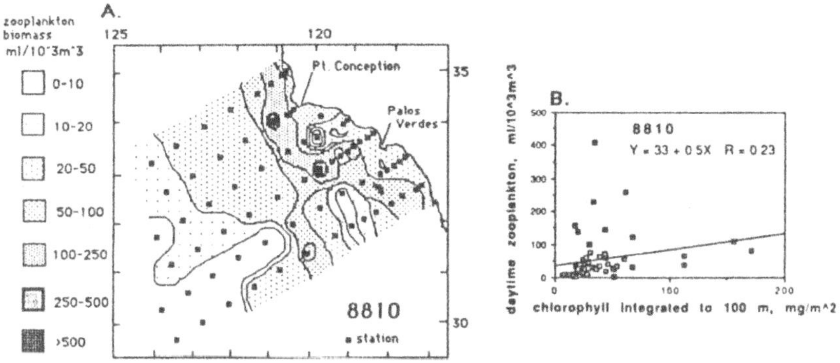

Figure 7.5. Biomass of zooplankton in the California Current off southern California in October 1988. (A) Displacement volume of zooplankton (large organisms removed, corrected to daytime values), ml(10^3 m^3)$^{-1}$. (B) Displacement volume of zooplankton versus integrated chlorophyll. Correlation is not significant ($0.05<P<0.1$).

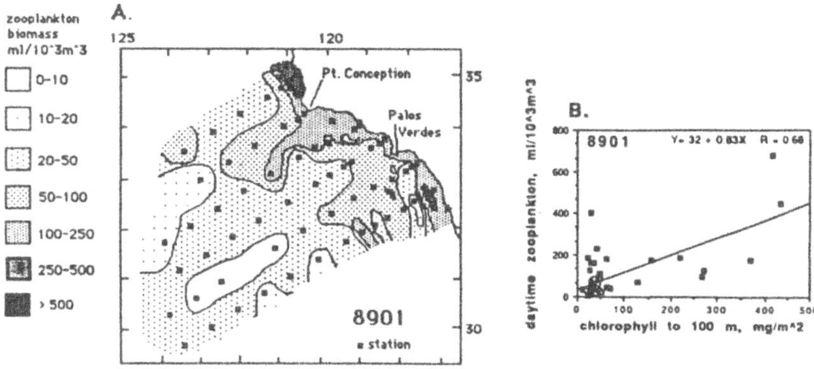

Figure 7.6. Biomass of zooplankton in the California Current off southern California in January 1989. (A) Displacement volume of zooplankton (large organisms removed, corrected to daytime values), ml(10^3 m^3)$^{-1}$. (B) Displacement volume of zooplankton versus integrated chlorophyll. Correlation is significant at $P<0.01$.

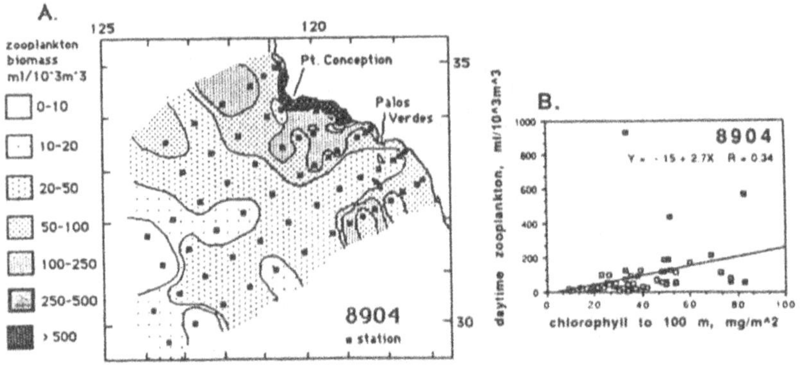

Figure 7.7. Biomass of zooplankton in the California Current off southern California in April 1989. (A) Displacement volume of zooplankton (large organisms removed, corrected to daytime values), ml(10^3 m^3)$^{-1}$. (B) Displacement volume of zooplankton versus integrated chlorophyll. Correlation is significant at P<=0.01.

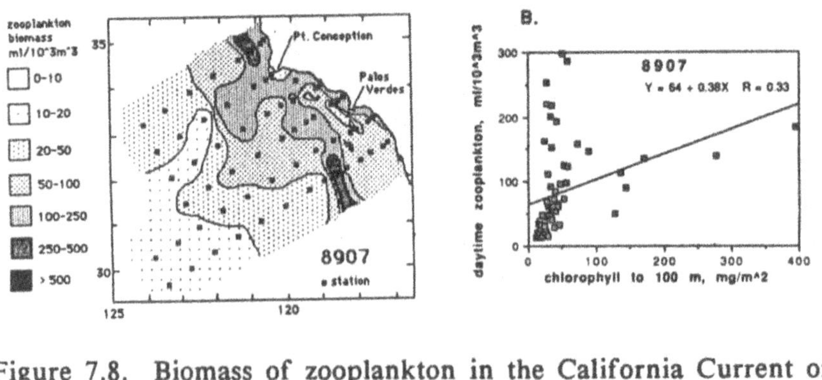

Figure 7.8. Biomass of zooplankton in the California Current off southern California in July 1989. (A) Displacement volume of zooplankton (large organisms removed, corrected to daytime values), ml(10^3 m^3)$^{-1}$. (B) Displacement volume of zooplankton versus integrated chlorophyll. Correlation is significant at P<=0.01.

northeast, where near-surface temperatures were about 5°C lower on each cruise than in the southwest, where biomass was low.

In terms of the total secondary productivity in the area surveyed (which is approximately 2.6×10^5 km^2), calculating secondary productivity in a simple, linear way from either the total chlorophyll (i.e., secondary production a constant fraction of primary production, primary production linearly related to chlorophyll by the same relation for all cruises) or the total biomass of zooplankton (i.e., a simple P/B ratio) would indicate that secondary production was greatest on cruise 8901 (first 3 columns of Table 7.1). More importantly, the unexplained variances in Figures 7.5-7.8, panels B, mean that spatial patterns of secondary production derived from zooplanktonic biomasses would differ on small- and meso-scales from those based on primary production.

The biomasses of zooplankton in Table 7.1 refer to that retained by 0.5 mm mesh. Based on comparisons reported by Ohman and Wilkinson (1989) in the CalCOFI area, the biomasses retained by 0.2 mm mesh average 1.3 times these values. The P/B ratio appropriate to the 0.2-0.5 mm zooplankters is probably greater than that for the larger animals (though Huntley and Boyd, 1984, concluded that, for copepods, maximal growth rate was not size-dependent).

Derivation of more biologically realistic, variable P/B ratios, dependent upon concentration of available food and bodily size of the animals as well as temperature, involves several assumptions. There is first the assumption that measurements at each station adequately represent the conditions for up to 40 km around it. To assign the relevant temperature and concentration of food to each station, one would need to weight the known vertical distributions of temperature and chlorophyll (again assuming that chlorophyll = food) by the vertical distribution of zooplanktonic (ideally, herbivorous) biomass, which is not known. That is, one should use the temperature and concentration of food which the "median unit of biomass" actually experiences. Further, one should know this for different size categories of zooplankters, whose vertical distributions may differ.

Lacking the information to do the calculation correctly, I will assume that the distribution of temperature at 10 m (given in Mullin, 1991) adequately preserves the pattern of relative temperature experienced by the zooplankton, if not the absolute values. I will further assume that, because maximal chlorophyll in the water column is strongly correlated with the integrated chlorophyll (Mullin, 1991) the maps of integrated chlorophyll (Figures 7.1-7.4, panels A) represent the distribution of food and, further, that the chlorophyll is uniformly distributed in the upper 100 m (or, equivalently, that the vertical distribution of zooplanktonic biomass is random with respect to that of chlorophyll). Finally, I will assume that the zooplanktonic biomass caught by 0.5 mm mesh in the upper 200 m (Figures 7.5-7.8, panels A) derives its food from the upper 100 m, and that the "typical" zooplankter weighs 0.1 mg (dry weight).

Table 7.1. Biomass and production in the southern California sector of the California Current in October 1988 (8810), January 1989 (8901), April 1989 (8904), and July 1989 (8907). The sector has an area of 2.6×10^5 km^2. Total biomasses of phytoplankton are based on chlorophyll and of zooplankton are based on displacement volume caught by 0.5-mm mesh in the upper 200 m. Primary production is calculated assuming a daily rate of twice the 6-h rate, and the relations to chlorophyll given in Figures 7.1-7.4, panels B. Secondary production is calculated as 15% of primary production, and as 50% (unlimited by food--see text) and 6% (maximal production of eggs by *Calanus*) of the biomasses of >0.5 mm zooplankton.

	Biomasses 10^2 K tons C		Production, K tons C d^{-1}			
			primary	secondary		
Cruise	phytopl.	zoopl.		15% of p.p.	50% of zoopl.	6% of zoopl.
8810	3.9	1.2	123	18	60	7
8901	6.0	1.7	215	32	85	10
8904	4.4	1.6	239	36	80	10
8907	4.8	0.71	230	32	36	4
annual total, M tons C yr^{-1}			74	11	24	2.8

Huntley and Boyd (1984) concluded that there is a positive correlation between temperature and the threshold concentration of food below which the rate of growth is limited by food (their Figure 5). The further assumption that a carbon:chlorophyll ratio of 50 is reasonable for phytoplankton (though the real ratio is vertically and horizontally variable) results in the conclusion that these threshold concentrations are equivalent to 200-250 mg chlorophyll m^{-2} in the southwest, and 350-400 mg m^{-2} around Point Conception. By this reckoning, secondary production was less than its (temperature-dependent) maximal rate over almost all of the area on all four cruises.

In fact, the assumptions summarized above result in the conclusion that only in a small part of the region was the concentration of phytoplankton sufficient for secondary production

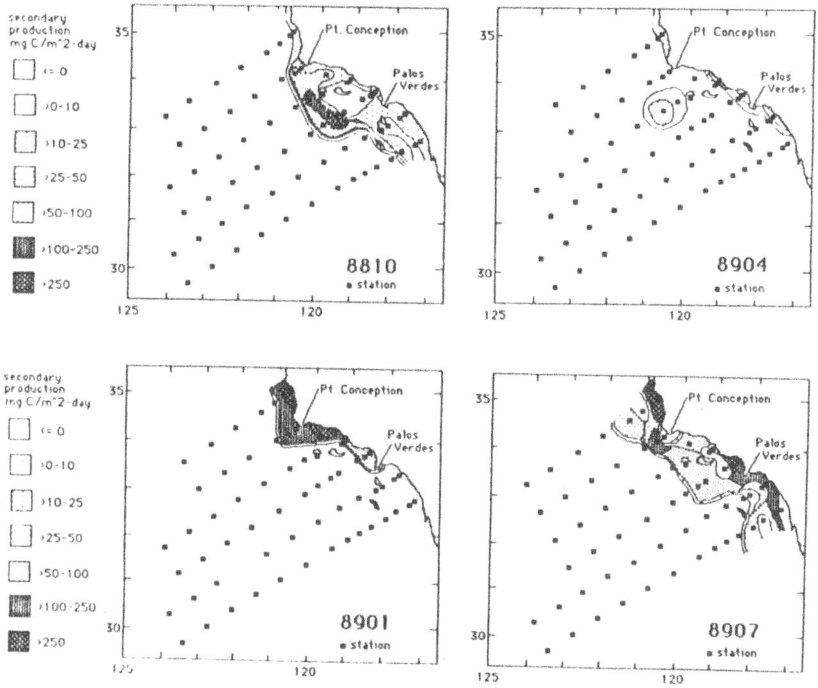

Figure 7.9. Rate of secondary production, in mg $C(m^2\ d)^{-1}$ in the California Current off southern California in October 1988 (8810), January 1989 (8901), April 1989 (8904), and July 1989 (8907), calculated following Huntley and Boyd (1984) from chlorophyll and zooplankton biomasses shown in Figures 7.1-7.8. Assumptions include: uniform distributions of phytoplankton to 100 m or the bottom, and zooplankton to 200 m or the bottom; phytoplankton C:chlorophyll = 50; zooplankton displacement volume:dry weight:C = 1:0.13:0.052; each zooplankter is 0.1 mg dry weight.

of zooplankton >0.5 mm (Figure 7.9). This conclusion is unreasonable on its face, though the areas where secondary production is great may resemble this figure. By altering one or more of the assumptions--particularly by increasing the C:chlorophyll ratio or altering the depth distribution of seston so as to increase the effective concentration of food at the depth where the zooplankton is considered to feed--one could alter the absolute values of secondary production considerably. In addition to this uncertainty, Figure 7.9 shares with the "Lindeman" approach the weakness that the suitability of this secondary production as food for particular species of carnivores is unknown.

Direct Measurement

The final approach which yields maps of secondary production in the southern Californian sector of the California Current is based on measured production of eggs by a particular species of zooplanktonic copepod--a species which often dominates the zooplanktonic biomass at some stations, but a single species nevertheless--*Calanus pacificus* (Figures 7.10-7.13). Actually, these are maps of secondary productivity per capita or per biomass unit, rather than the rate of production, and of the extent of food-limitation of this productivity. Details of methods, and the associated assumptions and problems, are given in Mullin (1991). Briefly, female *Calanus* spp. were incubated for 48 h in containers with ambient seawater and its seston or with cultured diatoms as a supplemental source of excess food, and eggs were collected after 24 and 48 h. The productivity by "fed" females represents potential, and by "unfed" (= ambient seston) the realized rate, or somewhat less than this (Mullin, 1991). To the extent that female *Calanus* feed in patches or layers of anomalously abundant food (a possibility which also affects the interpretation of the metabolic, P/B approach discussed above), the productivity *in situ* may approach that of the females incubated with supplemented food.

Panel A in each figure shows regions of high and low productivity of eggs; a rate of 10 eggs per female per day is equivalent to a daily P/B ratio of about 2% in terms of carbon. The regions where females in ambient seston exceed this rate (the vertical striping) in October (8810, Figure 7.10A) are near the coast, and there were several stations far offshore at which even females which had been fed for two days produced fewer than 10 eggs per day. In January (8901, Figure 7.11A), this rate was exceeded by unfed females over a much larger area, and in some regions the daily rate exceeded 6%. Rates exceeded 2% per day near the coast or in the northern portion of the sector in April (8904, Figure 7.12A), and also over much of the southeastern part of the area in July (8907, Figure 7.13A). [For comparison, daily P/B ratios of 3-16%, representing the most food-limited and the unlimited rates in the warmest season, were thought by Mullin and Brooks (1970, Table VII-3) to characterize the late copepodite stages of *Calanus pacificus* in the nearshore region of the Southern California Bight, and Peterson (1988) reported maximal egg production by *C. marshallae* at 10°C to be equivalent to a daily P/B of 7%.]

Ratio "b" is the ratio of production of eggs for two days by females in ambient seawater to those with supplemented food, and is one measure of short-term limitation by food. The ratio is relatively high (the shaded portions of Figures 7.10-7.13, panels B) when realized production approaches potential either because both are high

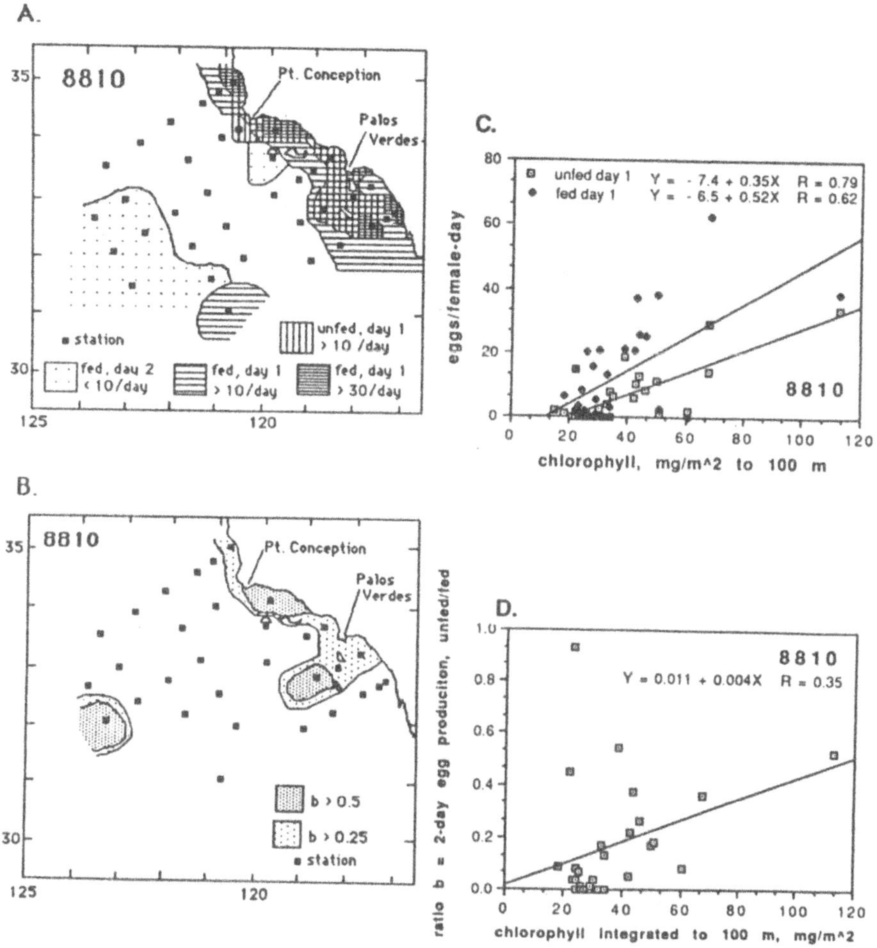

Figure 7.10. Production of eggs by *Calanus pacificus* in the California Current off southern California in October 1988. (A) Daily rates of females incubated for 2 d in seawater and ambient seston ("unfed") or with excess food supplied ("fed"). Poorest production is when females produce fewer than 10 eggs d^{-1} even on the second day with excess food. (B) Ratio ("b") of egg production over 2 d of unfed females to that of fed females. High ratios can indicate either little limitation by food (both rates of production high) or extreme limitation by food (even "fed" rate low). (C) Daily production by females in ambient seston or in supplemented food, versus chlorophyll in the upper 100 m. Both correlations are significant at $P<0.01$. (D) Ratio b versus chlorophyll. Correlation is significant at $P< =0.05$, 1-tailed.

A.

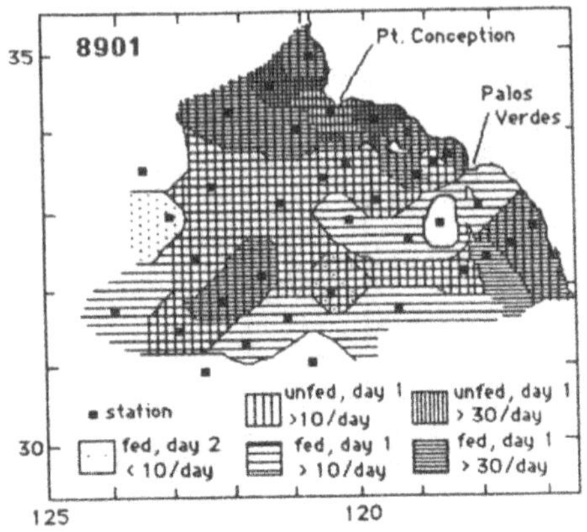

B.

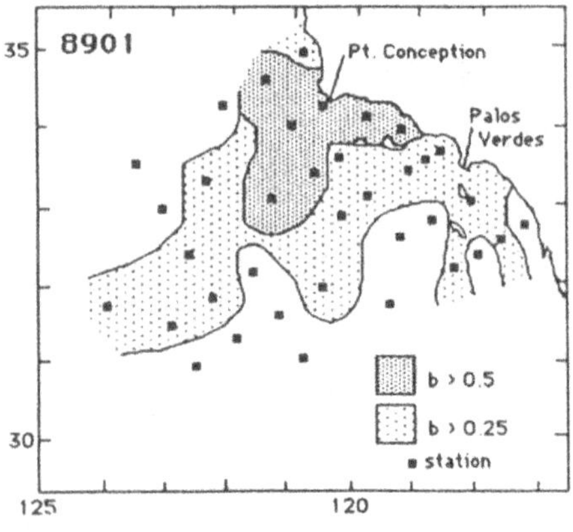

Figure 7.11. Production of eggs by *Calanus pacificus* in the California Current off southern California in January 1989. See Figure 7.10 legend for details. Panel C is shown as (a) All data, and (b) Data for <100 mg chlorophyll m^{-2}, where linearity is expected--correlation for unfed females is marginally significant ($0.025<P<0.05$, 1-tailed), and correlation for fed females is not significant ($P>0.05$, 1-tailed); in D, correlation is significant at $P<0.01$. Figure 7.11 continued on next page.

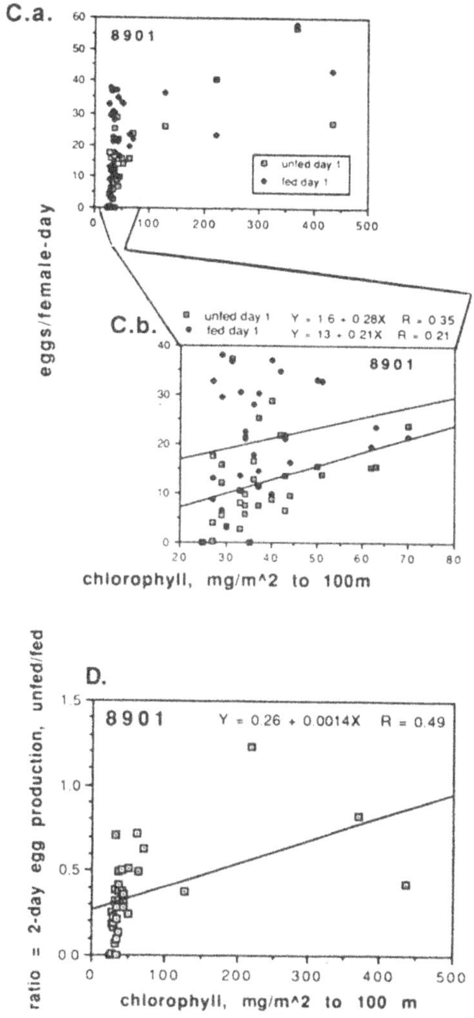

Figure 7.11 continued.

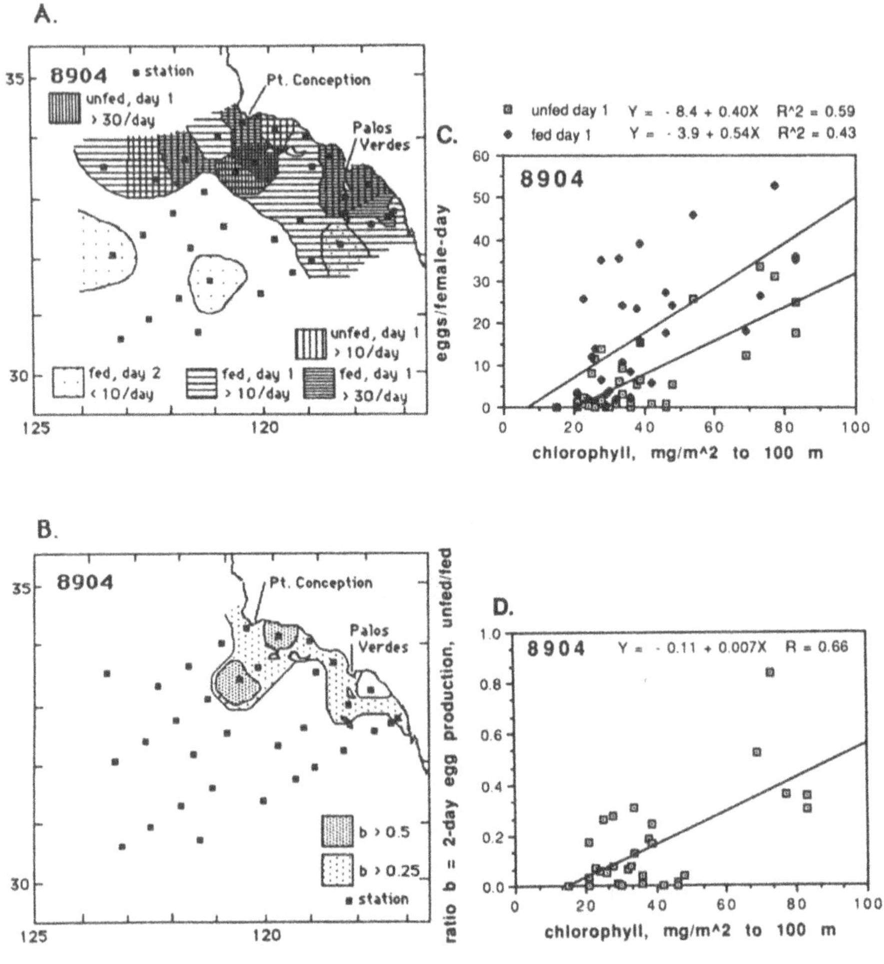

Figure 7.12. Production of eggs by *Calanus pacificus* in the California Current off southern California in April 1989. See Figure 7.10 legend for details. In C, both correlations are significant at P<0.01; in D, correlation is significant at P<0.01.

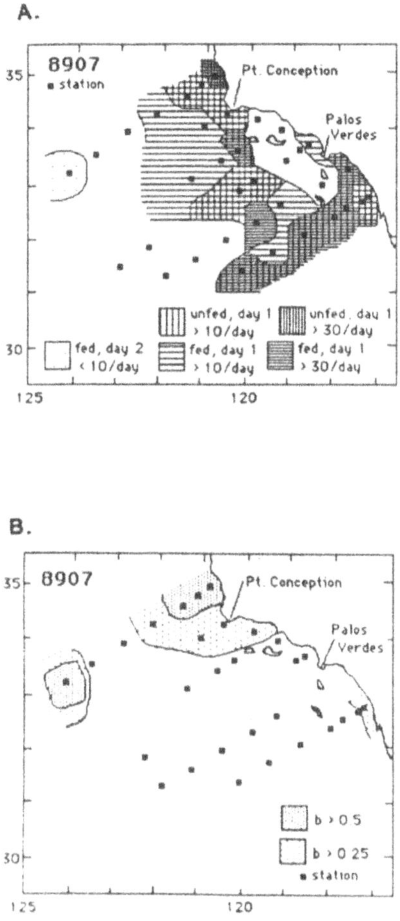

Figure 7.13. Production of eggs *Calanus pacificus* in the California Current off southern California in July 1989. See Figure 7.11 legend for details. In panel C.b., both correlations are non-significant ($P>0.1$); in D, the correlation is significant at $P<0.01$. Figure 7.13 continued on next page.

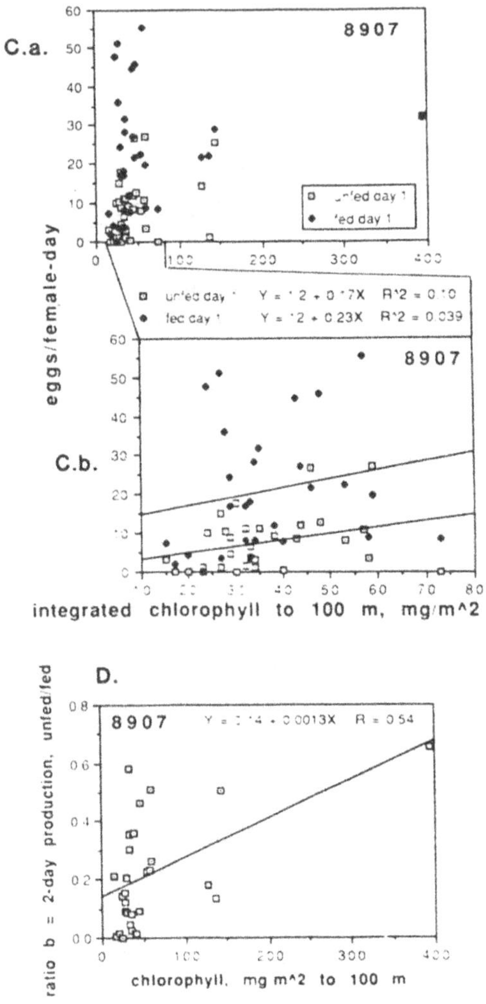

Figure 7.13 continued.

(good conditions) or because both are low (extremely poor conditions). By this standard, food-limitation of secondary production was much less in January than during the other three cruises.

On large scale, the patterns of productivity by female *Calanus* are similar to the distributions of chlorophyll (Figures 7.1-7.4), and in fact the degree of food-limitation indicated by the b ratios was significantly correlated with chlorophyll on all four cruises (Figures 7.10-7.13, panels D). However, the productivity of females in ambient seston (= "unfed") was positively correlated with chlorophyll on only three of the four cruises (Figures 7.10-7.13, panels C), and on all cruises there was considerable imprecision in these relations, meaning that on the scale of individual stations, the ability to predict secondary productivity from the biomass of chlorophyll could be quite poor.

The direct estimation of per capita secondary productivity by *Calanus* results in the impression that productivity of the sector greatest during 8901 (January) and least during 8810 (October). [8904 appears to be intermediate, but is difficult to compare because measurements were made on only one station along the northernmost transect line.] This agrees with the temporal ordering for total secondary production suggested by phytoplanktonic biomass (Table 7.1), but contrasts with the impression created by the results of the metabolic, P/B calculation (Figure 7.9). This must be due in part to the specific ecology of *Calanus*; it is doubtful that secondary production by all species is zero in the southwestern part of the sector, where the few *Calanus* present produced no eggs.

Results of several methods of estimating secondary production are compared in Table 7.1, in which I have given a "Lindeman" estimate based on primary production (for all herbivores, using the primary production to chlorophyll relations for each cruise separately), and, for zooplankton retained by 0.5 mm mesh, estimates assuming a constant P/B ratio, not limited by food, of 50% d^{-1}, and a constant P/B ratio of 6% d^{-1} derived from the maximal rates of production of eggs by *Calanus*.

The results for a daily P/B of 50% (i.e., not food-limited) are unrealistically high for pure herbivory, relative to primary production. This is unsurprising, since both the Huntley-Boyd calculation and the effect on egg production of adding food to the ambient seston indicate that food-limitation is in fact widespread. All three estimates have ambiguities. The "Lindeman" estimate applies conceptually to herbivores of all sizes, and their biomass and suitability as food is unknown. The other two estimates are based on measured biomasses of >0.5 mm zooplankton, which includes carnivorous as well as herbivorous biomass.

The largest unknown, in terms of the overall trophodynamic balance, is the secondary production of herbivorous protozoa and zooplankton <0.5 mm. Many of these smaller forms are, of course, major food sources of many larval fishes.

Conclusions

What is a fisheries recruitment trophodynamicist (if anyone would describe their profession so cacophonously) to make of all this? First, I think it remains to be shown empirically that variation in total secondary production has much effect on variation in recruitment, in spite of the large body of evidence indicating that larval fish are often hard-put to get enough to eat. If the conceptual attractiveness of this notion remains compelling nevertheless, I recommend that either one expend the effort to measure, more or less directly, the secondary production by those zooplankters (or at least those size categories) actually used as food by the young fish of particular species, or that one reason directly from primary production to potential recruitment of generic young fish without interposing the "black box" of secondary production at all. In the latter case, one should realize that, though meso- and small-scale details in primary production are attainable, there are too many sources of error to preserve this detail in the final prediction of recruitment. If knowledge of small- and mesoscale pattern in secondary production is required, direct estimation is likely to be the only valid means of attaining it. This is also likely to be the best means of establishing unambiguously the degree to which variability in secondary production affects recruitment of commercially important fishes.

References

Berggreen, V., B. Hansen, and T. Kiørboe. 1988. Food size spectra, ingestion, and the growth of the copepod *Acartia tonsa* during development: Implications for determination of copepod production. Mar. Biol. 99:341-352.

Boyd, C. 1985. Is secondary production in the Gulf of Maine limited by the availability of food? Arch. Hydrobiol. Beih. Ergebn. Limnol. 21:57-65.

Brinton, E. 1976. Population biology of *Euphausia pacifica* off Southern California. Fish. Bull., U.S. 74:733-762.

Checkley, D. M. 1980. Food limitation of egg production by a marine, planktonic copepod in the sea off Southern California. Limnol. Oceanogr. 25:991-998.

Chelton, D. B., P. A. Bernal, and J. A. McGowan. 1982. Large-scale interannual physical and biological interaction in the California Current. J. Mar. Res. 40:1095-1125.

Cox, J. L., L. R. Haury, and J. J. Simpson. 1982. Spatial patterns of grazing-related parameters in California coastal surface waters, July 1979. J. Mar. Res. 40:1127-1153.

Cox, J. L., S. Willason, and L. Harding. 1983. Consequences of distributional heterogeneity of *Calanus pacificus* grazing. Bull. Mar. Sci. 33:213-226.

Craig, H., and T. Hayward. 1987. Oxygen supersaturation in the ocean: biological versus physical contributions. Science 235: 199-202.

Dagg, M. J., and J. Littlepage. 1972. Relationships between growth rate and RNA, DNA, protein, and dry weight in *Artemia salina* and *Euchaeta elongata*. Mar. Biol. 17:162-170.

Dickey, T. D. 1988. Recent advances and future directions in multi-disciplinary *in situ* oceanographic measurement systems. pp. 555-598 *In* B. J. Rothschild (Ed.), Toward a theory on biological-physical interactions in the world ocean. Kluwer Academic Press, Dordrecht.

Eppley, R. W., E. Stewart, M. R. Abbott, and U. Heyman. 1985. Estimating ocean primary production from satellite chlorophyll. Introduction to regional differences and statistics for the Southern California Bight. J. Plankton Res. 7:57-70.

Greene, C. H., P. H. Wiebe, and J. Burczynski. 1989. Analyzing zooplankton size distributions using high-frequency sound. Limnol. Oceanogr. 34:129-139.

Hakanson, J. L. 1984. The long and short term feeding condition in field-caught *Calanus pacificus*, as determined from the lipid content. Limnol. Oceanogr. 29:794-809.

Hakanson, J. L. 1987. The feeding condition of *Calanus pacificus* and other zooplankton in relation to phytoplankton pigments in the California Current. Limnol. Oceanogr. 32:881-894.

Heinle, D. R. 1966. Production of a calanoid copepod, *Acartia tonsa*, in the Patuxent River Estuary. Chesapeake Sci. 7:59-74.

Huntley, M. 1985. A method for estimating food-limitation and potential production of zooplankton communities. Arch. Hydrobiol. Beih. Ergebn. Limnol. 21:41-55.

Huntley, M., and C. Boyd. 1984. Food-limited growth of marine zooplankton. Am. Nat. 124:455-478.

Huntley, M. E., V. Marin, and F. Escritor. 1987. Zooplankton grazers as transformers of ocean optics: A dynamic model. J. Mar. Res. 45:911-945.

Isaacs, J. D. 1972. Unstructured marine food webs and "pollutant analogs." Fish. Bull., U.S. 70:1053-1059.

Kimmerer, W. J., and A. D. McKinnon. 1987. Growth, mortality, and secondary production of the copepod *Acartia tranteri* in Westernport Bay, Australia. Limnol. Oceanogr. 32:14-28.

Lindeman, R. L. 1942. The trophic-dynamic aspect of ecology. Ecol. 23:399-418.

Mackas, D., and R. Bohrer. 1976. Fluorescence analysis of zooplankton gut contents and an investigation of diel feeding patterns. J. Exp. Mar. Biol. Ecol. 25:77-85.

Mann, K. H. 1969. The dynamics of aquatic ecosystems. Adv. Ecol. Res. 6:1-81.

McClatchie, S. 1987. Experimental test of an allometric method for estimating potential copepod production. Mar. Biol. 94:597-603.

Mullin, M. M. 1969. Production of zooplankton in the ocean: The present status and problems. Oceanogr. Mar. Biol. Ann. Rev. 7:293-314.

Mullin, M. M. 1982. How can enclosing seawater liberate biological oceanographers? Pp. 399-410 *In* G. D. Grice and M. R. Reeve (Eds.) Marine mesocosms. Springer-Verlag, New York.

Mullin, M. M. 1986. Spatial and temporal scales and patterns. Pp. 216-273 *In* R. W. Eppley (Ed.) Plankton dynamics of the Southern California Bight. Springer-Verlag, New York.

Mullin, M. M. 1991. Production of eggs by the copepod, *Calanus pacificus*, and its variability in the southern California sector of the California Current System. CalCOFI Rep. (in press).

Mullin, M. M., and E. R. Brooks. 1970. Production of the planktonic copepod, *Calanus helgolandicus*. Part VII of J. D. H. Strickland (Ed.) The ecology of the plankton off La Jolla, California, in the period April through September, 1967. Bull. Scripps Inst. Oceanogr. 17:89-103.

Mullin, M. M., and E. R. Brooks. 1976. Some consequences of distributional heterogeneity of phytoplankton and zooplankton. Limnol. Oceanogr. 21:784-796.

Ohman, M. D., and J. R. Wilkinson. 1989. Comparative standing stocks of meso- and macrozooplankton in the southern sector of the California Current System. Fish. Bull., U.S. 87:967-976.

Peterson, B. J. 1980. Aquatic primary productivity and the $^{14}C\text{-}CO^2$ method: A history of the productivity problem. Ann. Rev. Ecol. System. 11:359-386.

Peterson, W. T. 1988. Rates of egg production by the copepod *Calanus marshallae* in the laboratory and in the sea off Oregon, USA. Mar. Ecol Prog. Ser. 47:229-237.

Peterson, W. T., P. Tiselius, and T. Kiørboe. 1991. Copepod egg production, moulting and growth rates, and secondary production, in the Skagerrak in August 1988. J. Plankton Res. 13:131-154.

Richter, K. E. 1985. Acoustic scattering at 1.2 MHz from individual zooplankters and copepod populations. Deep-Sea Res. 32:149-161.

Ryther, J. H. 1969. Photosynthesis and fish production in the sea. Science 166:72-76.

Steele, J. H. 1965. Some problems in the study of marine resources. Int. Comm. Northw. Atlan. Fish., Spec. Publ. No. 6, Pp. 463-476.

V. M. Borisov

8. The State of the Main Commercial Species of Fish in the Changeable Barents Sea Ecosystem

Introduction

The Barents Sea was traditionally believed to be one of the most productive areas of the World Ocean. Its fish productivity (720 kg/km^2) was more than thrice the mean of the world ocean (225 kg/km^2) (Moiseev, 1989). As far back as the 1970s, the total catch of fish, crustaceans, molluscs, and algae taken by fishing countries averaged nearly 2 million metric tons (mmt). The record catch amounted to 2.8 mt in 1977. Later, the catches dropped noticeably due to the decline in the stocks of Gadidae, the main component of catches in the Barents Sea (Table 8.1). The decline was most pronounced in the early 1980s. Later, when the fishery for capelin was discontinued, the catch of all fishing countries was little more than 350,000 tons (t), in 1987. At present, the Barents Sea has lost its significance as a main fishing region.

The retrospective and present analyses of the situation indicate two principal reasons for the negative development of the process.

The first reason is connected with a worsening of hydrological conditions in the ecosystem owing to the regular cyclic decrease of the inflow of warm Atlantic water. In recent years additional stress has been imposed on the system due to increased anthropogenic pollution. It is promoted by the Gulf Stream, which transports refuse into the Barents Sea ecosystem from thousands of vessels and dozens of coastal countries.

The second reason is linked with the fact that the period of worsening hydrological conditions coincided with the heaviest stress from excessive fishery mortality that impeded the normal reproduction of stocks.

Table 8.1. Nominal Catches (Tons) of Fish, Crustaceans, Molluscs, in the Barents Sea.

	Countries							
Years	Bulgaria	U.K.	Denmark	Norway	Germany, F.R.	U.S.S.R.	Others	Total
1975	3,288	107,341	18,951	1,040,164	23,780	748,711	84,076	2,026,311
1976	-	74,886	5,071	1,569,403	35,440	885,456	25,971	2,596,227
1977	3,207	58,013	3,957	1,630,222	6,851	1,089,664	13,983	2,805,897
1978	-	26,108	13,542	1,044,837	-	859,256	2,835	1,946,578
1979	6,956	15,988	3,074	569,511	-	617,087	1,939	1,214,561
1980	1,539	7,728	9,618	445,242	-	584,448	3,139	1,051,714
1981	779	3,446	17,683	557,633	406	851,629	2,834	1,434,410
1982	385	3,703	11,488	548,478	136	592,427	615	1,157,232
1983	265	4,251	924	347,822	-	400,934	8,320	762,516
1984	-	1,413	5,103	295,741	21	360,716	-	662,994
1985	27	724	-	349,698	440	473,023	-	823,912
1986	-	1,730	370	211,521	220	220,781	-	434,622
1987	-	1,014	-	157,573	-	197,675	-	356,262

In light of this, an attempt will be made to review the state of the main commercial species of fish from the Barents Sea starting with a brief analysis of the hydrological situation.

Water Exchange and Hydroclimatic Fluctuations in the Region

A great deal of data on oceanographic and faunal characteristics has been accumulated in the history of the study of the Northeast Atlantic. Since the times of N. M. Knipovich (Knipovich, 1906) and F. Nansen (Nansen, 1906), the attention of scientists has been almost always focused on the dynamics of the environment and response of the species inhabiting the area adjacent to it.

Based on numerous scientific papers, it may be concluded that the state of the Barents Sea Ecosystem depends, primarily, on the intensity of the water exchange between the relatively warm Atlantic and the cooler Arctic Barents Sea. The speed and volume of the oppositely directed and interrelated flows are dependent upon both hydro-atmospheric conditions formed in the Atlantic and hydrocryo-atmospheric processes occurring in the Barents Sea, itself, and beyond it.

The mechanism of the water exchange can be presented as follows. In the winter long-term cooling season, water masses of high density are accumulated in the off-bottom layer of Arctic water. This is assisted by ice formation when salts are transported from the surface to the depths. The sinking heavier water forms two principal off-bottom currents, the East Spitzbergen and Persey Currents, which transport cool Arctic water into the Norwegian Sea through the Bear Island Channel (Tantsyura, 1959). The intensity of the currents may change greatly from year to year because their sources are not stable between years. The outflow from the Barents Sea of cold Arctic water moving southwest, primarily in the Bear Island Current, is compensated with warm Atlantic flows into the southern and central parts of the Barents Sea as the Norwegian Coastal and North Cape Currents, the south branch of the latter carrying Atlantic water far to the east and supplying the Murman Current. The flows meeting in the Barents Sea form the polar front where the more saline Atlantic water sinks when cooled. Owing to the intensive vertical circulation, deep layers are vigorously aerated and the surface layers are enriched with biogenic elements from the depths (Figure 8.1).

The periodicity of obtaining high catches in the Barents Sea and adjacent areas connected with favorable or unfavorable conditions for the reproduction of fish is well known. It manifests itself roughly as 10-yr cycles. The recent decline in natural resources occurred during the periods of 1957-1966 and 1970-1973. After a significant decrease

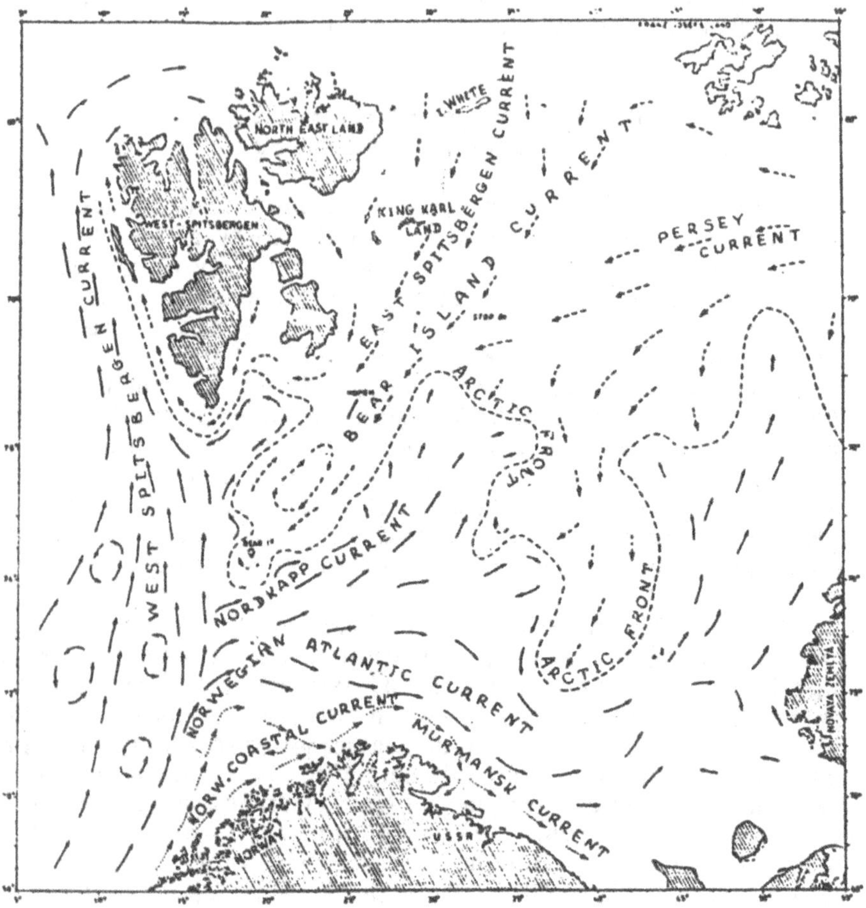

Figure 8.1. Atlantic (→), Arctic (-->), coastal (•••>)currents and polar front (—) in the Barents Sea. (Simplified after Tantsiura, 1959; and Novitskiy, 1961, by Midttun, 1989.)

in the stocks of Gadidae in the period of 1978-1985, their abundance recovered noticeably in 1986-1987. However, in 1985 the onset of the next in turn cool-period was recorded. So, in common with the preceding unfavorable periods, it affected the spawning rate, incubation of eggs, the feeding rate, and survival rate of juveniles, and, in the long run, the yield of year classes of almost all commercial species.

From the standpoint of natural conditions, the Barents Sea holds a most unique position. The close dependence of the reproduction of fish stocks on unsteady inflows of warm Atlantic water and on unstable interactions with counterflows of cool Arctic water, is responsible for vulnerability of populations.

The Hydrobiological Situation

The areas adjacent to the polar front may be considered as accumulators of organic matter maintaining the benthic biomass as high as 150-600 g/m^2, whereas it drops to 20-50 g/m^2 beyond the boundaries of the areas. The zooplankton biomass is assessed to be 150-250 mg/m^3, on the average, over a vast area between 20°-60°E and 70°-80°N, and the biomass of phytoplankton ranges from 40 mg/m^3 in the period of biological winter to 2,500 mg/m^3 in spring (Zenkevich, 1963).

In recent years the abundance of the main food species--euphausiids--has increased. The assessment made in PINRO shows that it was twice the long-term mean in winter of 1989, i.e., 668 specimens/1,000 m^3 as compared to the long-term mean of 332 specimens/1,000 m^3. The hydrobiological assessment shows that the annual primary production in the Barents Sea and adjacent waters is equal to 2×10^9 t. With a 15% ecological efficiency when it is converted from one trophic level to another, it promotes the yield of 3×10^8 t of phytoplankton-eaters, which can feed 4×10^7 t of plankton eaters (Drobysheva et al., 1988).

A considerable part of this abundant food resource that is a generous grant of nature, remains, however, unused because of an inadequate numerical strength of consumers. In comparison with the last period, in the second half of the 1980s, the abundance of the terminal link of the trophic web-predators was drastically declined, and moreover, plankton-eaters were also actually withdrawn.

The decline in the populations of herring, capelin, redfish, cod, haddock, polar cod, and shrimp has brought about some imbalance in the entire ecosystem, and a break in the trophic interrelations existing among the species.

Regulation of the Fishery

It should be admitted that, as mentioned above, the main or primary reason for the situation is the inadequately strict regulation of the fisheries, and secondarily the situation was driven by unfavorable hydrological conditions. The simultaneous effect of both factors manifests itself in an inadequate recruitment to the stocks of almost all commercial species of fish. It would be, undoubtedly, unfair to say that utter oblivion or disregard of protection measures exists in the region. The fishery is regulated in five main ways: (1) issue of quotas for the annual catch, (2) the size of a mesh in cod ends of trawls, (3) the minimum commercial size of fish, (4) permitted by-catch of the young, and (5) banned areas for fishing.

National quotas are annually set up on the basis of the total admissible catch of each species within the framework of the Joint Soviet-Norwegian Commission on Fisheries. It should be stressed, however, that the last word is often said not by scientists, but by fishermen of both Parties who are privately interested in obtaining as great quotas as possible. Such "mutually beneficial" tactics, contrary to the opinion of scientists, got the upper hand even in cases when any increase in current catches could deplete the stocks in future. This was, at least, one of the principal reasons of the overfishing in the fishery for capelin. This holds true for the fishery for cod, as well.

The State of the Main Commercial Species

Capelin

Owing to the high fishing intensity of the Soviet Union and Norway in the mid-1970s, the total catch of capelin reached nearly 3 mt in 1977. The maximum quotas set up in the framework of the Joint Soviet-Norwegian Commission on Fisheries (JSNCF) were systematically surpassed in the 1980s (Table 8.2). As a result of such irrational fisheries, the annual mortality rate exceeded the annual recruitment rate in the period of 1979-1986 in spite of the fact that this period is known for the yield of many highly abundant year classes. In 1986 the autumn fishery for capelin had to be stopped. According to the recommendations issued by ICES and the JSNCF, the ban should be extended until the spawning stock is recovered, at least to 1.5 mt. In 1986 it was as low as 60,000 t, and in 1989 it increased to 180,000 t (the assessment made at PINRO).

Atlanto-Scandian Herring

The stock is still in a state of deep depression; the source of which is marked by the 1960s. At that time the stock began declining while the intensity of the unregulated fishery got higher. The situation was aggravated to such an extent that since then, the large-scale fishery for the Atlanto-Scandian herring has no longer existed. From the late 1960s to the early 1980s the numerical strength of spawners was so low that they could not produce abundant year classes, even under favorable environmental conditions registered in 1969, 1973, 1976, 1979. It was not until 1983 that a rich year class

Table 8.2. TAC and actual catches of capelin.

Years	ICES TAC	Sov.-Norway Commiss. TAC	Actual Catches	Deviation from ICES TAC	Deviation from Sov.-Norway Comiss.
1979	1800	1800	1679	-122	-122
1980	1600	1600	1496	-104	-104
1981	1700-1900	1900	2041	+141	+141
1982	1400-1600	1700	1917	+317	+217
1983	2000-2200	2300	2336	+136	+36
1984	900-1120	1400	1463	+343	+63
1985	1100	1100	1026	-74	-74
1986	-	120	123	+123	+3
1987	No	fishery			

was produced. However, the expectations that the stock would recover were in vain, as in 1985-1986 this year class was subject to excessive fishing efforts and failed to produce an adequate size of recruitment to the spawning stock. In spite of a lack of spawners, the catch of mature herring of the 1983-year class reached 100,000-120,000 t in recent years as the subsequent year classes were poor.

Polar cod from the Barents Sea. The distribution, migrations, survival, and abundance of many species of fish, marine mammals, and birds in the Polar region depend on the abundance of the polar cod--their food species. With the decline in the abundance of capelin and north shrimp, the press of predators on polar cod increased. For example, in the absence of capelin, cod started migrating to the northeast, to Novaya Zemlya, feeding on polar cod exclusively. The numerical strength of polar cod has not increased noticeably under such conditions, despite the rich 1985- and 1986-year classes. The present state of the stock allows no hope for the renewal of a commercial fishery for polar cod in the near future.

Cod

The commercial stock of cod declined drastically in the early 1980s due to a very high fishing intensity and emergence of a series of poor year classes. It was expected that the stock would increase to 2.8 mmt, approaching the long-term mean, in 1988. However, owing to some depletion in food resources, a lower growth rate, a higher cannibalistic rate and disregard of the scientific recommendations issued by PINRO on the reduction in the TAC, this level of the numerical strength was not achieved. The commercial stock is on a low level, less than 1 mmt, and there is no hope that it would increase in the nearest future since the numerical strength of the rich 1983-year class has been already reduced and all the following generations (the 1985-, 1986-, 1987-, 1988-, 1989-year classes) were poor.

Haddock

All the aforesaid of cod holds true for haddock. The stock is on the level below the long-term mean. The fishery is based on the remainder of the rich 1982- and 1983-year classes, on the moderate 1984-year class and the poor 1985- and 1986-year classes. By analyzing the state of the stock of haddock, the JSNCF had to reduce the TAC to 25,000 t in 1990.

Deepwater Redfish, Greenland Halibut, and Other Species

Of other predators that together with Gadidae compose the upper trophic level of the bioproductive web, deepwater redfish and Greenland halibut should be mentioned. The stocks of these species which are important for the traditional trawl fishery, are also on a low level. Some increase in the catch of redfish is expected only in the first half of the 1990s when the rich 1982-year class attains the fishable age. The low abundance of the following year classes decreases still lower due to the grazing activity of cod in view of the lack of other fish food.

The stock of Greenland halibut is assessed to be only 100,000 mt. The increase in the stock to the level of the first half of the 1970s (200,000-300,000 mt) may be achieved if either the fishery would be temporarily banned for two or three years or the catch would be restricted to 4,000-7,000 mt instead of 20,000-22,000 mt taken nowadays (Nizovtsev, 1989).

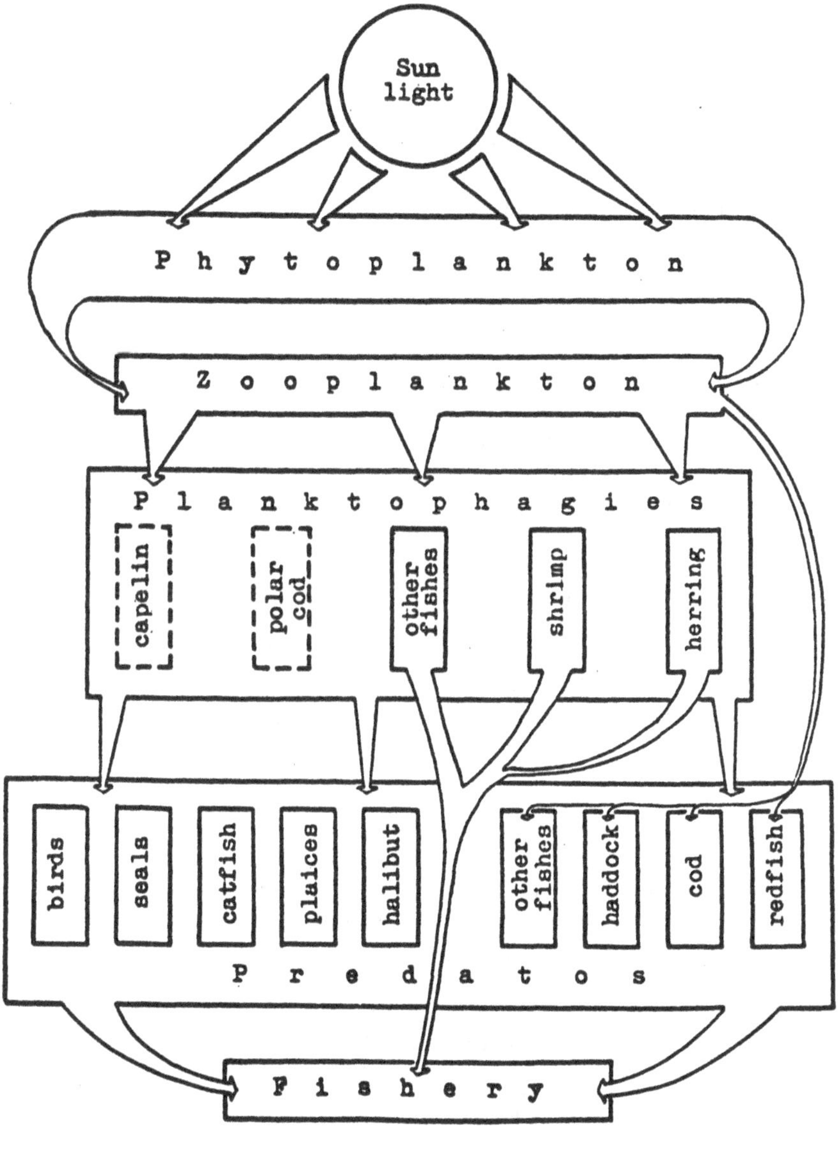

Figure 8.2. The general pattern of bioproduction use in the Barents Sea.

At present, the stocks of those species of fish that are not taken by specific fisheries, but caught as by-catch, are in a satisfactory state in the Barents Sea, including catfish, plaice, long rough dab. So the term "underutilization" of stocks can be referred, to some extent, only to these species.

Conclusion

The present state of the main commercial populations of fish in the Barents Sea indicates a serious upset in the ecosystem. It is believed to function in a normal way only to the level of zooplankton. The stream of organic matter to upper levels, that follows, is hampered owing to a lower abundance of consumers, such as herring, capelin, polar cod, and shrimp. The constrained consumption of zooplankton by predatory fish, omitting plankton-eaters, fails. In view of this fact, a great deal of bioproduction on lower energy levels remains intact, which leads to a certain reduction in the biological productivity, on the whole, in the sea (Figure 8.2).

A radical step under these conditions could be some reduction in the fishing effort, a transfer of effort to other areas of the World Ocean, a management regime for depleted stocks aimed at the restoration of them to an optimum level. The solution of the problem, as it seems, may be achieved through a quite realistic and effective measure, such as strict observation of scientific recommendations on the management of the fishery by all countries interested in the improvement of the ecological situation in the region.

References

Drobysheva, S. S., G. I. Nesvetova, V. N. Nesterova, V. M. Ryzhov, and A. N. Chepurnov. 1988. Regionalization of the Barents Sea with regard to formation of primary and planktonic production. ICES C.M.1988/L:7, 19 Pp.

Knipovich, N. M. 1970. Basis of the hydrology in the European Polar Ocean. St. Petersburgh, 1570 Pp. (In Russian)

Moiseev, P. A. 1989. Biological resources of the World Ocean. Agropromizdat, 368 Pp. (In Russian)

Nansen, F. 1906. Northern waters: Captain Roald Amundsen's oceanographic observations in the Arctic seas in 1901. Skr. Vidensk. Selsk Christiania Mat.-naturv. Pp. 1-145.

Nizovtsev, G. A. 1989. Recommendations on rational exploitation of the Norwegian-Barents Sea stock of Greenland halibut. PINRO. Murmansk., 93 Pp. (In Russian)

Novitsij, V. P. 1961. Permanent currents in the North Barents Sea. Trudy GOIN. Issue 64:3-32. (In Russian)
Tantsyura, A. I. 1959. On currents in the Barents Sea. Trudy PINRO. Issue II:35-53. (In Russian)
Midttun, L. 1989. Climatic fluctuations in the Barents Sea. Rapp. P.-v. Reun. Cons. int. Explor. Mer 188:23-35.

A. A. Rosenberg, M. Basson, and J. R. Beddington

9. Predictive Yield Models and Food Chain Theory

Abstract

This chapter considers the predictability of yield from single and multispecies fisheries. We have calculated confidence intervals for equilibrium yield and equilibrium spawning stock biomass given estimates of the variability in recruitment and estimates of other biological parameters for a range of stocks. Then, using the concept of a notional "safe" threshold level for spawning stock biomass, we calculate the probability that the stock goes below the threshold in any one year as a measure of the risk to sustainability.

In a single species concept we compare average and the coefficients of variation in yield and spawning stock biomass for eight species groups using data from the literature. The variability of yield increases with the mean, with slow-growing, long-lived species having lower yields and spawning stock biomasses, as a proportion of unexploited biomass, than fast-growing species. In other words, there is a clear tradeoff between uncertainty and average yield. This uncertainty is a measure of the predictability of the resource.

Two large marine ecosystem examples are examined in more detail, the South Georgia fishery for krill and Antarctic icefish and the Barents Sea fisheries for cod, capelin, and herring. In each case, the probability of maintaining the stock above a threshold level, the risk curve, is dependent on the harvest rates of all the species in the fishery. Predator-prey interactions in a multispecies fishery can buffer the increase in uncertainty in yield that normally accrues with increasing harvest rate.

A critical feature of predictive yield models is the quantification of both the uncertainty in yield and the risk to sustainability due to depletion of the spawning stock.

Introduction

The prediction of the potential yield of a fishery given biological information on a particular stock has always been a primary task of fishery scientists. In a single species context, this work is well developed and the pattern of equilibrium yield versus the fishing mortality rate on the stock is familiar (Beddington and Cooke, 1983). In recent years, increasing attention has been given to multispecies fisheries, accounting for both biological (predator-prey) interactions between target species and technological interaction due to common susceptibility to fishing gear of two or more species. A recent International Council of Exploration of the Sea Symposium in The Hague has summarized much of the recent work (Sissenwine and Daan, 1990).

In this paper we chose to focus on the predictability of yield from both single and multispecies fishery models. A dominant source of variability in the abundance of exploited fish stocks is the interannual variation in reproductive success of the stock, i.e., in the recruitment of new animals to the harvestable stock. Variation in abundance translates into annual variability in yield and hence the predictability of the sustainable yield level over time.

In the models presented below we have considered the predictability of long term yield in two ways. Firstly, we have calculated confidence intervals for equilibrium yield and equilibrium spawning stock biomass given the estimated variation in recruitment and estimates of other relevant biological parameters. Secondly, the probability that the spawning biomass of a stock goes below some notional "safe" threshold level in any year under various exploitation rates was calculated.

The idea of a notional threshold refers to a generalized concept of the relationship between stock and recruitment when there is substantial recruitment variability. While it is usually impossible to express recruitment of new individuals as a function of the size of the spawning stock, in general we assume that at low spawning stock abundance the risk of a poor recruitment (low reproductive success) is increased. This is illustrated schematically in Figure 9.1 where recruitment in the labelled "safe" area varies around some constant average. At low spawning stock levels we postulate a danger zone where poor recruitment is more likely and so the goal of management for a sustainable harvest is to maintain the stock in the safe zone.

Below we first consider a single species pattern of equilibrium yield and its predictability for a range of species groups. Then, two simple examples are considered in more detail in a multispecies context; the potential yield of Antarctic icefish and Antarctic krill around the island of South Georgia and the fisheries for cod, herring and capelin in the Barents Sea.

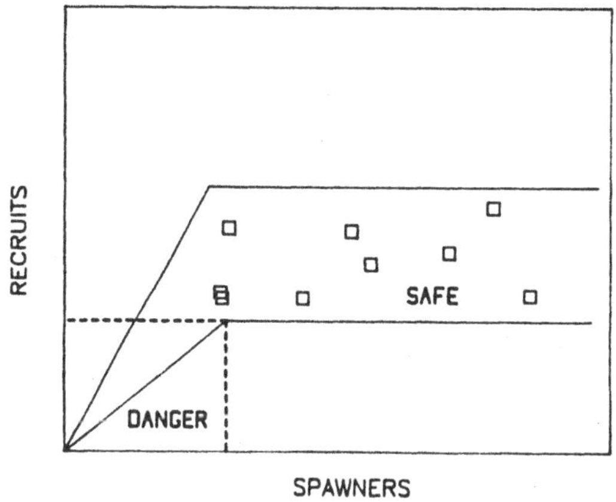

Figure 9.1. Schematic illustration of the relationship between stock and recruitment assumed in this analysis. Danger zone is defined by a notional threshold level of spawner abundance (spawning stock biomass). It is assumed that below this threshold, the chance of poor recruitment is greater.

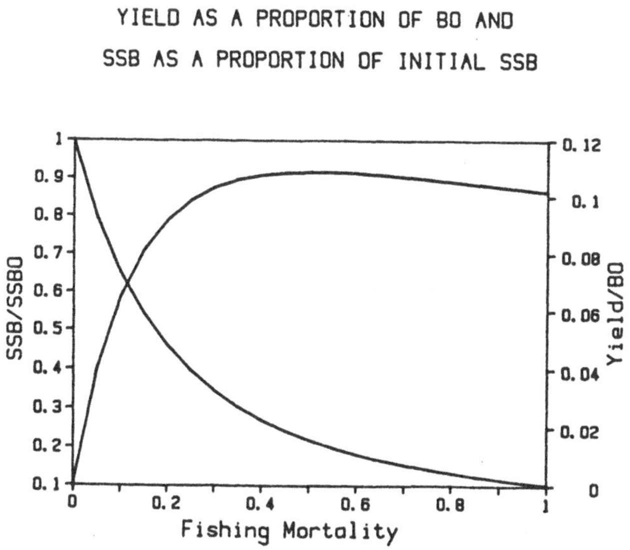

Figure 9.2. Basic pattern of equilibrium yield and spawning stock biomass versus fishing mortality rate from the Beverton and Holt model. Yield and biomass are expressed as proportions of the unexploited recruited biomass or unexploited spawning biomass respectively.

Yield and Its Predictability for Various Species Groups

We can use, as a basis for discussion, the well known pattern of equilibrium yield and spawning stock biomass (SSB) obtained from the dynamic pool models of Beverton and Holt (1957; Figure 9.2). For comparative purposes, it is simplest to consider the equilibrium yield as a proportion of the unexploited vulnerable biomass, the so-called recruited biomass. Similarly, the equilibrium spawning stock biomass is plotted as a proportion of the unexploited spawning biomass. For this model formulation, the shape of the yield and SSB curves with increasing instantaneous fishing mortality rate are determined by the natural mortality rate, m, the growth coefficient from the von Bertalanffy growth function, k, and the ages of animals in the stock which are exploited and mature.

Short-lived, fast growing species have equilibrium yield curves which peak at relatively high fishing mortality rates and there is little decrease in yield with increasing fishing mortality beyond the maximum yield level. Long-lived, slower growers have more domed equilibrium yield curves with maximum yield at quite low harvest rates. The degree of spawning stock biomass depletion coincides with the yield curve. As a general guide, the spawning stock is in the region of 20-30% of its unexploited level when the fishing mortality rate is approximately the level giving the maximum on the yield curve (Figure 9.2).

Beddington and Cooke (1983) give estimates of m, k, and ages at first capture and maturity for 61 stocks gleaned from the literature. Forming eight species groups for these stocks and calculating the average maximum equilibrium yield for each group using the equations given in their paper shows that proportionate yield varies between 5% and 10% in general, with some tendency for pelagic species such as sardine, anchovy and mackerel to have higher yields than groups higher up the food chain (Figure 9.3a).

The proportionate spawning stock biomass for the species groups shows a much stronger pattern (Figure 9.3b). There is an increasing level of depletion at the fishing mortality rate giving the maximum yield (F_{max}) for the demersal slow growing species compared to the pelagic fast growing stocks at the base of the food chain.

A key feature in consideration of the predictability of yield and the sustainability of the stock is the uncertainty or variability in yield and spawning stock biomass. To examine this variability we assumed that recruitment was gamma distributed with probability density function:

$$f(x) = [x^{a-1} e^{-a/b}]/[b^{a}G(a)]$$

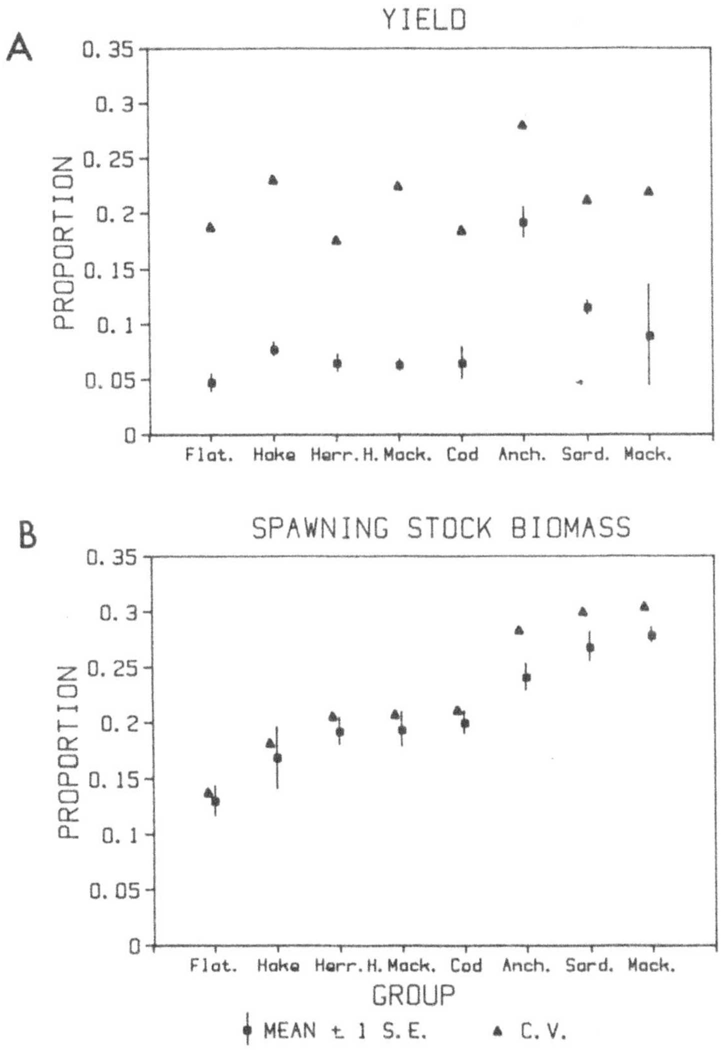

Figure 9.3A. Average maximum sustainable yield (squares) and coefficient of variation of the yield (triangles) for eight family groups of fishes. The groups are flatfishes, hakes, herrings, horse mackerels, cods, anchovies, sardines, and mackerels. The bars indicate one standard error around the means. B. Average spawning stock biomass at F_{max} and the coefficient of variation of the SSB for eight family groups as described above.

where G(a) is the gamma function and a and b are parameters.

The parameters of the distribution for each of the 61 stocks were estimated using the data in Beddington and Cooke (1983) using the relationships:

$$E(R) = ab$$

$$V(R) = ab^2$$

where E and V denote expected value and variance respectively. We then calculated the coefficient of variation (C.V.) of yield and SSB at the F_{max} using:

$$E(Y) = C_1(ab)$$

$$V(Y) = C_2(ab^2)$$

where

$$C_1 = F \int e^{-tM - (t-tr)F}\ W(t)\ dt$$

$$C_2 = F^2 \int e^{-2tM - 2(t-tr)F}\ W(t)^2\ dt$$

W(t) is the familiar von Bertalanffy growth function

$$W(t) = W_\infty\ (1\ e^{-kt})3$$

with parameters k and W_∞. The integrals are from the age at recruitment to the fishery (tr) to infinity.

There is an increase in C.V. as the mean proportionate yield increases and as the mean proportionate SSB at F_{max} increases. In other words, there is some tradeoff in uncertainty for the increase in yield. Variability in equilibrium yield is a measure of predictability under a long term management policy and this predictability is lower for pelagic, fast growing species. Variability is greatest for anchovies and lower for flatfish, cods and herring.

The variability in equilibrium SSB is some measure of the risk to the stock of harvesting in accordance with the long term strategy. This essentially implies that the productivity of the stock, i.e., its stock-recruitment function, has a notional threshold level below which the risk of poor recruitment is high (Figure 9.1). Clearly, the precise threshold level between the danger and safe areas of the relationship is unknown, but can be approximated to give a notional value for the existing data.

For a given level of variability in the equilibrium SSB it is appropriate to ask what the probability is that the spawning stock will be below this threshold in any one year and so have a high risk of poor recruitment in that year. Calculating this probability for different harvesting policies gives a measure of risk to the sustainability of the fishery of choosing a harvest rate. Of course, if the fishery is for more than one species and the species interact then the risk to sustainability will be affected by harvest policy for the ecosystem as a whole.

Here, two simple examples of exploited ecosystems are considered; the fisheries around the island of South Georgia in the Antarctic for Antarctic icefish (*Champsocephalus gunneri*) and for krill (*Euphausia superba*) and the fisheries in the Barents Sea for cod (*Gadus morhua*), capelin (*Mallotus virens*) and herring (*Clupea harengus*). In each case, two or three species at different levels of the food chain are harvested and there are either technological interactions between the species due to harvesting or predator-prey interactions. While many large marine ecosystems are far more complex than these two examples, they serve to illustrate the basic principles involved in developing measures of risk for multispecies fisheries.

Yield and Predictability for South Georgia Fisheries

Information on the biological parameters which characterize the sustainable yield and SSB of krill are given in Rosenberg et al. (1986) and in Basson (1988). These parameters are the von Bertallanffy growth equation parameter estimates, the instantaneous rate of natural mortality, the ages of recruitment and maturity and the level of recruitment variability in terms of the parameters of the gamma distribution. These latter were estimated from the age composition data given in Basson (1988) using the method described by Bravington (in preparation). For Antarctic icefish, the parameter estimates were obtained from Basson et al. (1989).

Because the focus is on predictability the 90% confidence interval of the yield for the two species was calculated. The shape of the yield curve is familiar and similar for the two species (Figure 9.4). The confidence interval is slightly wider for krill than for icefish because of the slightly higher recruitment variability for the euphausiid. Note that the confidence band is not symmetrical around the mean yield but, like the gamma distribution, is skewed. The overall pattern is of a strong increase in the width of the confidence interval as the harvesting rate increases, i.e., yield becomes less predictable when the stock is under heavier pressure--a common sense result.

The confidence bands for SSB tend to narrow at higher fishing mortality rates as the spawning stock is severely depleted (Figure 9.5).

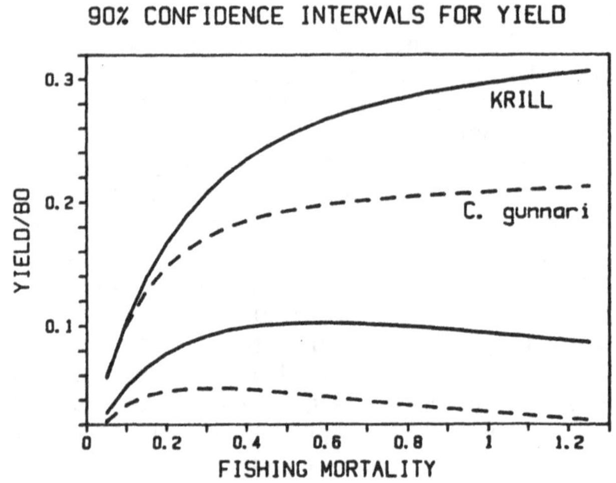

Figure 9.4. Confidence bands for equilibrium yield of Antarctic krill and icefish.

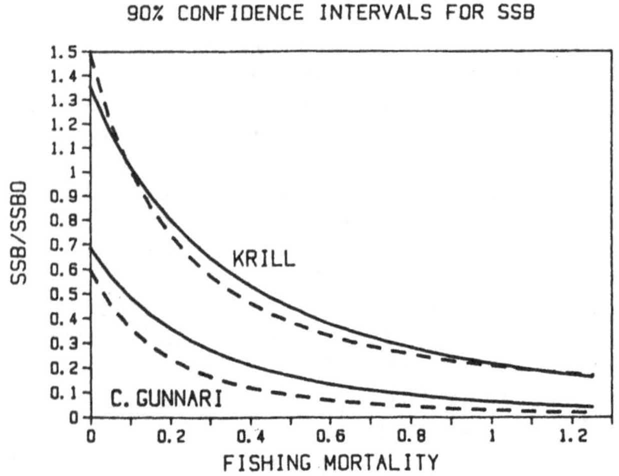

Figure 9.5. Confidence bands for SSB at F_{max} for Antarctic krill and icefish.

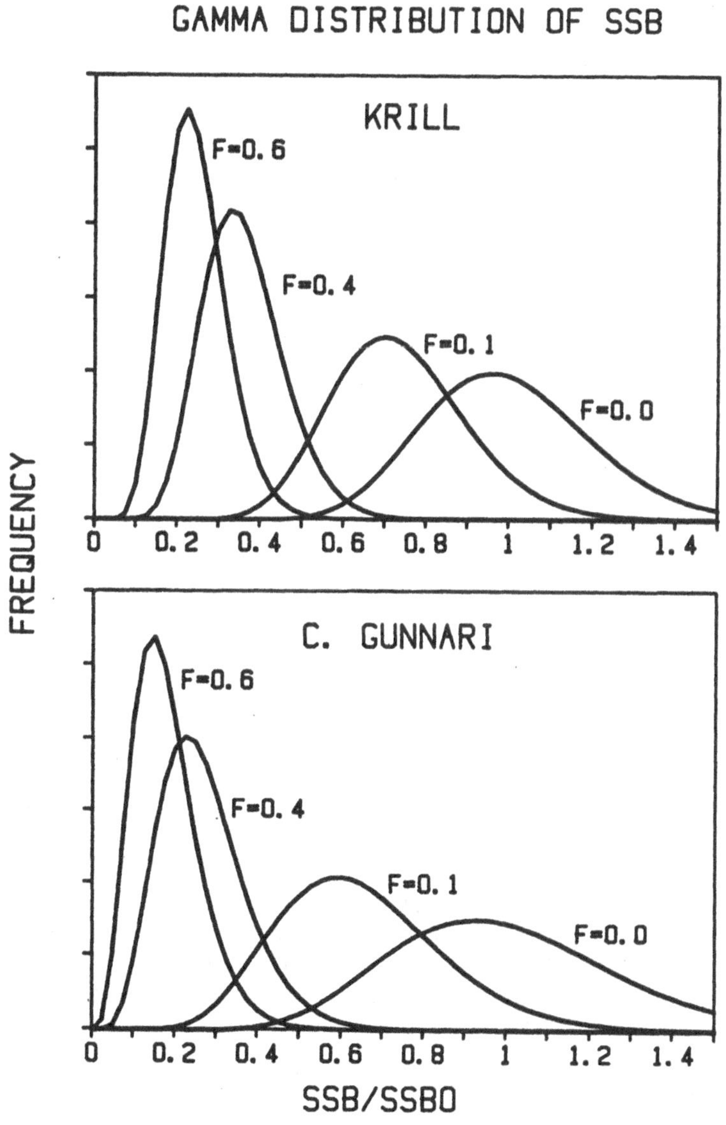

Figure 9.6. Distribution of SSB for krill and icefish under different harvest rates. The distribution results for the gamma distribution of recruitment which gives an approximately gamma distribution for SSB.

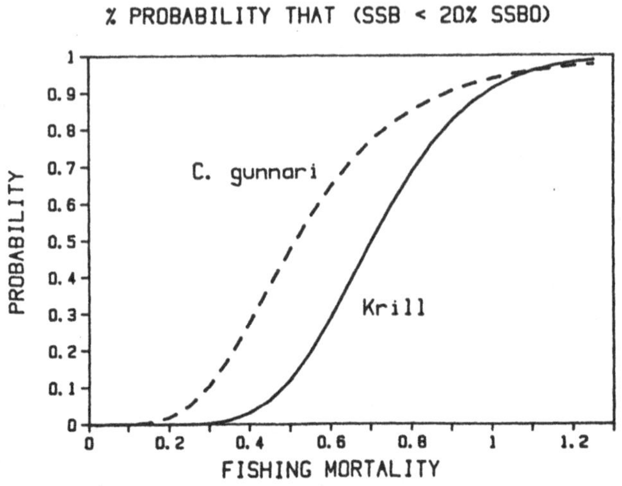

Figure 9.7. Risk curve for krill and icefish. The curves give the probability that the SSB of each stock goes below a 20% of the unexploited SSB in any one year. There is no interaction between the stocks for these runs.

This narrowing, of course, is due to a shift to the left as well as truncation of the righthand side of the distribution of SSB (Figure 9.6). The patterns are very similar for krill and icefish. Because of this depletion, SSB becomes more predictable as harvesting rate increases, in contrast to yield.

A notional SSB threshold where the risk of poor recruitment is likely to be high was chosen for both these species as 20% of the unexploited spawning stock biomass level. In the absence of interactions between the species the probability of each stock going below the threshold in any one year under a range of harvest rates (Figure 9.7) indicates that icefish is more vulnerable to increasing the harvest rate than the more robust krill stocks. The probability or "risk" curve for icefish is steeper and to the left of the krill curve. So, icefish may have a slightly more predictable yield at F_{max} but there is a trade-off in the risk to sustainability, which increases rapidly with the harvest rate compared to krill.

The krill stock in around South Georgia is many times larger than the stock of icefish, so even though icefish prey upon krill the impact of this predation is probably insignificant. However, there is another type of interaction between the species due to the fishery. The fine mesh nets used to harvest krill may take juvenile icefish and so affect the sustainability of the icefish fishery. The probability of the icefish

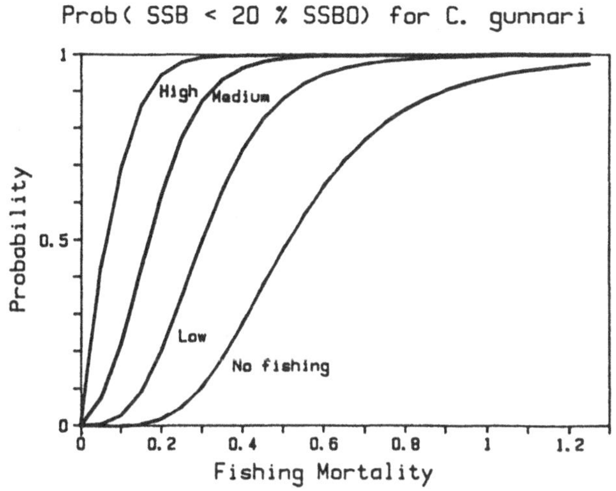

Figure 9.8. Risk curves for icefish as in Figure 9.7 except that the krill fishery has a by-catch of small icefish. The curves are for different levels of harvesting on the krill stocks as labelled (no fishing to high fishing).

stock being reduced below the 20% threshold value in any one year when krill are harvested as well is increased in line with the harvesting pressure on krill (Figure 9.8). A large krill fishery may eliminate the possibility for a sustainable icefish fishery because of the risk of poor recruitment would make such a fishery unpredictable and unviable.

The Barents Sea Fisheries

There are three main species harvested in the Barents Sea which are at different levels in the food chain. The confidence bands for equilibrium yield reflect the position of each species in the ecosystem (Figure 9.9). Capelin are small, short-lived pelagic planktivores with very variable recruitment and with a wide confidence band. They are the focus of harvest by Norwegian and Soviet fishermen, but the fishery has collapsed in recent years with the virtual disappearance of large schools of capelin from the fishing grounds (ICES, 1987).

The herring fishery in the Barents Sea has also varied substantially in the level of landings in recent years. The herring are longer-lived and larger than capelin and may prey upon them during the early life history (Hamre, 1988). The confidence interval for herring is consequently narrower than for capelin (Figure 9.9) as may

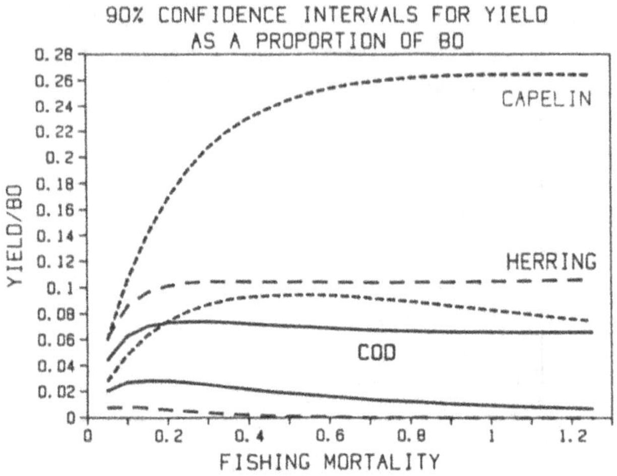

Figure 9.9. Confidence bands for equilibrium proportionate yield of Barents Sea capelin, herring, and cod.

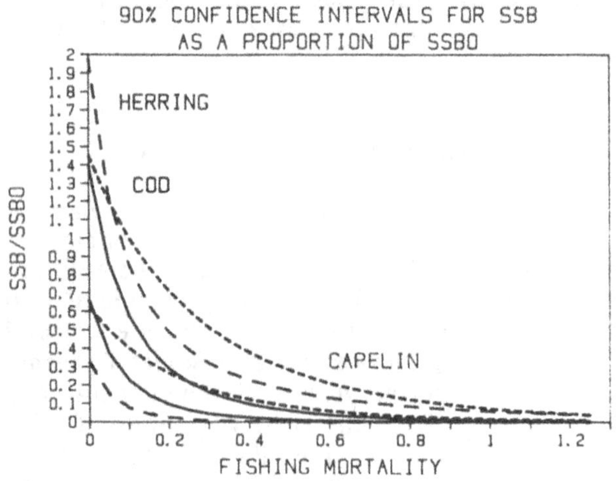

Figure 9.10. Confidence bands for SSB of Barents Sea capelin, herring, and cod.

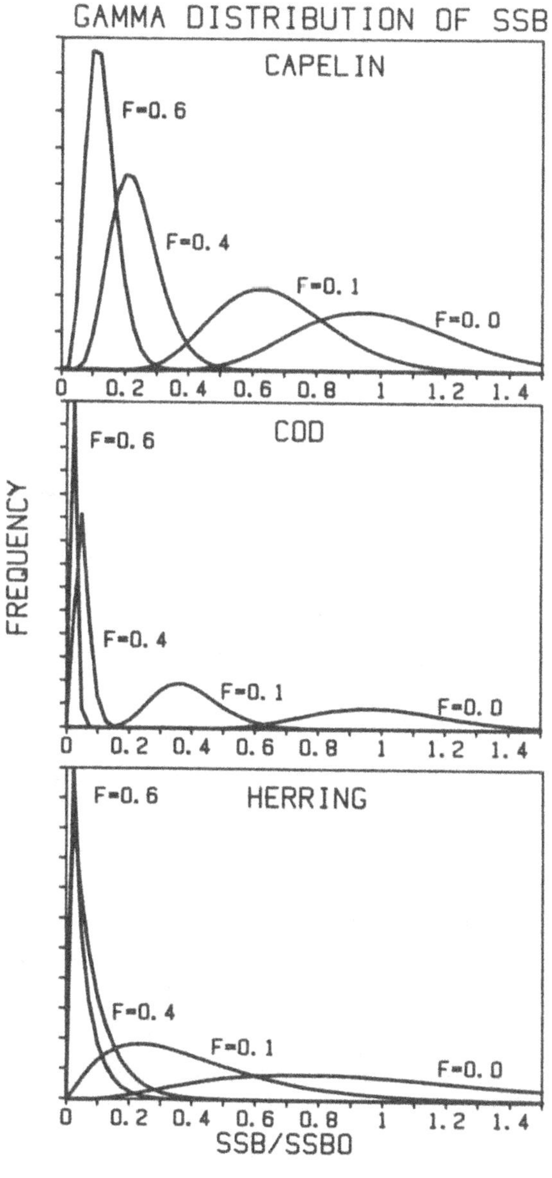

Figure 9.11. The distribution of SSB for capelin, herring, and cod under different harvest rates.

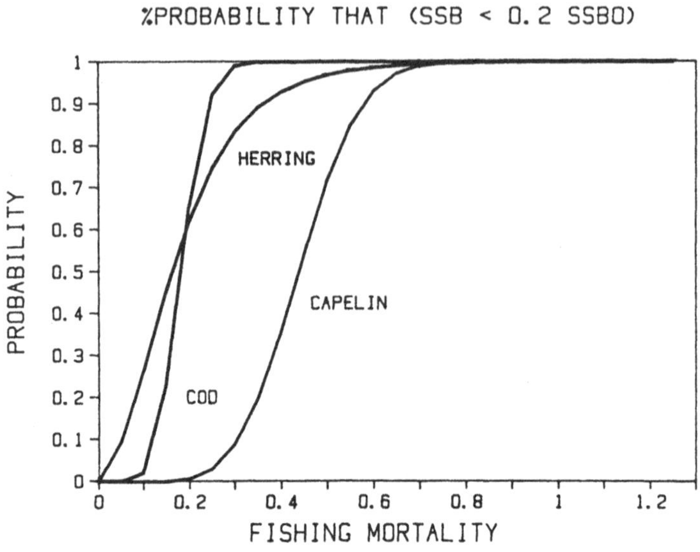

Figure 9.12. Risk curves for capelin, herring, and cod where there is no interaction between the species. The threshold value is assumed to be 20% of unexploited respective SSB.

be expected from the results of the analysis of different groups of species presented above (Figure 9.3).

The third species, cod, is thought to be the main predator in the system (Hamre, 1988) and is relatively long lived with a lower variation in recruitment than the other species. Consequently, the confidence interval for cod is substantially narrower than for the other two species (Figure 9.9). Note also that the yield curve for cod peaks at a lower fishing mortality rate than for herring which is lower than capelin.

The confidence bands for spawning stock biomass again reflect the position of the species in the food chain and the level of recruitment variability (Figure 9.10). The SSB bands narrow as harvest rate increases due to the compression of the distribution of spawning stock biomass to the left (Figure 9.11). Herring, in particular, has a very flat unexploited distribution of SSB which is skewed very strongly as harvest rate increases.

The threshold values between notional "safe" and "danger" areas of the recruitment curves were again taken as 20% of the unexploited SSB for each stock. The probability of the stock going into the danger area for each species reflects the risk to sustainability associated with different harvesting policies (Figure 9.12). The risk curve for cod rises very steeply at quite low fishing mortality rates. While the maximum yield for herring is at a similar fishing mortality

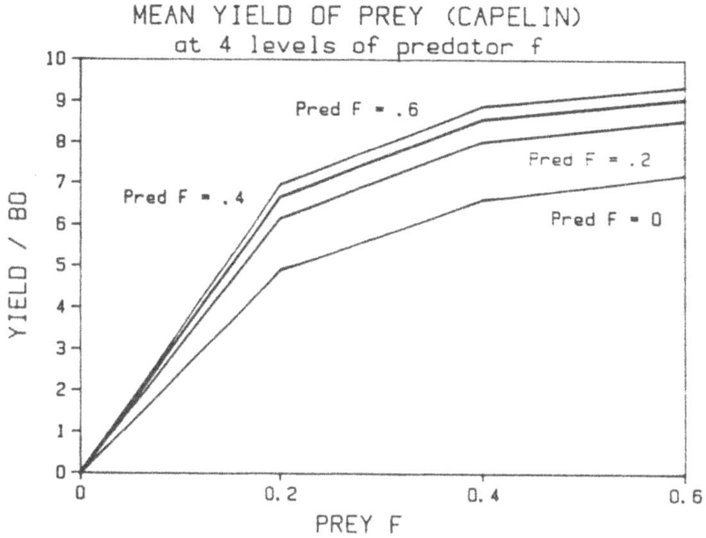

Figure 9.13. Yield curves for capelin when cod is the major predator for diferent levels of harvest on both predator and prey.

rate as for cod, its risk curve is much less steep, reflecting the longer tail of the distribution of SSB which results from the high variability of year-class size for this species. Capelin, at the base of the food chain, has a risk curve shifted well to the right because of its higher growth and natural mortality rate. This is again in accord with the results presented above for different groups of fishes. The slope of the capelin risk curve is similar to the herring curve. The slope is controlled by the level of recruitment variability, due to the buffering effect of occasional large year classes. The position of the curve on the fishing mortality axis is related to the rates of growth and natural mortality.

The interactions of these three species in the Barents Sea is complex. Cod is probably the major predator of capelin, focussing on the older ages. Cod also eat juvenile herring, but herring may take larval capelin (Hamre, 1988). Only two-way interactions are considered here. There are probably alternative prey for cod, so its yield profile and the predictability of yield is assumed not to be strongly affected by the harvest of either capelin or herring.

As the cod is harvested down, the yield curve for capelin increases in height nonlinearly (Figure 9.13) but the shape of the curve remains fairly constant. The variation in the yield of capelin, as measured by the C.V., decreases as the cod is harvested (Figure 9.14), offsetting the usual pattern of increasing variability with an increasing fishing mortality on the capelin itself.

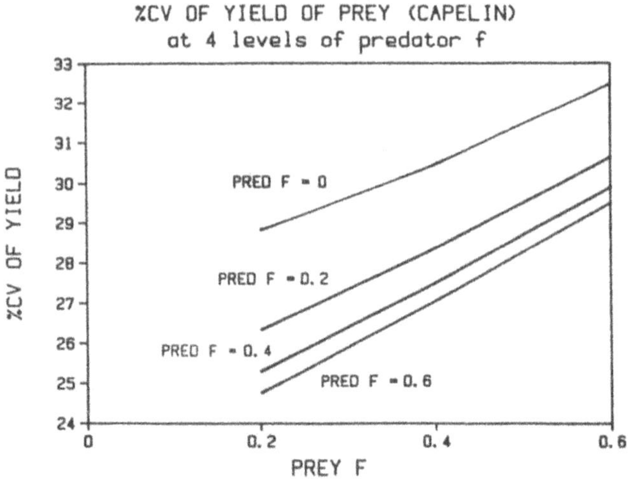

Figure 9.14. The coefficient of variation in yield for interacting populations of cod and capelin under different harvesting pressures.

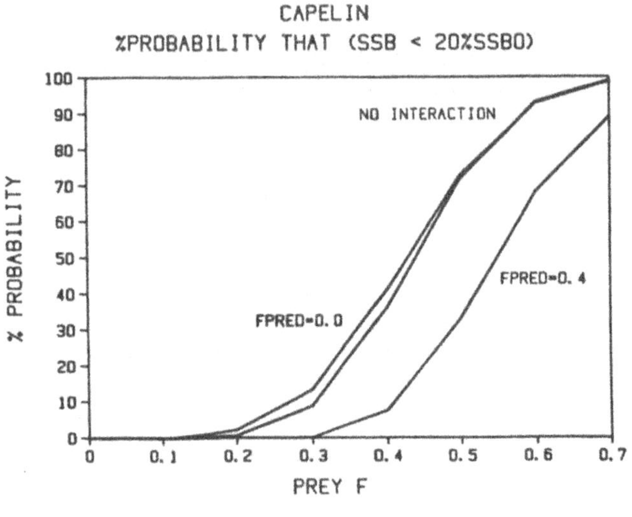

Figure 9.15. Risk curves for capelin preyed upon by cod which are harvested at two different levels (denoted by FPRED).

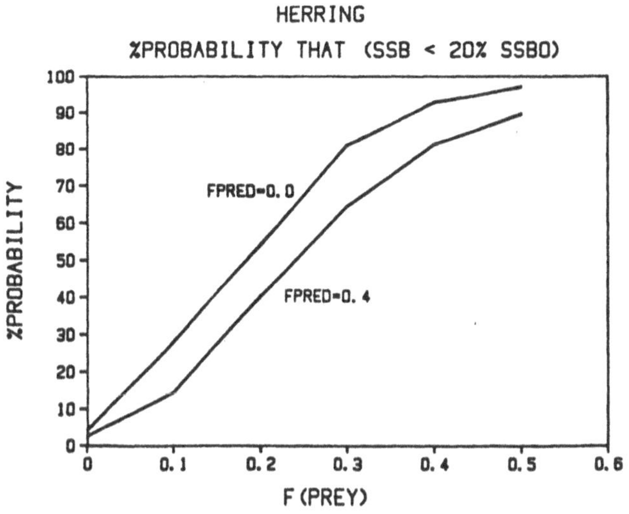

Figure 9.16. Risk curves for herring preyed upon by cod which are harvested at two different levels (denoted by FPRED).

In terms of the risk to the stock, the harvesting of the predator shifts the probability of going below the threshold SSB to higher F values without substantially changing the slope of the risk curve (Figure 9.15). This is a reflection of the decreasing natural mortality rate on the capelin as the predator is fished down. Since the growth rate of the capelin remains constant, with decreased natural mortality, a higher yield can be taken without depleting the spawning stock.

For the cod-herring interaction, the effect is similar both for the level of equilibrium yield and its variability. The risk curve again shifts to higher fishing mortality rates (Figure 9.16). There is little change in the slope of the risk curve because the distribution of SSB is not compressed by harvesting of the predator and so the buffering effect of a long tail on the distribution remains (Figure 9.17).

Discussion

As harvesting effort increases on a stock, the predictability of the equilibrium yield decreases. This predictability of yield is lower for fast-growing, shorter-lived species at the base of the food chain which tend to have highly variable recruitment.

In contrast, high variability in recruitment is reflected in a high variability in spawning stock biomass which can buffer the risk of overfishing because of the presence of occasional strong year class.

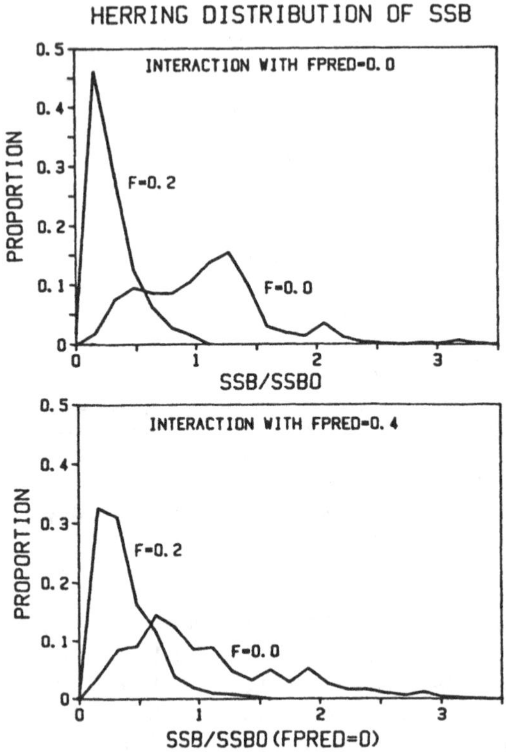

Figure 9.17. The distribution of herring spawning stock biomass where the herring is subjected to either no harvesting or with an F of 0.2 and the cod predators are either not harvested or are harvested with a FPRED of 0.4.

This means that the risk to the sustainability of the stock in terms of spawning stock biomass depression increases more rapidly with the harvest rate for higher level predators than for those species at the base of the food chain.

Predator-prey interactions in a multispecies fishery can buffer the increase in uncertainty in yield because variability decreases as the predator stock is harvested down. The risk curve shifts to the right on the prey harvest rate axis as the harvest rate on the predator increases without changing slope. The position of the risk curve on the prey harvest rate axis depends on the growth rate of the fish and on the non-fishing sources of mortality. The slope of the risk curve depends on the spread and skewness of the distribution of spawning stock biomass. Of course, these conclusions, from the simplified systems considered here, may be altered by inclusion of other biological effects of predator-prey interactions. For example, the assumption of consistent growth rates for capelin and herring over a range of stock sizes may not hold. As a result, there may be additional feedback to spawning biomass in response to cod predation.

A critical feature of predictive yield models is the quantification of both the uncertainty in yield and the risk to sustainability as expressed by the depletion of the spawning stock under different long term management scenarios. This risk can only be determined for a given notional threshold level of spawning stock biomass. Hopefully, policy can adapt to increasing information on threshold levels to better describe risk. Just this sort of approach of quantifying the risk to the spawning stock is currently being applied in the fisheries around the Falkland Islands in the South Atlantic to guide management decisions (Beddington et al., 1990).

References

Basson, M. 1988. Population dynamics of krill interactions with its major predators. Ph.D. Thesis, Imperial College, University of London. 364 Pp.

Basson, M., J. R. Beddington, and W. Slosarczyk. 1989. The status of the *Champsocephalus gunnari* stock in the South Georgia area. Working Group on Fish Stock Assessment, Hobart, Australia, 25 Oct. to 2 Nov. 1989. WG-FSA-89/8 Rev. 1. Report of the Eighth Meeting of the Scientific Committee, Scientific Committee for the Conservation of Antarctic Marine Living Resources, Secretariat, Hobart, Australia. SC-CAMLR-VIII.

Beddington, J. R., A. A. Rosenberg, J. A. Crombie, and G. P. Kirkwood. 1990. Stock assessment and the provision of management advice for the short fin squid fishery in Falkland Islands waters. Fish. Research. 8:351-365.

Beddington, J. R., and J. G. Cooke. 1983. The potential yield of fish stocks. FAO Fish Tech. Pap. 242, 47 Pp.

Beverton, R.J.H., and S. J. Holt. 1957. On the dynamics of exploited fish population. U.K. Min. Agr. Fish. Fish. Invest. (Ser. 2) 19, 533 Pp.

Bravington, M. In prep. Estimating mortality and recruitment variability from age composition data. (to be submitted to Biometrics).

Hamre, J. 1988. Some aspects of the interrelation between the herring in the Norwegian Sea and the stocks of capelin and cod in the Barents Sea. ICES C.M. 1988/H:42; 15 Pp.

International Council for the Exploration of the Sea [ICES]. 1987. Report of the Atlanto-Scandian herring and capelin working group. ICES C.M. Assess 8:57 Pp.

Rosenberg, A. A., J. R. Beddington, and M. Basson. 1986. The growth and longevity of krill during the first decade of pelagic whaling. Nature 324:152-154.

Jeremy S. Collie

10. Adaptive Strategies for Management of Fisheries Resources in Large Marine Ecosystems

Abstract

The adaptive approach to fisheries management involves experimentally manipulating fish populations to learn about the processes regulating fish population size. Although the development and application of adaptive management has focused on Pacific salmon, adaptive strategies have also been formulated for groundfish such as the yellowtail flounder. Application of adaptive management strategies to large marine ecosystems such as the eastern Bering Sea (EBS) is problematic because of the lack of spatial replicates. The biggest resource issue in the EBS is management of walleye pollock. Some of the most important questions about pollock are whether discrete stocks exist in the EBS, the role of cannibalism on recruitment, and the reliance of sea birds and marine mammals on pollock as a food source. In the absence of discrete pollock stocks in the EBS, it may be necessary to compare the EBS stock with the Gulf of Alaska stock, which exhibits coherence with EBS recruitment. The North Pacific Fisheries Management Council currently manages groundfish on a single-species basis without considering multispecies interactions. Adaptive management could be implemented under the existing regulatory framework but it will require a philosophical shift toward recognizing and testing alternative hypotheses.

Introduction

Of the many factors affecting the abundance of marine fish populations, fishing is the only one over which managers exert control. The key to adaptive management is to recognize and to identify alternative hypotheses about factors affecting fish production. The passive adaptive approach seeks the best management strategy averaged across the different hypotheses, or states of nature. The active adaptive approach deliberately manipulates population size to test alternative hypotheses. In the context of large marine ecosystems, hypotheses that could be tested with adaptive management are those depending on the density of harvested species.

Within the scope of this symposium on large marine ecosystems, the main hypothesis to be tested is: H(1) food chain dynamics are important in regulating the abundance of exploited fish populations over decadal time scales. This general hypothesis can be divided into at least two sub-hypotheses:

H(1a) Competitive Replacement Hypothesis. Food available from lower trophic levels limits total fish production and thereby the harvestable surplus.

H(1b) Predation Hypothesis. Predator-prey interactions among the exploited fish regulate the abundance of individual species and thereby the harvestable surplus.

Some experimental approaches to test these hypotheses are:

1. Nutrient enrichment of coastal waters to increase food availability (This may not be a deliberate "experiment" but the outcome is uncertain.);

2. Manage for total fish biomass, recognizing the limitation of food availability;

3. Manipulate the levels of individual species to test multispecies interactions and to develop a desirable species mix.

Adaptive management is needed because of the limitations of other approaches. Process-oriented studies are limited by the inability to study all potentially important mechanisms on a scale relevant to the fishery resource. Empirical data analyses, like process-oriented studies, are useful for hypothesis generation but often do not provide powerful tests of hypotheses. Another reason for adaptive management is that marine ecosystems may exhibit higher-order properties that can not be deduced or predicted from single-species or pair-wise models (Tyler et al., 1982). As long as status quo management strategies are pursued there may be insufficient natural variability to distinguish species interactions and density-dependent effects from environmental perturbations (Walters and Collie, 1988).

In this paper I first briefly review the theory of adaptive management as it applies to large marine ecosystems. Second, I present examples of evaluating alternative harvest policies: one for yellowtail flounder in the northwest Atlantic and one for pollock in the eastern Bering Sea. In the latter case the major questions relating to food chain dynamics are identified and experimental strategies for answering these questions proposed. Finally I discuss how adaptive management can be implemented under existing fisheries management plans.

The Adaptive Approach to Fisheries Management

The essence of adaptive management is to recognize and to confront uncertainties about demographic parameters of harvested populations (Walters, 1986). The best adaptive policy is robust to all possible future states of the fish population and is often different than the optimal policy for the most likely state of the fish population. The actively adaptive approach comes from recognizing that some policies are more informative about future states than others, and that more informative policies may lead to higher long-term yields. Dual control involves finding the best current management policy while accounting for the longer-term benefits of learning. Adaptive management theory allows the value of information to be expressed in the same units as the yield from the fishery.

The seminal work on adaptive management was done by Walters and Hilborn (1976) and applied mostly to Pacific salmon. Recently, adaptive management plans have been developed for groundfish in continental shelf ecosystems (e.g. Sainsbury, 1988). If fishing experiments are conducted on a single population, without replicates or controls, the effect of fishing may be confounded with uncontrolled environmental factors (see Walters and Collie [1988] for an example). To account for potentially confounded responses, Walters et al. (1988) applied the principles of experimental design to the management of spatially replicated populations. To the extent that geographically adjacent stocks share demographic parameters, learning about these parameters can be accelerated with an adaptive approach. In addition, catch can be traded off among stocks to keep the total catch relatively constant.

A key issue that a defensible experimental design should address is local uniqueness in behavior among experimental units, which implies the need for replication and randomization of treatments. A second issue is that experimental units may exhibit shared temporal patterns due to large-scale factors such as ocean climate; this implies the need for treatment/control comparisons. We applied these design principles to the analysis of mesoscale management experiments (Walters et al., 1988).

"Replicate" subpopulations are not quantitatively similar: there is a conflict between increasing the experiment size and having to admit increasingly dissimilar replicates into the experiment. The larger the experimental units in question, the fewer and more dissimilar will be the replicates. In the case of large marine ecosystems, there may be no suitable "replicate" ecosystem, in which case a controlled experiment is impossible. Some unique fishery resources, such as Pacific halibut (*Hippoglossus stenolepis*), may be too valuable to experiment with at all. We should admit that there are questions about some such populations that we can't expect to answer (e.g. what causes recruitment variation?).

Adaptive Management of Yellowtail Flounder

In this section a worked example of an adaptive management strategy applied to a spatially replicated groundfish population (Collie and Walters, 1991) is presented. This plan has not been implemented; its purpose is to illustrate the steps in formulating and evaluating the potential performance of an adaptive strategy. There are six major stocks of yellowtail flounder (*Limanda ferruginea*) off the east coast of North America of which the Grand Banks, Georges Bank and Southern New England are most important. These stocks are more-or-less discrete and can thus be considered "experimental units". For each stock there exist commercial catch, CPUE and research survey data, to which we fit a production-type model. There are basically three unknown parameters of the model: a--productivity at low stock size, b--density dependence, and q--a catchability coefficient. The model fit the data but there was considerable parameter uncertainty. The three-dimensional confidence region showed that most of the parameter uncertainty lay along one major axis. Only certain combinations of the three parameters were consistent with the data. If a was fixed the two other parameters (b and q) were more tightly constrained. We defined alternative models, or hypotheses about yellowtail flounder productivity, by choosing a discrete set of a values and refitting the production model.

The value of different harvest rates was then estimated for each alternative model by Monte Carlo simulation. The population dynamics routine simulated the six subpopulations according to the production model, with random process and measurement errors included. The simulated manager collected data from the fishery and calculated the likelihood of each model being the correct one. If the relative likelihood of a model being correct exceeded a threshold (0.9) the manager switched to the optimal policy for that model.

The simulations evaluated each combination of model and initial harvest rate. The values in Figure 10.1 are cumulative 50-year yields from all six yellowtail flounder stocks summed. The initial harvest rates correspond to the harvest rate for maximum sustainable yield for

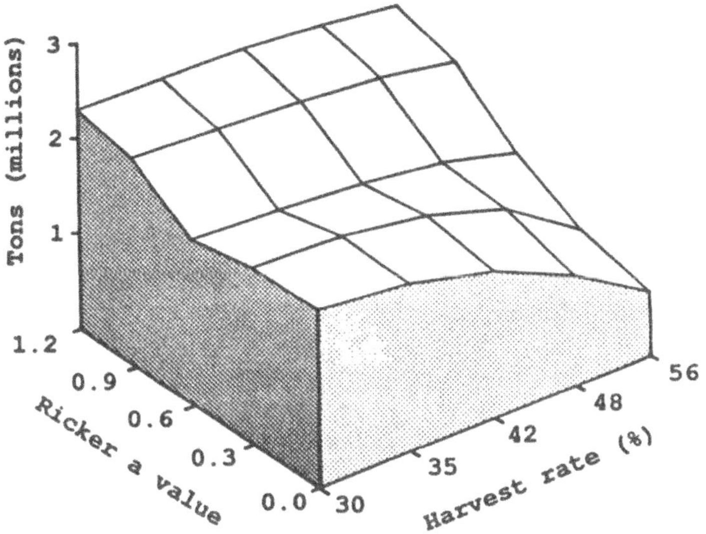

Figure 10.1. Simulated 50-year yields of six yellowtail flounder stocks in the northwest Atlantic. Total cumulative yield was simulated with five different models, represented by the Ricker *a* value, and five initial harvest rates. See text for details of the simulation. From Collie and Walters (1991).

each of the alternative models; thus, one diagonal corresponds to initially applying the correct harvest rate. The off-diagonal cells result from initially applying the wrong harvest rate and trying to identify the correct model. To the extent that learning is possible, the yields for each model tend to be the same regardless of the initial harvest rate.

The value of learning is emphasized by the difference in yield between simulations with and without adaptive learning (Figure 10.2). As one deviates from applying the correct initial harvest rate there is an increasing value in recognizing one's mistake. The surprise was the asymmetry of this surface: if a low productivity model was correct and a high harvest rate applied, the stock did not respond enough to distinguish the correct model. Currently the yellowtail flounder stocks probably lie on this part of the plane. The purpose of the Monte Carlo simulations is not to predict the future, but to estimate the value of adaptive management. This single-species example

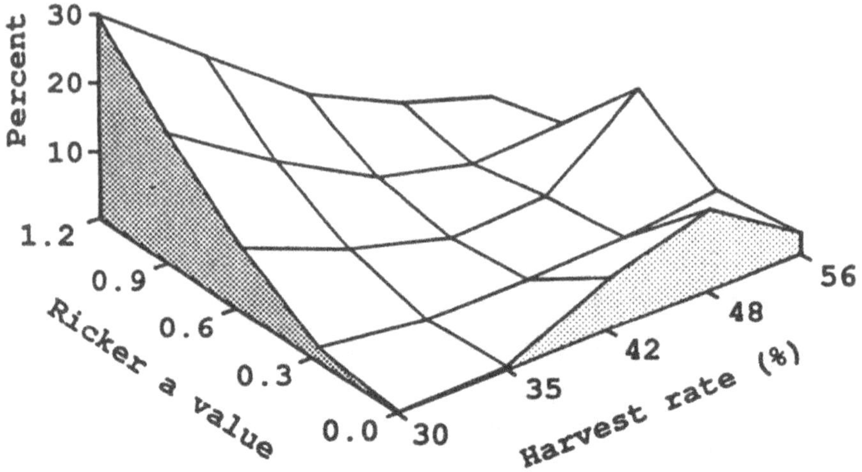

Figure 10.2. Difference in yield between simulations with and without adaptive learning. A positive difference indicates the value of learning. Details of the simulations are as described for Figure 10.1. From Collie and Walters (1990).

showed that the value of information was always positive and could be expressed in units of yield from the fishery.

Trophic Interactions Among Eastern Bering Sea Groundfish

Management issues important at the ecosystem level included: (1) Is there a biological basis for an all-species optimum yield cap? and (2) Do multispecies interactions limit the abundance of individual species? The key species is walleye pollock (*Theragra chalcogramma*) for which there is an additional question of whether discrete stocks exist. Survey-based approaches to these questions are hampered by the vast size of the EBS shelf (1.2×10^6 km^2) and the fact that much of the shelf is ice covered for much of the year. Experimental management could thus supplement ongoing surveys of the eastern Bering Sea.

Evidence of trophic interactions among eastern Bering Sea groundfish species comes from time series of abundances and food habits studies. Species supporting the largest fisheries are pollock,

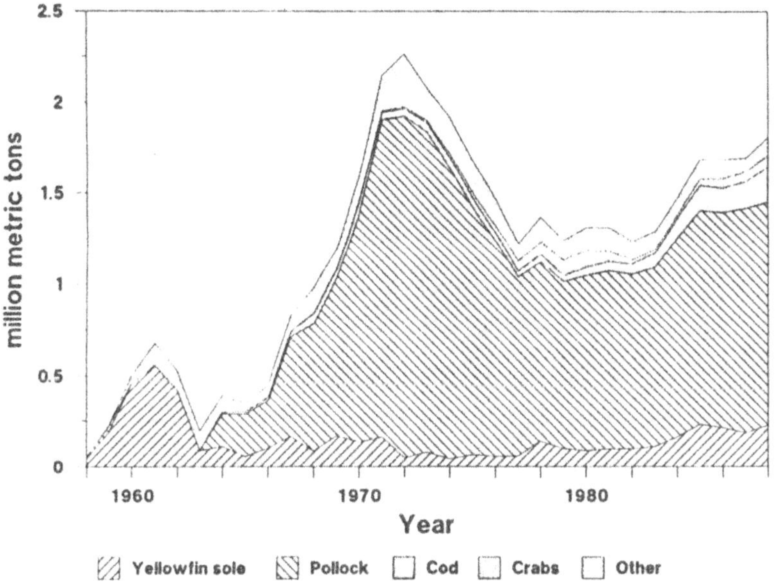

Figure 10.3. Eastern Bering Sea groundfish and crab catches, 1958-1988. The other species catch consists mainly of rockfish in the early years and other flatfishes in more recent years. The crab catch statistics start in 1966 and include king and Tanner crabs.

yellowfin sole (*Limanda aspera*), Pacific cod (*Gadus macrocephalus*), Greenland turbot (*Reinhardtius hippoglosoides*), arrowtooth flounder (*Atheresthes stomias*), and sablefish (*Anoplopoma fimbria*) (Figure 10.3). Pacific salmon, Pacific herring and other pelagic fishes are not included in Figure 10.3 or 10.4. The trawl fishery intensified in the 1960s, and changes in total yield since 1970 reflect changes in abundance and management restrictions. Optimum yield of the multispecies fishery is currently constrained between 1.4 and 2.0 million metric tons (mmt) by the North Pacific Fisheries Management Council (NPFMC). Long-term biomass estimates are available for pollock and yellowfin sole and more recently for Pacific cod and king and Tanner crabs (Figure 10.4). Biomass was estimated from commercial catch and survey data and recent estimates are the most reliable. In general the biomasses of major groundfish species fluctuate in unison as if responding in common to some external influence. An exception to this generalization is the apparent replacement of herring by pollock in the 1960s (Wespestad and Fried, 1983). The apparent interaction could be due to competition for food or predation on herring by pollock but there is scant evidence to test either hypothesis.

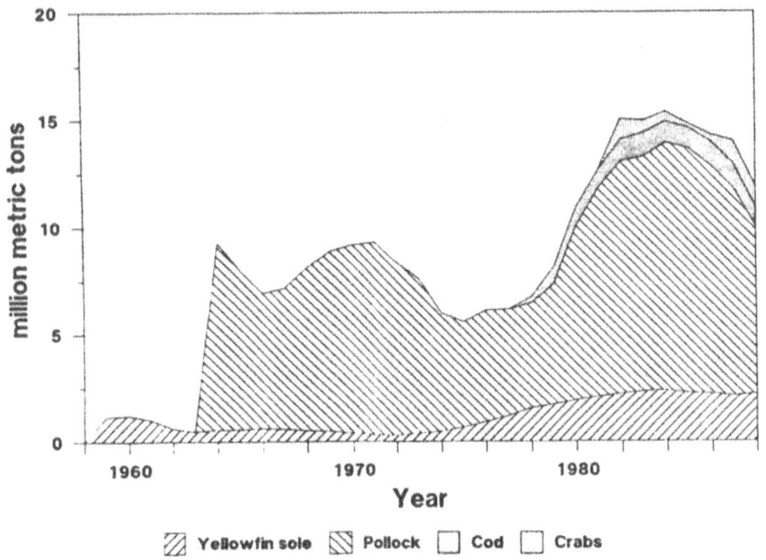

Figure 10.4. Biomass of selected groundfish and crabs in the eastern Bering Sea. Yellowfin sole biomass was estimated with cohort analysis (Bakkala and Wespestad, 1986). Pollock biomass was estimated with catch-age analysis (Quinn and Collie, 1989) starting in 1964. Cod and crab biomass were estimated from trawl survey data starting in 1978 for cod (Thompson, 1989) and 1982 for crabs (NMFS, Kodiak, AK).

The major groundfish species appear to have partitioned the eastern Bering Sea spatially (Wespestad, 1989) and by prey type (Livingston et al., 1986). Pollock occur primarily over the outer shelf and oceanic domains and feed primarily upon copepods, euphausids and juvenile pollock (Figure 10.5). Pacific cod live primarily on the middle and outer shelf and have a varied diet including shrimp, crabs and other crustaceans, juvenile pollock and pleuronectids. Yellowfin sole occur chiefly on the middle shelf and coastal domains and feed primarily on bivalve and other benthic invertebrates. There is some among-species diet overlap for polychaetes, shrimp and euphausids but the prey causing most of the diet overlap is juvenile pollock (Livingston et al., 1986). There is little direct evidence of competition among groundfish species, and food limitation; if it is important, it must be indirect. There is no compelling biological reason for constraining the optimal yield of all groundfish between 1.4 and 2.0 mmt, although there are economic reasons for keeping total yield stable. The productivity of individual species remains

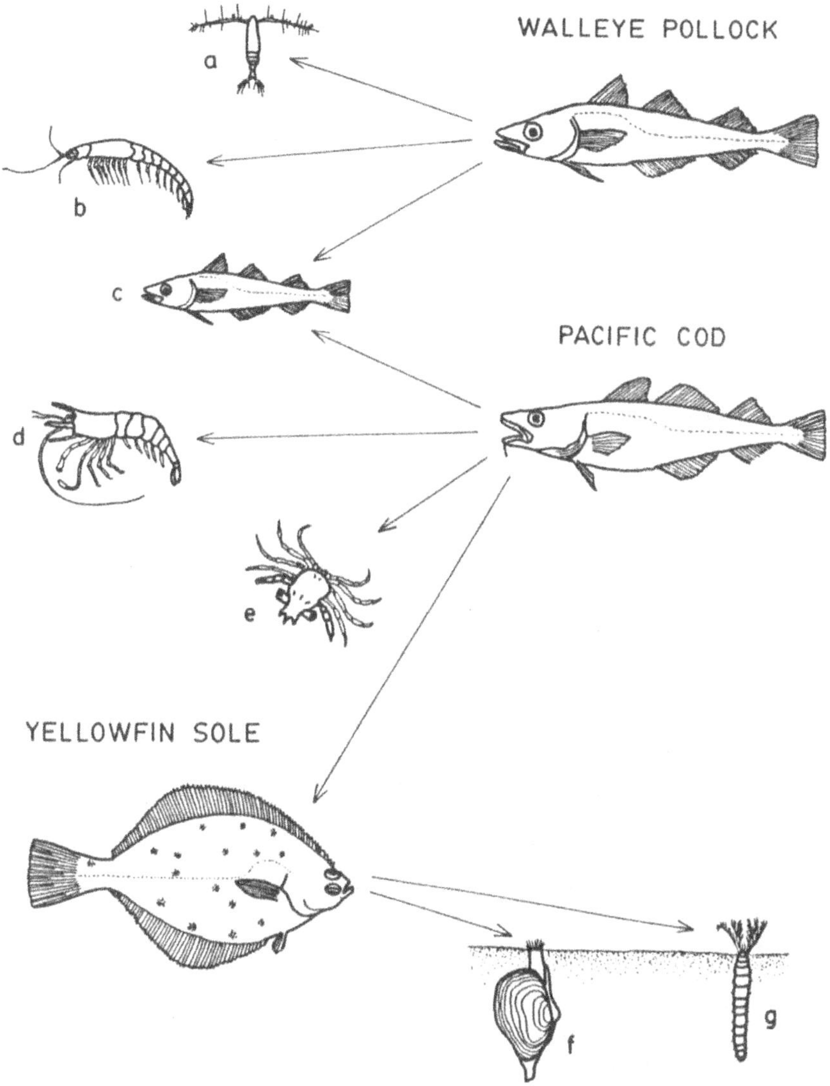

Figure 10.5. Principal prey of walleye pollock, Pacific cod, and yellowfin sole as described by Livingston et al. (1986): a--copepod, b--euphausiid, c--juvenile pollock, d--shrimp, e--juvenile crabs, f--bivalve, g--polychaete.

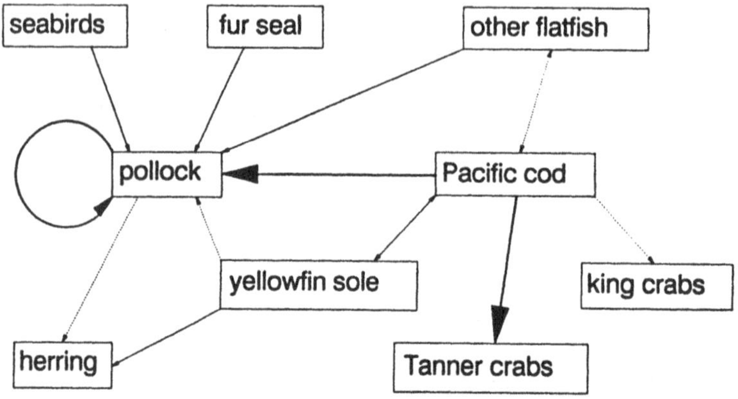

Figure 10.6. Partial food web of eastern Bering Sea groundfish, based on the food habits studies of Livingston (1989), and Livingston et al. (1986). The line types subjectively indicate the strength of the interactions: broken line--weak; solid line--intermediate; bold line--strong.

uncertain and, as long as biomass remains high, the optimum yield cap provides a conservative margin of error.

A partial food web (Figure 10.6) was constructed based largely on stomach contents data (Livingston et al., 1986) and other feeding habits studies. In most instances only the diet composition has been determined; in others consumption rates have been calculated (Livingston, 1989). The strongest apparent trophic interactions involve pollock as prey and Pacific cod as predator. Most of the predation on pollock (including cannibalism) is on the juvenile stage prior to recruitment to the fishery. Black-legged kittiwakes (*Rissa tridactyla*) and murres (*Uria* spp.) feed on one-year-old pollock, especially during the breeding season in the St. Matthew Island and the Pribilof Islands region (Springer and Byrd, 1989). Fur seals (*Callorhinus ursinus*) pup and breed on the Pribilof Islands and also feed on juvenile pollock (Lowry et al., 1989).

Apart from food habits studies there is little empirical evidence that trophic interactions affect the abundance of predators or prey. Available abundance estimates permit only crude tests of correlation because the time series are short and the spatial resolution very coarse. However, this crude level of analysis is appropriate for considering adaptive management experiments because it reflects the resolution to which groundfish populations are managed and can be manipulated. For a management experiment to be successful, the response should be apparent even with coarse measurement precision.

One of the most convincing predator-prey relationships is that of large pollock eating juvenile pollock. The spawner-recruit data

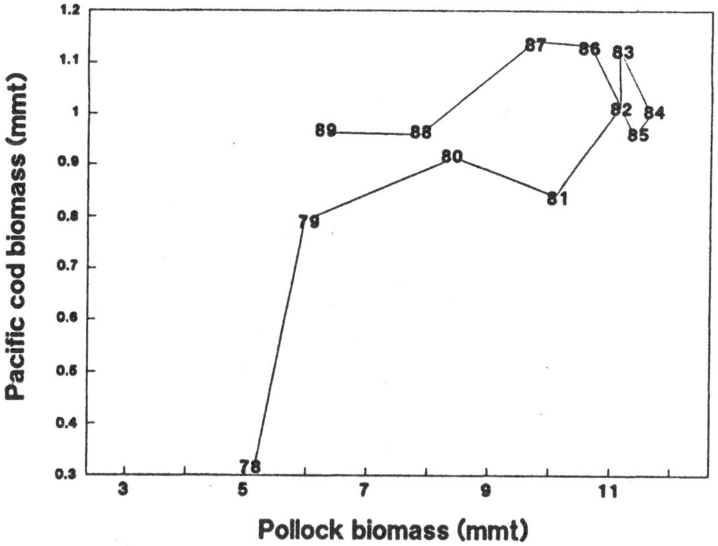

Figure 10.7. Relationship between pollock and Pacific cod biomass, 1978-1988.

(Quinn and Collie, 1990) indicate strong density-dependent recruitment and are consistent with the hypothesis that cannibalism is the major biotic regulator of pollock recruitment. It is therefore unlikely that other predators affect pollock recruitment but pollock recruitment may affect predator abundance. Pacific cod biomass appears to be positively correlated with pollock biomass (Figure 10.7) which could indicate a beneficial effect of pollock recruitment on cod feeding or a common response to a third factor. The only other empirical relationship is a negative correlation between yellowfin sole recruitment and herring abundance (Wespestad, 1989); yellowfin sole are known to eat herring roe.

Experimental Approaches to Multispecies Management

Examination of the eastern Bering Sea food web (Figure 10.6) indicates that the experimental options are restricted to a few possibilities. Sea birds and fur seals are beyond the jurisdiction of fisheries management and the main issue is the effect of pollock recruitment on these predators, not the reverse. A possible experiment would be to manipulate the abundance of pollock and to observe potential effects on its predators. Because the predators act on a pollock year class before the fishery, the abundance of pollock as prey could only be manipulated indirectly via the effect of spawning stock on recruitment. Variations in pollock recruitment appear to be caused more by uncontrollable factors than by stock size and therefore it is not feasible to manipulate the level of pollock recruitment. To ensure an adequate forage base for sea birds and marine mammals fisheries managers can maintain the pollock population within the range that has historically generated average recruitments, and impose mesh restrictions to minimize the capture of young fish.

Recent assessments of eastern Bering Sea pollock (Quinn and Collie, 1990; Wespestad et al., 1990) indicate that recruitment might be maximized on the average at stock sizes less than those prevalent during the 1980s. A possible experiment would be to fish down the pollock stock to intermediate levels and monitor recruitment. Instead of experimenting with such a large and important resource it may be preferable to contrast pollock recruitment between the eastern Bering Sea and other stocks such as the Gulf of Alaska. Although different population regulation processes are thought to operate in these different environments, there is coherence in the pollock recruitment data (Figure 10.8). The two areas have some (e.g., 1978) but not all (e.g., 1982) strong year classes in common. It is unclear whether the Gulf of Alaska stock is currently at a relatively low or high abundance level (Hallowed and Megrey, 1990). It may be desirable to deliberately maintain a contrast between the Bering Sea and Gulf of Alaska to test the effect of high or low stock size on recruitment under similar environmental conditions.

The pollock fishery in the Aleutian Basin, which has intensified since 1985 (Wespestad et al., 1990), can be considered an unplanned experiment. A primary question is whether fish harvested in the Aleutian Basin belong to the same stock as the shelf pollock. The answer to this question depends on the rate of mixing between the two areas at different stages of the pollock life history. A secondary question, conditional on the degree of mixing, is whether harvesting pollock in the Aleutian Basin affects recruitment to the shelf component of the population. If the basin and shelf are linked, the first apparent response to overfishing in the Aleutian Basin would be a decline in recruitment of young pollock to the EBS shelf. Any potential decline in recruitment rate would need to be measured against the background of natural recruitment variability, implying the need for an experimental control. In a broad sense Gulf of Alaska pollock, which are not harvested by the international fishery, provide an experimental control (Figure 10.8) and western Bering Sea pollock, which are also harvested in the Aleutian Basin, provide psuedoreplication of the treatment. Admittedly, this "experiment" has a low statistical power of detecting the recruitment response to overfishing the Aleutian Basin, but it indicates the level of resolution possible on the spatial scale of large marine ecosystems.

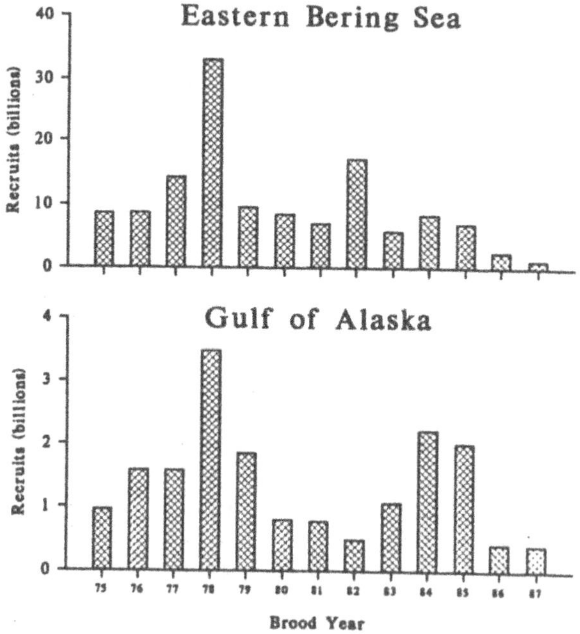

Figure 10.8. Recruitment of walleye pollock expressed as the number of age two recruits from 1975 to 1987 brood years. The eastern Bering Sea estimates are from Quinn and Collie (1990); the Gulf of Alaska estimates are from Hallowed and Megrey (1990, Table 12).

Probably the most feasible management experiment would be to manipulate the abundance of Pacific cod and to monitor the effect on its prey species, particularly Tanner crabs. Although reliable estimates of cod biomass do not extend very far back in time, abundance seems to have been highest during the late 1980s (Figure 10.4) while harvest rates were generally at or below the target rates for managing cod on a single-species basis (Thompson, 1990). There appears to be considerable scope for experimentally reducing cod abundance. The reason for the relatively light exploitation of cod is that large cod quotas would have exceeded the 2 mmt yield cap, forcing a reduction in the quota of some other more desirable species. Although crab populations could potentially benefit from reduced cod predation, many would be caught as by-catch resulting from increased bottom trawling. It might, therefore, be necessary to limit the experimental fishery to longline gear. Potential biological and economic impacts of increased cod quotas were considered in an amendment proposal to increase the optimum yield cap for Bering Sea groundfish above 2 mmt (NPFMC, 1988). Extensive simulation modelling of predator-prey interactions would be required to estimate the overall value of reducing cod abundance.

A common argument against experimentally manipulating one species of a multispecies complex is that recruitment variability causes natural contrasts in stock size without the need to disrupt the fishery. The recent lack of strong year classes of Pacific cod and pollock suggest that cod and pollock biomass will decline without the help of fishing (Figure 10.7). There are two reasons why experimental manipulations may still be desirable. Firstly, data sets from the past few decades indicate that natural trends are insufficient to empirically demonstrate multispecies interactions (e.g., Sissenwine et al., 1982). Secondly, because abundance of several groundfish species is positively covariant, deliberate manipulation is required to induce informative contrasts in abundance.

Eastern Bering Sea groundfish have been managed since 1982 under the Bering Sea/Aleutian Islands Groundfish Fishery Management Plan (FMP) which was developed by the North Pacific Fisheries Management Council (NPFMC) under the Magnuson Fishery Conservation and Management Act. Each year the NPFMC sets quotas for each species and stock. Proposed amendments to the FMP are reviewed annually. Amendment proposals and appropriate alternatives are evaluated against the status quo policies for their efficacy and for their potential biological and socioeconomic impacts. These potential impacts are typically very uncertain and the missing ingredient for a truly adaptive framework is to evaluate the management alternatives across a set of hypotheses or states of nature.

There is sufficient latitude in setting quotas for an adaptive or experimental strategy to be implemented without amending the FMP, provided that the total yield falls within the optimal yield range. A 1991 amendment proposal would establish NPFMC guidelines for authorizing fishing experiments. During the annual status of the stocks assessment there is considerable uncertainty in the current estimates of stock size and productivity. Different survey and assessment methods give competing estimates of stock size (e.g., Hallowed and Megrey, 1990). The NPFMC Plan Teams spend much time and energy deriving a "best estimate" of current stock size. In evaluating different policy options, the Plan Teams consider only a single column of Table 10.1. Instead, the Plan Teams should entertain the range of estimates and seek the harvest strategy (quota) which is most robust to different hypotheses about stock size. In Table 10.1 this means evaluating each policy across all columns to find the overall value.

Table 10.1. Expected value (V) of alternative policy options (u) for a set of hypotheses (H) about the response of a population to harvesting. The overall value of a harvest policy is the mean, weighted by the prior probability (p) assigned to each hypothesis. Bold type indicates the maximum yield for each hypothesis and the maximum overall.

	State of nature					
	H_1	H_2	H_3	H_4	...	Weighted
Prior probability	P_1	P_2	P_3	P_4	...	mean
Status quo (u_1)	V_{11}	V_{12}	**V_{13}**	V_{14}		$V(u_1)$
Alternative (u_2)	**V_{21}**	**V_{22}**	**V_{23}**	**V_{24}**		$V(u_2)$
Alternative (u_3)	V_{31}	**V_{32}**	V_{33}	**V_{34}**		**$V(u_3)$**
Alternative (u_4)	**V_{41}**	V_{42}	**V_{43}**	**V_{44}**		$V(u_4)$
.						
.						
.						

Conclusions

The unit of fisheries management should continue to be the species stock. Although fish production is ultimately limited by food availability and most animal production is ultimately eaten by some predator, there is little empirical evidence (at least for eastern Bering Sea groundfish) of competitive replacements on decadal time scales. The price differential among groundfish species necessitates continued species-by-species management even though it is complicated by by-catch. Multispecies management should concentrate on key predator-prey interactions as identified from food habits studies. Multispecies fisheries models incorporating the entire suite of trophic relationships have never been useful management tools because many of the model components remain unmeasured.

Food habits studies indicate that predator-prey interactions are potentially important but there is little empirical evidence that trophic interactions influence the abundance or yield of the species involved. Manipulative experiments are probably the only definitive way to measure the importance of species interactions. Manipulative experiments should generally be conducted on predators and not prey, because most of the predation is on pre-recruit fish (Sissenwine, 1984). Fisheries managers can therefore only control the abundance of prey species indirectly thorough the highly variable effect of stock size on recruitment. Finally, an adaptive framework should be adopted by fisheries management councils, if only passively. Even if deliberate manipulations are not feasible or desirable, managers should entertain alternative hypotheses and evaluate harvest policies across the alternatives.

Acknowledgments

I thank the following people for providing fisheries data or for steering me to the appropriate report: Richard Bakkala, Nic Bax, and Peggy Murphy. Patricia Livingston, Vidar Wespestad, and an anonymous reviewer improved the manuscript with their comments. This study was supported by the University of Alaska, Fairbanks.

References

Bakkala, R. G., and V. G. Wespestad. 1986. Pp. 49-62. *In* R. G. Bakkala and L.-L. Low (Eds.) Condition of groundfish resources of the eastern Berning Sea and Aleutian Islands region in 1985. U.S. Dep. Commer., NOAA Tech. Mem. NMFS F/NWC-104.

Collie, J. S., and C. J. Walters. 1991. Adaptive management of spatially replicated groundfish populations. Can. J. Fish. Squat. Sci. 48(7):in press.

Hollowed, A. B., and B. A. Megrey. 1990. Walleye pollock, Pp. 22-89. *In* Stock assessment and fishery evaluation report for the 1991 Gulf of Alaska groundfish fishery. North Pacific Fishery Management Council, Anchorage, AK.

Livingston, P. A. 1989. Interannual trends in Pacific cod (*Gadus macrocephalus*) predation of three commercially important crab species in the eastern Bering Sea. Fish. Bull., U.S. 87:807-827.

Livingston, P. A., D. A. Dwyer, D. L. Wencker, M. S. Yang, and G. M. Lang. 1986. Tropic interactions of the key fish species in the eastern Bering Sea. Int. N. Pac. Fish. Comm. Bull. 47:49-65.

Lowry, L.F., K. J. Frost, and T. R. Loughlin. 1989. Importance of walleye pollock in the diets of marine mammals in the Gulf of

Alaska and Bering Sea, and implications for fishery management. Pp. 701-726. *In* Proceedings of the international symposium on the biology and management of walleye pollock. Alaska Sea Grant Rep. No. 89-1.

North Pacific Fisheries Management Council [NPFMC]. 1988. Draft supplemental environmental impact statement for a proposal to increase the optimum yield range in the Fishery Management Plan for the groundfish fishery of the Bering Sea and Aleutian Islands. North Pacific Fisheries Management Council, Anchorage, AK. 192 Pp.

Quinn, T. J., and J. S. Collie. 1989. Alternative population models for eastern Bering Sea pollock. Int. N. Pac. Fish. Comm. Bull. 50:243-257.

Sainsbury, K. J. 1988. The ecological basis of multispecies fisheries, and management of a demersal fishery in tropical Australia. Pp. 349-382. *In* J. A. Gulland (Ed.) Fish population dynamics. 2nd ed. Wiley, New York.

Sissenwine, M. P. 1985. Why do fish populations vary? Pp. 59-94. *In* R. M. May (Ed.) Exploitation of marine communities. Dahlem Conference, Life Sciences Research Report 32, Springer-Verlag, New York.

Sissenwine, M. P., B. E. Brown, J. E. Palmer, R. J. Essig, and W. K. Smith. 1982. An empirical examination of population interactions for the fishery resources off the northeastern USA. Pp. 149-156. *In* M. C. Mercer (Ed.) Multispecies approaches to fisheries management advice. Can. Spec. Publ. Fish. Aquat. Sci. 59.

Springer, A. W., and G. V. Byrd. 1989. Seabird dependence on walleye pollock in the southeastern Bering Sea. Pp. 667-677. *In* Proceedings of the international symposium on the biology and management of walleye pollock. Alaska Sea Grant Rep. No. 89-1.

Thompson, G. G. 1989. Pacific cod. Pp. 51-85. *In* Stock assessment and fishery evaluation document for groundfish resources in the Bering Sea/Aleutian Islands region as projected for 1991. North Pacific Fishery Management Council, Anchorage, AK.

Tyler, A. V., W. L. Gabriel and W. J. Overholtz. 1982. Adaptive management based on structure of fish assemblages of northern continental shelves. Pp. 149-156. *In* M. C. Mercer (Ed.) Multispecies approaches to fisheries management advice. Can. Spec. Publ. Fish. Aquat. Sci. 59.

Walters, C. J. 1986. Adaptive management of renewable resources. Macmillan, New York. 374 Pp.

Walters, C. J., and J. S. Collie. 1988. Is research on environmental factors useful to fisheries management? Can. J. Fish. Aquat. Sci. 45:1848-1854.

Walters, C. J., and R. Hilborn. 1976. Adaptive control of fishing systems. J. Fish. Res. Board Can. 33:145–159.

Walters, C. J., J. S. Collie, and T. Webb. 1988. Experimental designs for estimating transient responses to management disturbances. Can. J. Fish. Aquat. Sci. 45:1848–1854.

Wespestad, V. G. 1989. Physical oceanography of the eastern Bering Sea relative to interannual variation in biota. Unpublished ms. NMFS, Alaska Fisheries Science Center, Seattle, WA.

Wespestad, V. G., and S. M. Fried. 1983. Review of the biology and abundance of Pacific herring (*Clupea harengus pallasi*). Pp. 17-29. *In* W. S. Wooster (Ed.) From year to year, Interannual variability of the environment and fisheries of the Gulf of Alaska and the eastern Bering Sea. Washington Sea Grant, WSG-WO 83-3.

Wespestad, V. G., R. G. Bakkala, and P. Dawson. 1990. Walleye pollock. Pp. 27–50. *In* Stock assessment and fishery evaluation document for groundfish resources in the Bering Sea/Aleutian Islands region as projected for 1991. North Pacific Fishery Management Council, Anchorage, AK.

Marc Mangel

11. Empirical and Theoretical Aspects of Fisheries Yield Models for Large Marine Ecosystems

Abstract

Fisheries yield models are based on a number of (usually unstated) assumptions. These include stationarity of fluctuations and an independence of history, one-way linkages between the stock and ecosystem (the ecosystem affects the stock, but not vice versa), and genetic homogeneity. I reviewed empirical evidence concerning these assumptions and concluded that there is little support for them. New kinds of models are needed to deal with yield in large marine ecosystems.

Introduction

> The problem is perhaps that ecologically, fish are not a unit.
>
> Ursin (1982)

Fish are *not* an ecological unit. Yet most of our management schemes and plans are based on catching a certain number of "tons of fish." Many of these plans have evolved from the use of yield models that may actually not apply to multiple species in large ecosystems. In this chapter, I attempt to identify the main assumptions that underlie yield models and then examine the empirical support for these assumptions. The next section defines and gives examples of yield models in which the future of the stock is, modified by random factors, determined solely by the current size of the stock. I then review assumptions that underlie yield models and discuss the empirical basis for these assumptions. This includes measurements of time and space scales, estimates of the effect of the stock on the ecosystem as well as the ecosystem on the stock, and genetic structure of stocks. Theoretical concepts that may be helpful in the development of new models are discussed. I argue that new kinds of models are needed.

What Is a Yield Model?

For all populations, the basic components of production are recruitment, growth, natural mortality (predation, disease), and anthropogenic mortality (harvesting, habitat destruction). Models used in fisheries management deal with these processes at different levels of aggregation. Generally, two kinds of models are used (e.g., Clark, 1976, 1985; Ricker, 1977; Rothschild, 1986). The first type is *yield-per-recruit* theory, which describes how the biomass of a cohort changes over time, after it has entered the ecosystem. This theory deals with the components of growth and mortality, as they affect production. The second type of model is *production-model* or *stock-recruitment* theory, which is used to describe the relationship between the rate of change of biomass and biomass. Although these two branches can be combined to define a relatively complete model of production, often they are not integrated, leading to assertions such as stocks can be either "recruitment overfished" or "stock overfished." But as Rothschild points out, "yield-per-recruit is a function of biomass, and biomass is a function of growth and mortality. As growth increases, egg production increases, thus modifying the recruitment-stock function [Rothschild, 1986]." The two theories are inexorably linked.

In this chapter, a "yield model" is any model in which population size at a given time completely determines population size at future times, modified by random factors. Typical examples are:

Schaefer Model

$$dX/dt = rX(1-X/K) - qEX \tag{1}$$

where X is biomass, r and K are logistic growth rate parameters, q is catchability, and E is harvesting effort;

Discrete Time Models of the Form

$$X_t = \rho(X_t - C_t) + R_t \tag{2}$$

where ρ is survivorship, C_t is catch in year t, and R_t is the recruitment given by, for example,

$$R_t = \alpha(X_t - C_t)\exp(-\beta[X_t - C_t]) \quad \text{Ricker model)}$$

$$R_t = \alpha(X_t - C_t)/(1+\beta[X_t - C_t]) \quad \text{(Beverton-Holt model)} \tag{3}$$

These models can be elaborated in many ways, such as including age structure (Clark, 1976, 1985), or stochastic effects (Clark, 1985; Mangel, 1985; Walters, 1986), but retain the general feature that population size at future times is a clear and definite (if not deterministic) function of population size at the current time. Getz and Haight (1989) provide a good summary of the new level of technical sophistication to which one can take these models.

Models such as (1) and (2) have lead to "one number" management, particularly when the models are coupled to optimization criteria. Such "point targets" are not particularly good, but are, unfortunately, too common (Soule, 1987). Examples of the use of these models for management are given by Garrod (1977) and Bannister (1977). Garrod suggests that there exists sufficient data on Atlantic cod stocks to provide detailed analyses of yield per recruit. Bannister, on the other hand, finds that the models for plaice do not work well.

Yield models are based on a number of implicit, and usually unstated, assumptions. In this chapter I try to clearly identify these assumptions and to evaluate the evidence to support them. By doing so, I hope to highlight areas that require further research.

Assumptions That Underlie Yield Models

Among the most important assumptions underlying yield models are the following.

Nature of Interactions

Interactions with the environment are assumed to be only a function of population size. Alternatively put: population size itself is a surrogate for interactions between the young and the ecosystem. This assumption has two components.

(1) ***Stationarity***. The assumption of stationarity of fluctuations (a "white" spectrum) is what allows one to draw a "stock-recruitment" curve without paying attention to how the stock got to where it is. The alternative assumption is that the "noise" is colored with persistent effects on the stock. (The assumption of stationarity, which allows relative ease in analysis of time-series data, is used in most ecological models.)

(2) ***Linkage***. The assumption is that there are only one-way effects of the environment on the stock and no history effects. The assumption is equivalent to the following scenario: At stock level S_1, we observe a distribution of stock and recruitment f_1. Then the stock

moves to a new level, S_2, and finally returns to S_1. The assumption is that the distribution of stock and recruitment after the return will be f_1. Multispecies models (e.g., May, 1984) usually make this assumption as well, but with a focus on more than one stock.

Time Scales and Space Scales

Only one time scale is sufficiently important to be explicitly included in the model. Space enters only in a nominal way; for example, as "reaction-diffusion" (Murray, 1989) extensions of (1).

Genetics

The population is composed of genetically identical (or at least unimportant) subunits (cf. Gauldie, 1991).

It is true that these assumptions are typical of most ecological models, not just the ones used in fisheries management. But if we want to use such models for management of valuable resources, it is imperative to understand the kind of insight that they provide and the foundations of the assumptions. What kind of empirical support exists for these assumptions?

Time Scales and Space Scales

In general, the target organism in a fishery may have time and space scales that differ from both its predators and its prey. The predator may have a characteristic time scale (e.g., mean lifetime) and space scale (a measure of motion) larger than the target organism, which itself has time and space scales larger than the plankton or smaller fish that it eats. Thus, the target organism is both a producer (for its predator) and a consumer (of its prey). Most yield models capture only one set of time/space scales and ignore the others.

An example of disparate spatial scales is shown in Figure 11.1, modified from Smith et al. (1989a). These data suggest that at least four spatial scales are important for the dynamics of a target organism and its prey. The nature of these spatial scales is determined by the foraging ecology of the target organism, the behavioral response of its prey, and the behavioral response of its predator.

Smith and Moser (1988) used scale deposition rates to study historical biomass of stocks. They found great disparity in deposition indices at sites separated by order of 1,000 km; this suggests that the critical spatial scale of the events that affect biomass is less than 1,000 km. It also suggests that changes in biomass over a large spatial scale may have properties strongly independent of total biomass. They also

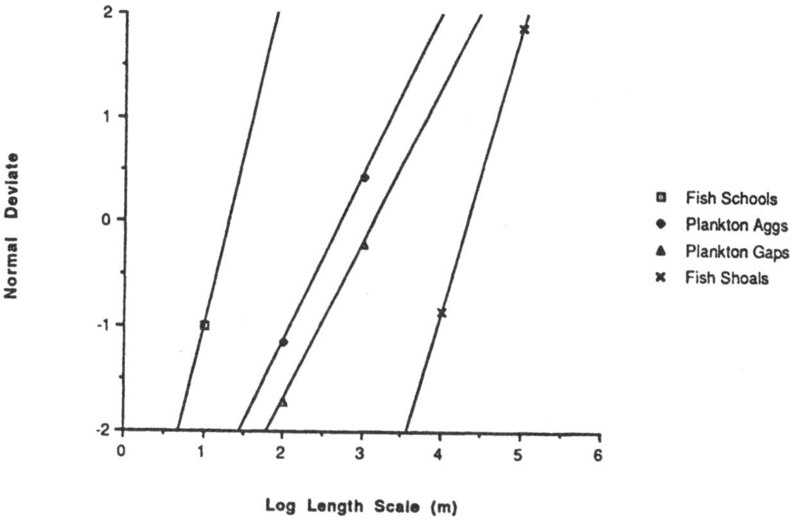

Figure 11.1. Characteristic patterns of length scales of fish schools (square), zooplankton aggregations (diamond), gaps in zooplankton aggregations (triangle), and fish shoal groups (cross), after Smith et al. (1989a). The normal deviate is a particular length minus the mean, divided by the standard deviation. Thus, a length scale with normal deviate = 0 corresponds to 50% of the subjects less than this length scale. The fish shoal groups may represent long-term (evolutionary) responses to hydrographic patterns. The plankton patterns are responses to patterns in phytoplankton spacing, subject to mortality pressures from fish and turbulent diffusion. The fish schools respond to the zooplankton patterns.

discovered two time scales of change. One is on the order of 3-5 yr and one is of the order of 3-6 decades. They concluded that "Spectral analysis of the fish-scale record indicates long-lasting population trends in sardine and anchovy and does not suggest equilibrium phenomena as population controls."

In general, we must conclude that a large marine ecosystem involves a number of temporal and spatial scales that may be important and the models should take this into account.

Does the Stock Affect the Ecosystem?

If the assumption of one-way interactions is valid, then changes in stock size should have little effect on the ecosystem. Testing this idea--either with experiments or using natural disturbance--in pelagic ecosystems is very difficult. However, the seminal work of R. Paine over the last 20 yr does shed considerable light on this question, at least for sessile organisms.

Paine (1984, Figure 5) describes the interspecific relationships in a guild of five coralline algae (producers) with the same grazer (consumer) under three different "ecological conditions." These were, with the grazer on a natural surface, without the grazer on a natural surface, and with the grazer on an artificial surface. When two guild members come into contact, one overgrows the other and thus "wins" the competition. Paine shows that the results of competition are in general not hierarchial (except when the grazer is present on the artificial surface), not transitive, and are changed significantly when the grazer is removed.

The question of removal and return is very important because very often in fisheries management we face the question of recovery of a stock after severe depletion. The assumption that the stock can recover to its previous value is equivalent to the assumption that the absence of the stock had no effect on the ecosystem. Riffenburgh (1969) needed a *nonstationary* Markov chain to describe successfully a system composed of Pacific sardine, anchovy, their predators, and prey. This model showed that sardines could not be reinstated as the most abundant species as the result of selective fishing (i.e., that recovery does indeed depend upon how the stock affects the ecosystem and, thus, on how changes in the stock lead to changes in the ecosystem).

Paine et al. (1985) describe an experiment which is an analogy to the sequence of depletion and recovery. In the experiment, for about 1 yr the starfish *Stichaster australis* was removed from an exposed shoreline. The initial treatment also involved removal of the giant brown alga *Durvillaea antarctica* from a band beneath the pre-existing bed of mussel *Perna canaliculus*. After 1 yr of exclusion, the starfish was allowed to return to the site. The results, including the percentage cover by the alga in a control site are shown below (Paine et al., 1985):

Date	% Cover		
	Mussel	Alga	(Alga in Control)
9/1968	0	85	(85)
3/1969	20	15	(100)
5/1969	50	0	(100)
9/1969	98	2	(95)
10/1970	85	5	(90)
2/1971	65	25	(100)
8/1976	85	10	(80)
1/1983	60	35	(100)

The conclusion is striking and forceful: After nearly 15 yr of "recovery" the ecosystem had not returned to its previous state. The 1 yr removal of the predator and one competitor had a profound effect on the state of the ecosystem.

Walters (1987b) addresses many of these same issues. In particular, he analyzes stock-surplus production relationships for six Northeast Atlantic herring stocks. Three of these stocks show "hooks" and "loops" (i.e., hysteresis in the stock-production relationship that suggests a two-way interaction between the stock and the ecosystem). The starfish and herring data are consistent in their support that the stock affects the ecosystem. The way that this effect is propagated through different trophic levels is not clear (Bradbury and Mundy, 1989), but it does appear that the effect is nonlinear.

MacCall (1986) describes the interaction between the brown pelican and northern anchovy and sardine. The biomass of anchovy or sardine affects the population dynamics of pelican (e.g., the disappearance of sardine affected the breeding success of pelican near Monterey), although it is not as clear that sea birds in general are sufficiently numerous to greatly influence the dynamics of sardine or anchovy.

In general, we must conclude that the stock affects the ecosystem, and that yield models should take this into account.

How Does the Ecosystem Affect the Stock?

MacCall (1984) developed a recruitment model for the northern anchovy. The three terms bearing on recruitment strength are fecundity, density-independent mortality, and density-dependent mortality. He estimated the contribution of each term and concluded:

> These values [of the parameters] indicate that the ratio of density independent to density dependent factors (neglecting density dependent factors on fecundity) is 10 to 1. The preponderance of density independent influences tends to mask the relatively weak regulation of stock abundance. This is consistent with the large variability characteristic of clupeid and engraulid stocks. The large contribution of density independent variability does not allow the population to equilibrate at a stable level. Rather, equilibrium takes the form of a probability distribution about a mean value . . . . A production or equilibrium yield curve may be derived, but it should be used with caution: It must be interpreted in a probabilistic rather than deterministic sense.

Smith et al. (1989b), in a study of sardine and anchovy, find that 30% of mortality of sardine and anchovy eggs is due to cannibalism; this means that 70% of the mortality is due to factors independent of the adult spawning stock. They conclude that:

> . . . while cannibalism is sufficiently important as a source of egg mortality to control the size of populations like sardine and anchovy over the long term, it is unlikely that the interannual variation in rates of mortality due to cannibalism are sufficient to cause major recruitment variations.

Smith and Moser (1988) conclude from a study of recruitment variation in northern anchovy that recruitment variation far exceeds variation in specific reproductive parameters. This suggests a variation enhancing effect of the environment. Recruitment variation is common and typically very large (Sissenwine et al., 1988). It appears that much of this recruitment variation is caused by the environment and not by the stock. Sousa (1984) reviews the role of natural disturbance in ecological systems. Our knowledge and understanding of the role of disturbance concentrates on sessile organisms. Very little is known about the role of natural disturbance in pelagic systems (or at least very little is mentioned by Sousa).

Paine's work also suggests a protocol for the study of the role of disturbance. In experimental studies of the interactions of sessile marine invertebrates, a "reference state" is achieved by removing all consumers. One can then add consumers, one at a time, and study the effects on the producers. In particular, one can ask how the consumers influence the system at natural densities. Paine (1984) proposes that this kind of experiment can be used to distinguish between weak interactions among trophic levels and strong interactions among trophic levels. For example, one might remove consumers from a system in which producer P_1 dominates, and find that P_2 dominates. Then consumers could be added one at a time. We would ask: does the system return to a situation in which P_1 dominates and how does this return, or non-return, depend upon the C_1? Figure 8.1 in Slobodkin (1980) is in this spirit and Paine and Levin (1981) suggest a general theoretical formulation. Of course, experiments in large marine ecosystems are much harder to do than in the rocky intertidal. On the other hand, natural disturbance is continually providing perturbations which, if we are sufficiently clever, may take the place of experimental perturbations.

In conclusion, our models must be able to treat the profound effects that the ecosystem has on the stock dynamics.

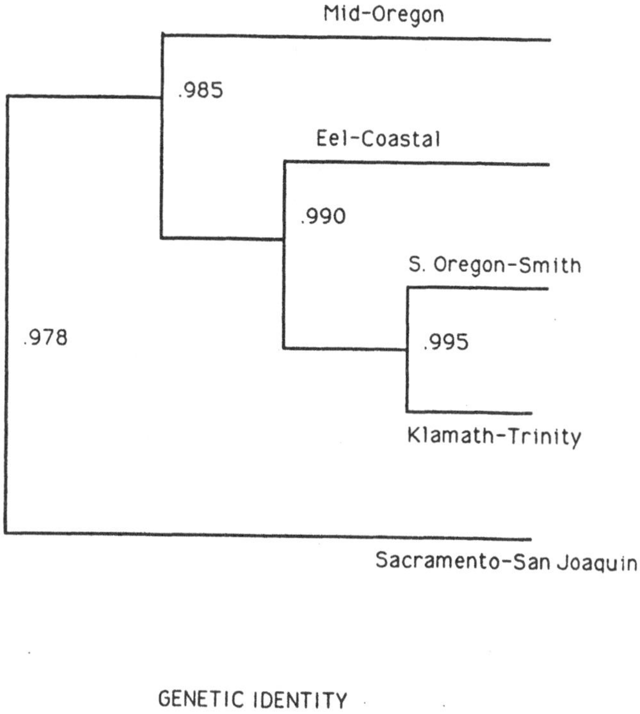

Figure 11.2. A dendrogram depicting genetic similarity among the genetic groups of chinook salmon used in studies by Gall et al. (1989).

Genetics

The study of allozyme variation, particularly in the last 10 or 15 yr, has shown that populations are not composed of genetically homogeneous stocks and that variability can be considerable, even over a modest geographic scale (however, see Gauldie, 1991). Figure 11.2 provides an example of the dendrogram for west coast chinook salmon stocks. Details are given in Gall et al., 1989. The west coast chinook stocks are not genetically homogeneous and management of "California chinook salmon" as a group may submerge important genetic aspects. It is possible to use such genetic information for identification of the composition of mixtures of stocks, but we do not know how genetic variability affects survival and/or population dynamics. The assumption commonly used in the management of large marine ecosystems of pan-mictic populations is not consistent with these data.

At the current time, discussions about the importance of maintaining genetic diversity abound (e.g., Soule, 1987). One issue of particular concern is the way in which a fishery may affect the genetic composition of a stock. It is, in fact, not clear how genetic diversity really affects conservation or management of species. Until such information is known, our models should take a conservative approach in which genetically distinct stocks are assumed (e.g., based on geographic discreteness or morphology).

Some Concepts That May Be Helpful

In yield models, ecological interactions of exploited species are implicitly treated as stock parameters, yet the results described above suggest that a more explicit treatment of interactions and natural disturbance is required. Some of the concepts that may be useful for the conceptualization of ecological interactions include:

Trophic cascades. Natural change at one trophic level often affects two or more trophic levels.

Alternative community states. Many of the common models in population and community ecology can produce multiple steady states. These need not be stable, just persistent over the correct time scales, to have profound effects on ecosystems. An example from neurobiology is the Fitzhugh-Nagumo (Rinzel, 1981) equations, which are a model for the Hodgkin-Huxley equations of nerve activation. The Fitzhugh-Nagumo equations have a single, stable steady state, but if perturbations from this state are sufficiently large, the system exhibits an enormous quasi-cycle before returning to the stable steady state. It is highly likely that many ecosystems exhibit the same kinds of properties. An example, drawn from insect ecology, is the spruce budworm system (Ludwig et al., 1978).

Ursin (1982) argues to the contrary, that marine ecosystems are fundamentally stable. This argument is based on the following example: between 1965 and 1975 cod year classes varied by a factor of 6, which Ursin says is "remarkably small variation considering that a cod spawns between half a million and five million eggs per year." Similar data on recruitment variability are quoted for plaice (a factor of 5) and haddock (a factor of 100). Ursin's argument raises again a fundamental question put by Rothschild (1986): Of all the vertebrates only fish have evolved the system of egg production that is enormously larger than adult population size. The elucidation of this pattern remains a central problem in the evolutionary ecology of fish.

Approaches. Should we build models that are "bottom up" (from primary productivity . . .) versus "top down" (from consumers . . .)? The currently existing data suggest that both kinds of models should be built and their results compared. For example, Mills and Forney (1988) describe top down predator control and bottom up producer induced feedback as important aspects of the food dynamics in lakes. Bartell et al. (1988) show that both top down and bottom up control of phytoplankton production is possible. As part of the modelling exercise, one must explicate the causes of interactions and linkages and their time dependence.

Multiple time and space scales. The literature of applied mathematics is filled with examples of multiple time and space scales (Nayfeh, 1973). A number of biological examples are discussed by Murray (1989). Our models need to take multiple time and space scales into account in a consistent manner.

Multispecies management and constraints. One approach to multispecies management is to maximize the harvest from a given stock, subject to certain constraints on other stocks. For example, one might maximize herring and squid catch in the North Atlantic, subject to the constraint that pilot whale stocks increase at a rate no smaller than a given fraction of the rate in the absence of harvest. Development of such models, in which additional stocks provide constraints, is an important, open area for future work.

Food Chains, Assemblies of Organisms and Biotic Interactions

Do food chains need to be modeled explicitly for an understanding of large marine ecosystems, and if so, what does such modelling offer for prediction and management? Ursin (1982) asks what properties of food webs might help stabilize marine ecosystems and suggests replacing the standard food chain

$$\text{algae} \rightarrow \text{copepods} \rightarrow \text{herring} \rightarrow \text{cod}$$

by "a food web of triangular meshes each containing a prey species, a predator species, and a species which is predator to the first and prey to the latter" (Ursin 1982).

$$\begin{array}{ccc} \text{small prey} & \rightarrow & \text{big predator} \\ \downarrow & & \downarrow \\ \text{small predator/big prey} & & \end{array}$$

Schoener (1989) compares the standard HSS model, with a 4-level model and shows that the nature of hypothesized interactions can be changed completely by adding an additional trophic level. Schoener's argument, in the marine setting, looks as follows:

HSS model	Interaction	4 level model
Piscivores	Competition	Medium Piscivores
Planktivores	Predation	Small Piscivores
Producers	Competition	Small Planktivores
	Predation	Producers

By adding a fourth level, the nature of interactions is completely reversed. In fact, it is *both* competition and predation that matter.

The study of food webs over the last 20 yr has led to a number of generalizations, summarized by Fenchel (1988), Lawton (1989), and Schoener (1989). The most important ones are:

- The typical length ratios between prey and predator are 1:10, very rarely less than 1:100 or more than 1:1.

- Interactions in food webs are trophic, not competitive or mutualistic (e.g., sardine and anchovy interactions are difficult to explicitly incorporate into current food web models).

- On average, each species feeds on a constant number of others. That is, the ratio of prey (trophic species eaten) to predators (those eating) is constant.

- Top species have no predators, intermediate species serve as predators and prey, and basal species are prey. In general, the proportion of links between basal, intermediate, and top species is relatively constant and independent of the number of species in the web. In addition, the fraction of trophic species in a web belonging to top, intermediate, and basal species is roughly constant.

- Omnivores (feeding at more than one level in the food chain) are rarer than expected. This may be very important for pelagic systems, in which fish may use more than one kind of feeding (e.g., biting or filtering) on different kinds of prey.

• Marine Pelagic webs have the greatest complexity, at least measured by Cohen's maximum chain length.

• Webs from fluctuating and constant environments differ. Webs from constant environments have proportionately more links between basal and top species.

• In general, food chains are short and habitat structure influences the length of food chains--which are shorter in 2-dimensional habitats than in 3-dimensional habitats.

• There appear to be two "constants" of webs. These are web vulnerability = average number of predators per prey and web generalization = average number of prey eaten per predator, which are constant in that they do not vary with web size.

• Possible explanations for structure of food chains include models based on (a) energy flow, (b) Lotka-Volterra dynamic models with dynamic constraints, (c) donor controlled interactions--victim populations significantly influence predators, but predators have no significant impact on victims, and (d) a cascade model involving body size.

Even though the theoretical work involving food webs has been enormous, there are a number of difficulties that limit their usefulness for both understanding and managing large marine ecosystems. Schoener (1989) points out one of the main problems:

> I see as probably the major problem with web description the decision to draw a link or not. Many species have broad ranges of prey types included in their diet but concentrate on only a few. At what percent occurrence should a prey no longer be counted?

This is not a trivial matter. For example, pelagic marine species have egg, larval, young fish, and mature fish stages. Is each one of these a "quasi-species" in the food chain? If yes, then the chain cannot be very informative about population dynamics (and thus management issues). If no, then we essentially don't know how to draw food chains in which a species enters at four trophic levels. This is not a purely theoretical point; Smith and Moser (1988) point out that:

> Analysis of fish species assemblages suggests that predation by major mesopelagic species of the California Current region could affect survival rates of ontogenetic states of

> epipelagic populations (e.g., sardine and anchovy) and thus influence recruitment variations.

Terazaki (1989) shows drawing such links is a real problem and that it is compounded by the immigration and emmigration of species. Crawford et al. (1989) note alteration of flows between different trophic levels in the same ecosystem and that many of the more abundant vertebrates are opportunistic (rather than selective) feeders. Opportunistic feeding makes links more difficult to draw.

Paine (1988) challenges the "manner in which theory has been linked to data on natural communities." Difficulties identified by Paine include connectance and the spatial scale of observation, choice of biological or "trophic" species:

> Natural systems are dynamic and continually changing, with distance from equilibrium being an important unknown. To me, this implies that focus on dynamic rather than static (= completely descriptive) properties would constitute a primary desideratum . . . Web metrics change in time and space [Paine, 1988].

Paine lists three criteria that should be satisfied. (1) Is the web under consideration a biologically realistic representation of the community? (2) Whenever possible, have the species been identified rather than aggregated? (3) Has the role of interaction strength been explored both empirically and theoretically? The background for this is the following: May (1973) identified three independent features of webs and their incumbent species that could influence trophic properties: the number of species (S), their average connectance (C), and mean interaction strength (I). Even though varying interaction strength is crucial for dynamics of food webs, it is still not clear how to characterize interaction strengths, and, perhaps for this reason, it is rarely done.

What Can Phenomenological Models Offer for Empirical Understanding and Management?

To be certain, theory has contributed and undoubtedly will continue to contribute to our understanding, particularly qualitative understanding and intuition, of large marine ecosystems. For example, simple Lotka-Volterra equations show how oscillations can be driven by population dynamics and bifurcations can occur as population parameters change. Adding economic terms to such models show how extinction of a harvested stock may be economically (if not biologically!) optimal.

In those real systems that have been studied with experimental (i.e., controlled) manipulations, we see the importance of both competitive and predative processes in population interactions (Paine, 1977). Abrams (1984) has given an example which links such dynamic models and food webs. This theory has three components: Resource (R), Consumer (C), and Predator (P). The equations of the theory are:

$$dR/dt = Rf(R) - g_c(R)T_cC$$

$$dC/dt = Cb_1g_c(R)T_c - g_p(C)h(T_c)P - (L(T_c) + \mu_c)C$$

$$dP/dt = Pb_2g_c(C)h(T_c) - \mu_pP \qquad (4)$$

Most of the terms in these equations are self explanatory, except that T_c = fraction of average consumer's life spent foraging; $L(T_c)$ = non-predator mortality related to foraging. This model "requires an explicit method for determining T_c as a function of other components of the model" (Abrams, 1984). Mangel and Clark (1988) have provided a protocol to do this. Within this theory, Abrams identifies five kinds of interactions: (1) effect of resource density on predator population growth rate, (2) effect of consumer density on predator population growth rate, (3) effect of predator density on predator population growth rate, (4) effect of predator density on resource population growth rate, and (5) effect of consumer density on resource population growth rate. A study of such interactions will help to elucidate the interaction strengths of the food chains.

Ludwig (1987) and Walters (1987a) describe an approach that may be particularly well suited for management of large marine ecosystems. They show that simple models may do better than more complex ones when we have to estimate parameters. This is troublesome to many colleagues (Botsford, 1987)--how can a model that is "less realistic" do better? But these simple phenomenological models can be used to study the efficacy of management actions that produce information as well as fish. The issue of simplicity versus complexity of models and interactions is illustrated by Walters and Collie (1988) who give an example involving Pacific halibut and cyclic environmental variables.

Conclusion: Remember Darwin's Chain

In summary, the empirical support for the currently used models of large marine ecosystems is relatively weak and a new generation of models is needed. In developing those models, we should keep in mind that a key role of theory is to help us to be prepared for the unanticipated, to provide for the possibility of being wrong, and to understand consequences. Using these as guidelines will force the theories to be linked more closely to empirical results. Gulland (1988) stresses that two of the most important issues facing fisheries management in the future are improved assessments and management under uncertainty. Our new theories should reflect these considerations.

Darwin (1859) recognized the importance of theoretical formulations that explicate links in ecosystems when he described the following chain in *Origin of the Species*: old maids keep cats, which eat mice, which eat bumblebees, which pollinate grass, which is eaten by cows. Thus, cow production is driven by the density of old maids.

Acknowledgments

I thank Paul Smith, Carl Walters, and Bob Paine for their time in discussing this matter with me. A referee provided a number of insightful comments which improved the manuscript. This work is supported in part by the National Science Foundation through grant OCE 91-16895 and by the Center for Population Biology, University of California, Davis.

References

Abrams, P. A. 1984. Foraging time optimization and interactions in food webs. Am. Nat. 124:80-96.

Bannister, R.C.A. 1977. North Sea plaice. Pp. *In* J. E. Gulland (Ed.) Fish population dynamics. Wiley, New York.

Bartell, S. M., A. L. Brenkert, R. V. O'Neill, and R. H. Gardner. 1988. Temporal variation of production in a pelagic food web model. Pp. 101-118. *In* S. R. Carpenter (Ed.) Complex interactions in lake communities. Springer-Verlag, New York.

Botsford, L. 1987. Participants' comment on Ludwig (1987). Lect. Notes Biomath. 72:137-138.

Bradbury, R. H., and C. N. Mundy. 1989. Large-scale shifts in biomass of the Great Barrier Reef ecosystem. Pp. 143-168. *In* K. Sherman and L. M. Alexander (Eds.) Biomass yields and geography of large marine ecosystems. AAAS Selected Symposium 111, Westview Press, Inc., Boulder. 493 Pp.

Clark, C. W. 1976. Mathematical bioeconomics. Wiley, New York.

Clark, C. W. 1985. Bioeconomic modelling and fisheries management. Wiley, New York.

Crawford, R.J.M., L. V. Shannon, and P. A. Shelton. 1989. Characteristics and management of the Benguela as a large marine ecosystem. Pp. 169-220. *In* K. Sherman and L. M. Alexander (Eds.) Biomass yields and geography of large marine ecosystems. AAAS Selected Symposium 111, Westview Press, Inc., Boulder. 493 Pp.

Darwin, C. 1859. The origin of the species. Modern Library ed., Viking Press, New York.

Fenchel, T. 1988. Marine plankton food chains. Ann. Rev. Ecol. Syst. 19:19-38.

Gall, G.A.E., B. Bentley, C. Panattoni, E. Childs, C-f. Qi, S. Fox, M. Mangel, J. Brodziak, and R. Gomulkiewicz. 1989. Chinook mixed fishery project, 1986-1989. Rep., Calif. Dep. Fish Game, Univ. Calif., Davis.

Garrod, D. J. 1977. The north Atlantic cod. Pp. 216-242. *In* J. E. Gulland (Ed.) Fish population dynamics. Wiley, New York.

Gauldie, R. W. 1991. Taking stock of genetic concepts in fisheries management. Can. J. Fish. Aquat. Sci. 48:722-731.

Getz, W. M., and R. G. Haight. 1989. Population harvesting. Princeton Univ. Press, Princeton, NJ.

Gulland, J. 1988. Fish population dynamics, 2nd ed. Wiley, New York.

Lawton, J. H. 1989. Food webs. *In* J. M. Cherrett (Ed.) Ecological concepts. Blackwell Scientific, Oxford.

Ludwing, D. 1987. Computer-intensive methods for fisheries stock assessment. Lect. Notes Biomath. 72:125-136.

Ludwig, D., D. D. Jones, and C. S. Holling. 1978. Qualitative analysis of insect outbreak systems: The spruce budworm and forest. J. Anim. Ecol. 47:315-332.

MacCall, A. D. 1984. Population models of habitat selection, with application to the northern anchovy. NOAA, National Marine Fisheries Service, Southwest Fisheries Center, La Jolla, CA. Southw. Fish. Ctr Admin. Rep. LJ-84-01.

MacCall, A. D. 1986. Changes in the biomass of the California Current ecosystem. Pp. 33-54. *In* K. Sherman and L. M. Alexander (Eds.) Variability and management of large marine ecosystems. AAAS Selected Symposium 99, Westview Press, Inc., Boulder. 319 Pp.

Mangel, M. 1985. Decision and control in uncertain resource systems. Academic Press, New York.

Mangel, M., and C. W. Clark. 1988. Dynamic modeling in behavioral ecology. Princeton Univ. Press, Princeton, NJ.

May, R. M. 1984. Exploitation of marine communities. Springer-Verlag, Berlin.

Mills, E. L., and J. L. Forney. 1988. Trophic dynamics and development of freshwater pelagic food webs. Pp. 11-30. *In* S.

R. Carpenter (Ed.) Complex interactions in lake communities. Springer-Verlag, New York.

Murray, J. D. 1989. Mathematical biology. Springer-Verlag, New York.

Nayfeh, A. H. 1973. Perturbation methods. Wiley, New York.

Paine, R. T. 1977. Controlled manipulations in the marine intertidal zone, and their contributions to ecological theory. Pp. 245-270. *In* The changing scenes in natural sciences, 1776-1976. Acad. Nat. Sci., Spec. Publ. 12, Philadelphia, PA.

Paine, R. T. 1984. Ecological determinism in the competition for space. Ecology 65:1339-1348.

Paine, R. T. 1988. Food webs: Road maps of interactions or grist for theoretical development? Ecology 69:1648-1654.

Paine, R. T., and S. A. Levin. 1981. Intertidal landscapes: Disturbance and the dynamics of pattern. Ecol. Monogr. 51:145-178.

Paine, R. T., J. C. Castillo, and J. Cancino. 1985. Perturbation and recovery patterns of starfish-dominated intertidal assemblages in Chile, New Zealand, and Washington State. Am. Nat. 125:679-691.

Ricker, W. E. 1977. The historical development. Pp. 1-26. *In* J. E. Gulland (Ed.) Fish population dynamics. Wiley, New York.

Riffenburgh, R. H. 1969. A stochastic model of interpopulation dynamics in marine ecology. J. Fish. Res. Board Can. 26:2843-2880.

Rinzel, J. 1981. Models in neurobiology. Pp. 345-367. *In* R. H. Enns, B. L. Jones, R. M. Miura, and S. S. Rangnekar (Eds.) Nonlinear phenomena in physics and biology. Plenum Press, New York.

Rothschild, B. J. 1986. Dynamics of marine fish populations. Harvard Univ. Press, Cambridge, MA.

Schoener, T. W. 1989. Food webs from the small to the large. Ecology 70:1559-1589.

Slobodkin, L. B. 1980. Growth and regulation of animal populations. Dover Publications, New York.

Sissenwine, M. P., M. J. Fogarty, and W. J. Overholtz. 1988. Some fisheries management implications of recruitment variability. Pp. 129-152. *In* J. A. Gulland (Ed.) Fish population dynamics, 2nd ed., Wiley, New York.

Smith, P. E., and H. G. Moser. 1988. CalCOFI time series: An overview of fishes. Calif. Coop. Oceanic Fish. Invest. Rep. 29:66-77.

Smith, P. E., M. D. Ohman, and L. E. Eber. 1989. Analysis of the patterns of distribution of zooplankton aggregations from an acoustic Doppler current profiler. Calif. Coop. Oceanic Fish. Invest. Rep. 30:89-103.

Smith, P. E., H. Santander, and J. Alheit. 1989. Comparison of the mortality rates of sardine (*Sardinops sagax*) and anchovy (*Engraulis ringens*) off Peru. Fish. Bull., U.S. 87:497-508.

Soule, M. 1987. Viable populations for conservation. Cambridge Univ. Press, Cambridge, UK.

Sousa, W. P. 1984. The role of disturbance in natural communities. Ann. Rev. Ecol. Syst. 15:353-391.

Terazaki, M. 1989. Recent large-scale changes in the biomass of the Kuroshio Current ecosystem. Pp. 37-66. *In* K. Sherman and L. M. Alexander (Eds.) Biomass yields and geography of large marine ecosystems. AAAS Selected Symposium 111, Westview Press, Inc., Boulder.

Ursin, E. 1982. Stability and variability in the marine ecosystem. Dana 2:51-67.

Walters, C. J. 1986. Adaptive management of renewable resources. Macmillan Co., New York.

Walters, C. J. 1987a. Approaches to adaptive policy design for harvest management. Lect. Notes Biomath. 72:114-125.

Walters, C. J. 1987b. Nonstationarity of production relationships in exploited populations. Can. J. Fish. Aquat. Sci. 44(Supp. II):156-165.

Walters, C. J., and J. S. Collie. 1988. Is research on environmental factors useful to fisheries management? Can. J. Fish. Aquat. Sci. 45:1848-1854.

Brian J. Rothschild

12. On the Causes for Variability of Fish Populations: The Linkage Between Large and Small Scales[1]

Abstract

Fluctuations in fish stock abundance are caused by fishing, modification of population-dynamic stabilizing mechanisms, and changes in the physical environment. The magnitude of each cause, in any particular case, is difficult to assess. The difficulty lies in the highly-dimensional "environment" and in the nonlinear population-dynamics response. Because of these facts, empirical correlations are difficult to interpret. While "spurious correlation" is well known, some correlations may appear spurious, even when they are not. To circumvent the difficulty, it is necessary to understand the relationship between the population-dynamics process and the physical environment at the most fundamental level. This chapter describes aspects of this interaction at the basin and microscale levels.

Introduction

Whether the observed fluctuations in fish-stock abundance result primarily from fishing or "the environment" has been a subject of continuing debate. When stock abundance is relatively depressed, advocates who identify fishing as a major cause of decline suggest that reducing fishing effort will result in an increase in stock abundance. In contrast, other advocates who identify "the environment" as a major cause of decline are confident that as soon as the environment changes, the stocks will increase again, more or less independent of the trajectory of fishing effort.

---

[1] Contribution No. 2122, Center for Environmental and Estuarine Studies of the University of Maryland System.

Of course, neither side is correct--both fishing and the environment interact to affect the abundance of stocks. We know this because fish stocks have been known to fluctuate in abundance for centuries, even before the onset of industrialized fishing (Uda, 1961; Cushing, 1982, 1988a; DeVries and Pearcy, 1982; Soutar and Isaacs, 1969).

Nevertheless, the problem remains. While there is a relatively good understanding of the effects of fishing, the interaction between the physical environment and stock abundance continues to be hardly understood from a cause-and-effect point of view and, at least in the Western literature, the question of how the environment drives fish-stock fluctuations has been generally suppressed. There appear to be three reasons. First, attempts to correlate fish-stock variabiltiy with the physical environment have generally been thought to be statistical "fishing expeditions"; fish-stock abundance causality has been difficult to discover (cf Shepherd et al., 1984); and such computations have generally had little predictive value. Second, the theory of fishing, burgeoning, and reaching a pinnacle with the works of Beverton and Holt (1957) and Ricker (1954), provided a concrete relation between fishing mortality and stock size lending to a view, often implicit, that difficult questions of environmental interactions could be treated as statistical noise (see Schaefer, 1954). And third, in the early 1970s many stocks, particularly clupeid stocks, had "collapsed" (Murphy, 1973) and the thought was that overfishing was heavily implicated in the collapses (there was even doubt as to the effect of the 1972 El Niño on the collapse of the Peruvian Anchovetta (Anonymous, 1974, see also Rothschild, 1986).

Concerns for "global change" have remarkably rejuvenated interest in the role of the environment in driving fish-stock variability (Anonymous, 1989). If we can understand the effects of the environment on fish-stock fluctuations, and how the environment interacts with fishing, either to accentuate or to dampen, the variability in fish stocks, then we can be scientifically positioned to address how global change--that is, changes or variability in the earth's temperature, irradiance, and wind-velocity fields, affects the world's fisheries.

This chapter brings together some ideas on the interactions between fish-stock fluctuations and basin-scale, long-term changes in the environment. In particular it concentrates on the linkage between the long-term, basin-scale, environmental events that are commonly thought of as "climate," and short-term microscale events that affect the dynamics of populations. The chapter focuses particularly on aspects of the kinetic structure of the environment. The kinetic structure of the environment may be the least known source of climatic variability influencing fish stocks, yet remarkably a more important source of variability in population fluctuation than temperature, for example. We consider (1) the nature of the problem of linking environmental and population-dynamic variability, (2) the dynamics of populations, and (3) the relation between basin and microscale physics in affecting the abundance of fish stocks.

The Nature of the Problem

This section shows how the high-dimensionality of ecosystems contributes to the difficulty of drawing inferences from only observations, in the absence of theoretical foundations (Rothschild, 1988). The high dimensionality of the ecosystem contributes to a long and complex causal chain mediating the relationship between climate and fish-stock fluctuation. Because of this it is difficult to take account, or even identify, all of the components in the causal chain, a situation that can lead to either apparently or actually spurious relationships. The difficulty can arise in two ways. The first has to do with the ecosystem's high-dimensionality. The second has to do with the selection of the "best" of all possible relationships in a highly dimensional system.

In the first instance, as a practical matter, because of high dimensionality, all correlations must be computed with a set of variables that represent a much smaller dimensionality, than that which actually exists. That is to say, the number of all possible variables is a very large number, while the number of variables in any practical correlational analysis is a very small number. Symbolically, let the large number of all variables be N and the small number of variables in the correlation analysis be n. Then we must assume that the $N-n$ variables are *conditions* which must either be unimportant or constant for the correlation of the n variables to hold into the future and thus have predictive utility. This assumption which is always implicit is often not made explicitly. Because the $N-n$ *conditions* is a relatively large number, there is a high probability that some of the conditions will not have a constant effect and that some that were not important will become important in the future according to chance alone. So when this happens, the original correlation does not hold and even though the relationship was meaningful, it appears to have been spurious (even if it was not) and in any event, it will not have a predictive utility unless the system returns to its original state (which is unlikely).

In the second instance, as an example of selecting the "best" relationship, from an examination of many possible relationships, consider that environmental variables consist of several time series from different geographical regions. Suppose there are 10 series or regions. Then there are at least 10! possible averages that one could use to represent the environment. We would expect that (0.05 ×

10!) = $0.05 \times 3.63 \times 10^6 \cong 4 \times 10^6$ of these averages would be "statistically significant" just by chance, alone. This means that from a computational view, it would be very easy to find significant relations in a table of random numbers organized as time series, provided that we could examine roughly 4×10^6 of the relationships.

The resolution of this dilemma cannot be obtained through more intense observation. Rather, it is necessary to proceed through an understanding of the fundamental causal processes and the way that the processes are linked together in the causal chain. To do this, it is necessary to link the *population-dynamics process* and the physical setting. First we review the population-dynamics process, then the physical setting, and finally their interrelationship.

Population-Dynamics Process

In searching for fundamental causal relationships between the variability in the physical environment and in the stocks of fish (Rothschild, 1986: Chapter 8), the population-dynamics process tends to be obscured by consideration of only the temporal trajectories of the biomass of the stocks, or recruitment of only the particular fish population of interest, in the absence of the predators and the prey of the population, rather than the process itself.

The population-dynamics process takes into account the life history of the generalized multi life-history stage organism. Each life-history stage is considered independently because in any population, considerable differences in vital rates are generally found among the various life-history stages. In the population-dynamics process, each life-history stage is considered as a density-dependent module (Figure 12.1). Each module consists of density-enhancing and density-depressing components. The density-enhancing component causes the population to increase when it is at a relatively low level of abundance and the density-depressing module causes the population to decrease when it is at a relatively high level of abundance. The enhancing component is turned on because, in general, when the population is at a low level of abundance its food resources increase, and when it is at a high level of abundance its food resources decrease.

Thus, each module tends to be stable and when all life-history stage modules are linked together as in Figure 12.2, they have a redundant stability so that breakdowns in the stabilizing mechanisms at any stage are repaired by an increased density-dependent reaction at the next stage.

Thus, the process is highly stable and links together all populations in the ecosystem. The highly stabilizing density-dependent components are generally only observable when the

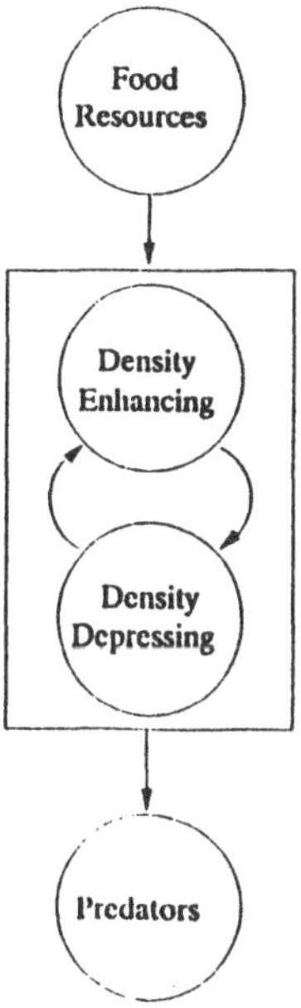

Figure 12.1. Nonlinear density-dependent population-stabilizing module. The module represents a single life-history stage. As an example, increased food resources generated by depressed population levels turns on reproduction and growth causing the population to increase. Increased predation generated by a relatively abundant population turns on predation causing population to decrease.

population in question is unusually large, or unusually small, so that they would only be easy to observe in time series that are relatively great in length.

Changes in the Physical Setting

The physical setting for the population-dynamics process is obviously quite complex and of a very large scope. To circumscribe the discussion, we will focus only on the relation between basin-scale climatic events and microscale events that interact with the population-dynamics process.

From a climate-change perspective, taking account of very long-term (thousands of years) and basin-scale events, we know that the physical setting has changed substantially with the waxing and waning of the glaciers and the accompanying probable changes in thermohaline circulation (Broecker et al., 1985). While it is difficult to portend the "big picture" direction of climate change, it is clear that

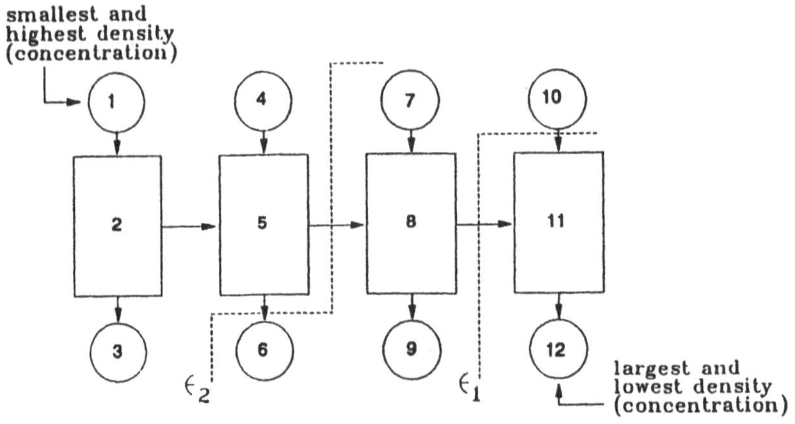

Figure 12.2. Linkages among nonlinear density-dependent population-stabilizing modules. The integers rank the size and hence the concentration or density of organisms. Block 2 is eggs; block 5, larvae; block 8, juveniles; and block 11, adults. Blocks 1, 4, 7, and 10 are food (obviously block 4 is transferred from the adult stage). Blocks 3, 6, 9, and 12 are predators. The integers correspond to the ranked size distribution of individuals. As density is correlated with size, the highest densities occur in circle 1, whereas the lowest densities occur in circle 12. The effects of turbulence measured at two different turbulent environments, one with a low turbulent dissipation rate, ϵ_1, and the other with a high turbulent dissipation rate, ϵ_2, are superimposed on the population-dynamics process model. Because apparent motion enhancement owes to the turbulent dissipation rate and $\epsilon_2 > \epsilon_1$, encounter velocities are enhanced under ϵ_2 for organisms having rank >5, whereas under a smaller value of ϵ, ϵ_1, only organisms with rank >10 have apparently enhanced encounter velocities, although for the larger organisms, the effect may be small depending upon swimming velocity.

there have been important basin-wide decadal changes during contemporaneous time, for example.

A number of sources reflect that in the North Atlantic, the years 1880 to 1920 were relatively cool years; from 1920 to 1960 were relatively warm; and then 1960 to 1980 were relatively cool (e.g., Cushing, 1982: Figure 41).

These stanzas of warming and cooling are evidently coupled with anomalies between surface pressure between warm and cool years. Dickson et al. (1988) compared the surface pressure during the "biologically active" months of March and April in the Northern Hemisphere for the 1970s (cool years) with the 1950s (warm years). They found a strong positive anomaly in pressure to the west of the

British Isles and the Iberian Peninsula. Using the data of Lamb and Weiss (1979) they pointed out there was not only an increase in a northerly index during the cool years, as deduced from surface pressure, but there was also a dramatic increase in the frequency of storms in the North Sea during these cool years. The cool years were also coupled with an evident breakdown in the conveyor belt as a mass of relatively cool water, the "great slug," appeared in the North Atlantic between 1976 and 1981.

The biological coupling with these events is recorded in various places. For example, Cushing (1982) reviews the northward movement of the benthos and other animals during the warm years; the spread of intertidal animals; the Russell cycle, and the rise and decline of the west Greenland cod fishery. Cushing (1988b) also shows the apparent effect of the great slug on many important commercial stocks of fish in the Northeast Atlantic. And finally, Dickson et al. (1988) show how the increased Northerly winds during the 1970s were coupled with increasing upwelling off of Portugal; a decline in zooplankton and phytoplankton in the North Atlantic, and declines in the catch of sardines off of Portugal.

The Linkage Between the Population Dynamics Process and the Physics

It appears that the physical conditions exert two types of effects. The first, and better known, are translocational. The translocational effects appear to be most related to changes in temperature. That is, the position of any particular fauna and flora seems to be coupled with or linked to particular isotherms, and as these isotherms move north or south (for example) the fauna and flora move north or south. In some instances the abundance of the organism does not seem to be affected, while in others, it does.

However, the maximum range in temperature that a population is exposed to, is often quite small, particularly given its capability to "track" isothermal movements, and it may be that variations in the kinetic structure of the water generate more population variability than changes in temperature.

Observation of such phenomena have been known for a long time in the phytoplankton literature. For example, Sverdrup (1953) showed that an onset of a bloom required that the mixing depth in a water column must be less than the critical depth. Dickson et al. (1988) in fact used this criterion to explain how the intensified northern winds in the eastern North Atlantic contributed to the decline in phytoplankton production and hence zooplankton production cited above. Further, the review of Cushing (1989) discusses a body of literature that implies that the structure of the phytoplankton community is related to the kinetic structure of the water in which the phytoplankton exist.

An early exemplification of the effects of kinetic structure on fish was made by Lasker (1975) who construed the relation between feeding conditions for anchovies in the California Current under stratified conditions and conditions where stratification was dissipated owing to increased winds.

Under stratified conditions the larval anchovies had ample food, but when wind destroyed the stratification, the food was dispersed, implying that the larvae might starve to death. Similar observations as to the effect of storms on the juxtaposition of predator and prey were made by Sissenwine (1984).

Keeping in mind the effects of variability of wind velocity on the kinetic structure of the environment, Rothschild and Osborn (1988) in a theoretical study drew upon the theory of homogeneous and isotropic turbulence to compute the uncorrelated, mean-square relative motion of a predator and a prey, a particular distance apart as a function of the turbulent energy dissipation rate which can be a function of wind strength.

The implications of their calculations can be seen relative to the population-dynamics process. As we can see from Figure 12.2, the various populations in the population-dynamics process vary over orders of magnitude in size. Mortality rates are also inversely correlated with the size of the organism. Now in the simplest case, as shown by Rothschild and Osborn, at relatively low wind velocities, the dissipation rate will be low, and so only contact rates of the less dense and largest organisms will be affected. But as the wind intensity (or other turbulence generating mechanism) increases, the uncorrelated mean-square velocity of the smaller, numerically more dense, organisms becomes affected. So with increasing intensity, encounters between predators and smaller, and therefore more densely distributed, organisms in the food chain have enhanced contact rates showing how wind velocity affects the efficiency of trophodynamic pathways, thereby changing the relative advantage and disadvantage of predator and prey.

Observations have been made which are consonant with the theory. For example, Sundby and Fossum (1989) have shown that the number of nauplii per-larval-cod stomach per-nauplii in the environment increases as a function of an index of turbulence adjusted for the effects of stratification, and Kiørobe et al. (1988) have shown that copepod egg production increases in stormy over calm conditions, implying that copepod feeding is enhanced even through there is no evident change in the chlorophyll content of the water. See also Marasé et al. (1990) and Costello et al. (1990).

Discussion

In this chapter, the fishery literature has been reviewed to examine the environmentally-induced variability in fish populations as opposed to fishing mortality-induced variability. Basically we have drawn attention to the difficult aspects of the correlation problem in the sense that because of high dimensionality, even meaningful relationships may seem "spurious." Resolving this problem will require examination of the most fundamental details of the causal chain. In this regard we have attempted to start at the beginning, considering the climatic aspects of thermohaline circulation and showing how these seemed to be coupled with the aspects of the thermohaline oceanographic and meteorological structure of the North Atlantic. A comparison of the cold and windy eras with the warm and calm eras reflects a considerable, perhaps decadal-scale, faunistic change, but shorter-term variability might be attributed to even the effects of changes in the mean and variance of the wind fields, taking into account, of course, the other variables that modify the effect of the wind on the sea surface. But the picture is hardly complete in the context of "global change" as these bits and pieces of observation and theory are not sufficiently coherent to achieve the understanding that society needs to make better management decisions. Knitting these fragments together is the challenge for understanding a major component of the earth system--the way that climate affects the sea--the way the sea affects the climate--and the way that both affect the biological metabolism of the fish stocks.

References

Anonymous. 1974. Report of the Fourth Session of the Panel of Experts on Stock Assessment on Peruvian Anchoveta. Instituto Del Mar Del Peru, 2(10):605-723.

Anonymous. 1989. Global ecosystem dynamics. EOS 70:82-85.

Beverton, R. J. H., and S. J. Holt. 1957. On the dynamics of exploited fish populations. Fish. Invest. Minist. Agric. Fish. Food (G.B.) Ser. II, 19:1-533.

Broecker, W. S., D. M. Peteet, and D. Rind. 1985. Does the ocean-atmosphere system have more than one stable mode of operation? Nature 315:21-26.

Costello, J. H., J. R. Strickler, C. Marasé, G. Trager, R. Zeller, and A. J. Freise. 1990. Grazing in a turbulent environment: Behavioral response of a calanoid copepod, *Centropages hamatus*. Proc. Nat. Acad. Sci., USA 87:1648-1652.

Cushing, D. H. 1982. Climate and fisheries. Academic Press, London.
Cushing, D. H. 1988a. The provident sea. Cambridge Univ. Press, Cambridge. 329 Pp.
Cushing, D. H. 1988b. The northerly wind. Pp. 235-244. *In* B. J. Rothschild (Ed.) Toward a theory on biological-physical interactions in the world ocean. Kluwer Academic Publishers, Dordrecht, The Netherlands. 650 Pp.
Cushing, D. H. 1989. Review--A difference in structure between ecosystems in strongly stratified waters and in those that are only weakly stratified. J. Plankton Res. 2(1):1-13.
DeVries, T. J., and W. G. Pearcy. 1982. Fish debris in sediments of the upwelling zone off central Peru: A late Quaternary record. Deep-Sea Res. 28:87-109.
Dickson, R. R., P. M. Kelly, J. M. Colebrook, W. S. Wooster, and D. H. Cushing. 1988. North winds and production in the eastern North Atlantic. J. Plankton Res. 10:151-169.
Kiørboe, T., P. Munk, K. Richardson, V. Christensen, and H. Paulsen. 1988. Plankton dynamics and larval herring growth, drift and survival in a frontal area. Mar. Ecol. Prog. Ser. 44:205-219.
Lamb, H. H., and I. Weiss. 1979. On recent changes of the wind and wave regime of the North Sea and the outlook. Fachl. Mitt. Amtl. Wehrgeophys. 194:1-108.
Lasker, R. 1975. Field criteria for survival of anchovy larvae: The relation between inshore chlorophyll layers and successful first feeding. Fish. Bull., U.S. 73:453-462.
Marasé, C., J. H. Costello, T. Granata, and J. R. Strickler. 1990. Grazing in a turbulent environment: Energy dissipation, encounter rates, and efficacy of feeding currents in *Centropages hamatus*. Proc. Nat. Acad. Sci., USA 87:1653-1657.
Murphy, G. I. 1973. Clupeoid fishes under exploitation with special reference to the Peruvian anchovy. Tech. Rep. No. 30, Hawaii Inst. Mar. Biol.
Ricker, W. E. 1954. Stock and recruitment. J. Fish. Res. Board Can. 11:559-623.
Rothschild, B. J. 1986. Variability in marine fish populations. Harvard Univ. Press, Cambridge, Mass. 271 Pp.
Rothschild, B. J. (Editor). 1988. Toward a theory on biological-physical interactions in the world ocean. Kluwer Academic Publishers, Dordrecht, The Netherlands. 650 Pp.
Rothschild, B. J., and T. R. Osborn. 1988. Small-scale turbulence and plankton contact rates. J. Plankton Res. 10:465-474.
Schaefer, M. B. 1954. Some aspects of the dynamics of populations important to the management of the commercial marine fisheries. Bull. Int.-Am. Trop. Tuna Comm. 1:27-56.

Shepherd, J. G., J. G. Pope, and R. D. Cousens. 1984. Variations in fish stocks and hypotheses concerning their links with climate. Rapp. P.-v. Reun. Cons. int. Explor. Mer 185:255-267.

Sissenwine, M. P. 1984. Why do fish populations vary? Pp. 59-94. *In* R. M. May (Ed.) Exploitation of marine communities. Springer-Verlag, New York.

Soutar, A., and J. D. Isaacs. 1969. History of fish populations inferred from fish scales in anaerobic sediments off California. Calif. Coop. Oceanic Fish. Invest. Rep. 13:63-70.

Sundby, S., and P. Fossum. 1989. Feeding conditions of north-east Arctic (Arcto-Norwegian) cod larvae compared to the Rothschild-Osborn theory on small-scale turbulence and plankton contact rates. ICES C.M. 1989/G:19, 11 pp.

Sverdrup, H. U. 1953. On the conditions for vernal blooming of phytoplankton. J. Cons. int. Explor. Mer 18:287-295.

Uda, M. 1961. Fisheries oceanography in Japan, especially on the principles of fish distribution, concentration, dispersal and fluctuation. Calif. Coop. Oceanic Fish. Invest. Rep. 8:25-31.

Theodore Smayda

13. Global Epidemic of Noxious Phytoplankton Blooms and Food Chain Consequences in Large Ecosystems

Abstract

The trophic consequences of the apparent global increase in the frequency of coastal red-tide blooms and spreading of toxic and benign phytoplankton species into new regions are considered. Various mechanisms are described by which blooms of red-tide species can negatively impact, lead to dysfunction in, and alter trophodynamics in all trophic compartments, from the microbial food web up to marine mammals. Reduced fecundity, survival and recruitment, and increased mortality of first feeding, juvenile, and adult stages can result from toxins produced by red-tide species, having either a direct impact upon or vectoring through the food web. The classical notion that red-tide blooms are insignificant to trophodynamics, and of interest primarily because of their impact on humans through paralytic shellfish poisoning, or economically from aquacultural dieoffs, is invalid. A trophic model showing the routing of toxins and inimical phytoplankton effects through the various trophic compartments of marine food webs is presented. Two major types of negative events can accompany nuisance blooms: water column events, usually associated with toxin production, and anoxic events which accompany benthic accumulation; and respiration of blooms which are underutilized by grazers. The potential consequences of the altered red-tide bloom characteristics in coastal waters to larval survival and recruitment of fish which have synchronized their spawning cycle strategies to phytoplankton bloom characteristics are touched upon. Assessments of the causes of variability at different life history stages of fish must include consideration of red-tide events and toxic species occurrences.

Introduction

In discussions of primary production and energy flow, the biomass-rich spring phytoplankton bloom is generally focused upon. What is commonly overlooked is that some species are always in bloom; this is the basis of succession (Smayda, 1980). Collectively, the production and biomass of these secondary bloom events, when summed over the annual cycle, most likely exceed spring bloom levels. In fact, there is sufficient evidence that pelagic food-web dynamics are primarily driven energetically by these so called minor blooms which are rarely studied *in situ*. The environmental carrying capacity for these secondary bloom events varies considerably: spatially, seasonally, and interannually. Periodically, some of the minor bloom species exhibit unusual, even extraordinary blooms termed red-tide events. In reality, red-,green-, yellow-, brown- or white-water discolorations may occur depending on the bloom species and its pigmentation. In this paper, red tide will be used as a collective term for such blooms. Historically, red-tide events generally have been treated as rogue blooms and anecdotally, such as whether fish and invertebrate kills occurred, whether H_2S production was evident, etc. The most quantified and scientifically studied aspect of red-tide blooms has been the frequent occurrence of Paralytic Shellfish Poisoning (PSP) accompanying certain dinoflagellate blooms.

Marine aquaculture of shellfish and fish farming increasingly has been accompanied by serious stock mortalities due to red-tide blooms. The causes of mortality and the species responsible for the dieoffs have varied, with both ichthyotoxin production and anoxia implicated. Table 13.1 lists some representative financial losses suffered by this industry. The magnitude of the economic losses per bloom episode, from about $4.5 million (U.S.) to >$60 million, reveals that such bloom events are not trivial. In the Seto Inland Sea, annual financial losses of cultured fish attributable to toxic blooms exceeded $0.7 million in 14 of the years between 1970-1987, with a cumulative loss >$100 million (Okaichi, 1989; Shirota, 1989). Between 1980-1984, 11 fish kills were recorded in Hong Kong where none had occurred previously, causing 35% mortality of the total cultured fish biomass (Lam and Ho, 1989; Wong and Wu, 1987). During the unprecedented 1988 bloom of the toxic flagellate *Chrysochromulina polylepis*, 200 caged salmon fish farms worth $200 million were towed to sea in a bloom avoidance measure; the total annual salmonid production value of this Norwegian activity is $750 million (Dundas et al., 1989). Such practical and commercial considerations require that the anecdotal approach to red-tide blooms be abandoned and scientific inquiry instituted.

Table 13.1. Examples of financial loss resulting from the mortality of aquacultured species accompanying red-tide blooms.

Date	Location	Species**	$U.S. Loss	Source
1989	Norway	salmon rainbow trout	4.5 M	(3)
1988	Norway Sweden	salmon	5.0 M	(2)
1987	Japan	yellowtail	15.0 M	(4)
1986	Chile	red salmon	21.0 M	(5)
1981	Korea	oyster	>60.0 M	(5)
1978	Korea	oyster	4.6 M	(1)
1978	Japan	yellowtail	~ 22.0 M	(4)
1977	Japan	yellowtail	~ 20.0 M	(4)
1972	Japan	yellowtail	~ 47.0 M	(4)

NOTE: (1) Cho, 1979; (2) Dundas et al., 1989; (3) Johnsen and Lien, 1989; (4) Okaichi, 1989; (5) Shirota, 1989; ** oyster (*Crassostrea gigas*); rainbow trout (*Onchorhyncus mykiss*); salmon (*Salmo salar*); yellowtail (*Seriola quinqueradiata*); ~ U.S. $ equivalent estimated from 150 yen per $1.0 (=1990 exchange rate); calculation may have exaggerated estimated loss value.

Is the mortality of cultured fish and molluscs during certain red-tide blooms an artifact of their enclosure in caged or raft culture farms? That is, do dieoffs caused by red tides occur in natural communities where behavioral avoidance, toxin dilution, and bloom species dispersal accompanying circulation processes attenuate the effects of such potentially lethal blooms? The evidence strongly indicates that they are not only natural events, but that mortality at selected, or all trophic levels can occur during a given bloom event (Figley et al., 1979; Smayda and Villareal, 1989; Smayda and Fofonoff, 1989; Underdal et al., 1989; Wardle et al., 1975; White, 1977, 1980, 1982). For example, the brown tide of *Aureococcus anophagefferens* during the 1985 summer in Narragansett Bay and

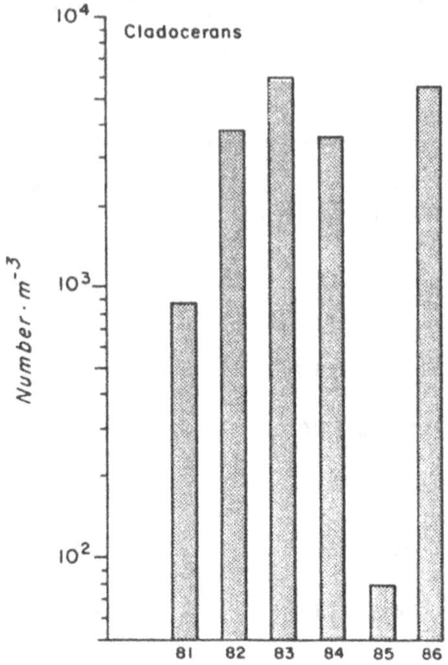

Figure 13.1. Mean summer abundances of cladocerans in lower Narragansett Bay from 1981-1986.

Long Island coastal embayments was accompanied by extensive mortality and recruitment failure of mussels and scallops (Tracey, 1988; Bricelj and Kuenstner, 1989); by a dieoff of macroalgae in Narragansett Bay and of eelgrass, *Zostera marina*, in Long Island embayments (Dennison et al., 1989); failure of the cladoceran community to appear (Figure 13.1) and the bay anchovy, *Anchoa mitchelli*, to spawn (Figure 13.2) in Narragansett Bay, as well as cessation of filtration by the mussel, *Mytilus edulis*, leading to starvation and mortality (Tracey, 1988). Thus, impaired physiological processes and trophic transfer as well as trophic dysfunction can also accompany red-tide events in natural communities, even in the absence of mortality.

There is increased reason to be concerned about red-tide blooms and the need to view them in a trophic context rather than anecdotally, or from the vantage point of aquacultural management needs. A remarkable spreading and increased bloom frequency of coastal red-tide species and events appear to be in progress globally (Anderson, 1989; Smayda, 1989a,b). Regional spreading phenomena

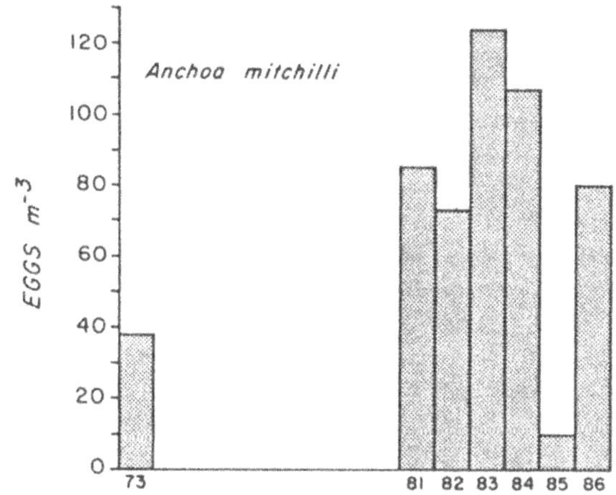

Figure 13.2. Mean annual summer egg abundance of bay anchovy, *Anchoa mitchelli*, in lower Narragansett Bay during 1985 brown-tide event relative to other years.

have been described for *Pyrodinium bahamense* var. *compressum* within Indo-Pacific waters (McLean, 1984), *Gymnodinium catenata* and *G. catenella* into Tasmanian waters (Hallegraeff et al., 1988), *Gyrodinium aureolum* within the North Sea (see Partensky and Sournia, 1986), and *Alexandrium tamarense* within Northern European, Northeast U.S. coastal, and South American waters (Anderson, 1989). Increased frequency in red-tide bloom events appears to have occurred in Tolo Harbour, Hong Kong; along the Romanian coast of the Black Sea; within sub-regions of the North Sea, the Kattegat, and the Baltic Sea (Smayda, 1989a,b). Similar long term increases have occurred in Japanese coastal waters (Okaichi, 1989). The apparent spreading and increased bloom frequency of both toxic and benign species represent two different phenomena. Spreading into new regions is not necessarily accompanied by red-tide blooms.

Smayda (1989a,b) has established that the increase in red-tide blooms regionally within the North Sea, in Scandinavian coastal waters, the Baltic Sea, and the Black Sea has been accompanied by increased concentrations of inorganic nitrogen and phosphorus, and reductions in their ionic ratios with silica. The reduced Si:P and Si:N ratios are hypothesized to have altered the chemical milieu and phytoplankton niche structure favoring the selection of flagellates over that of diatoms. The coastal enrichment of nutrients, also manifested in increased annual primary production rates (Smayda, 1989b), stimulates red-tide blooms of selected species. Figure 13.3 shows the sequence of novel bloom events exhibited by 24 species

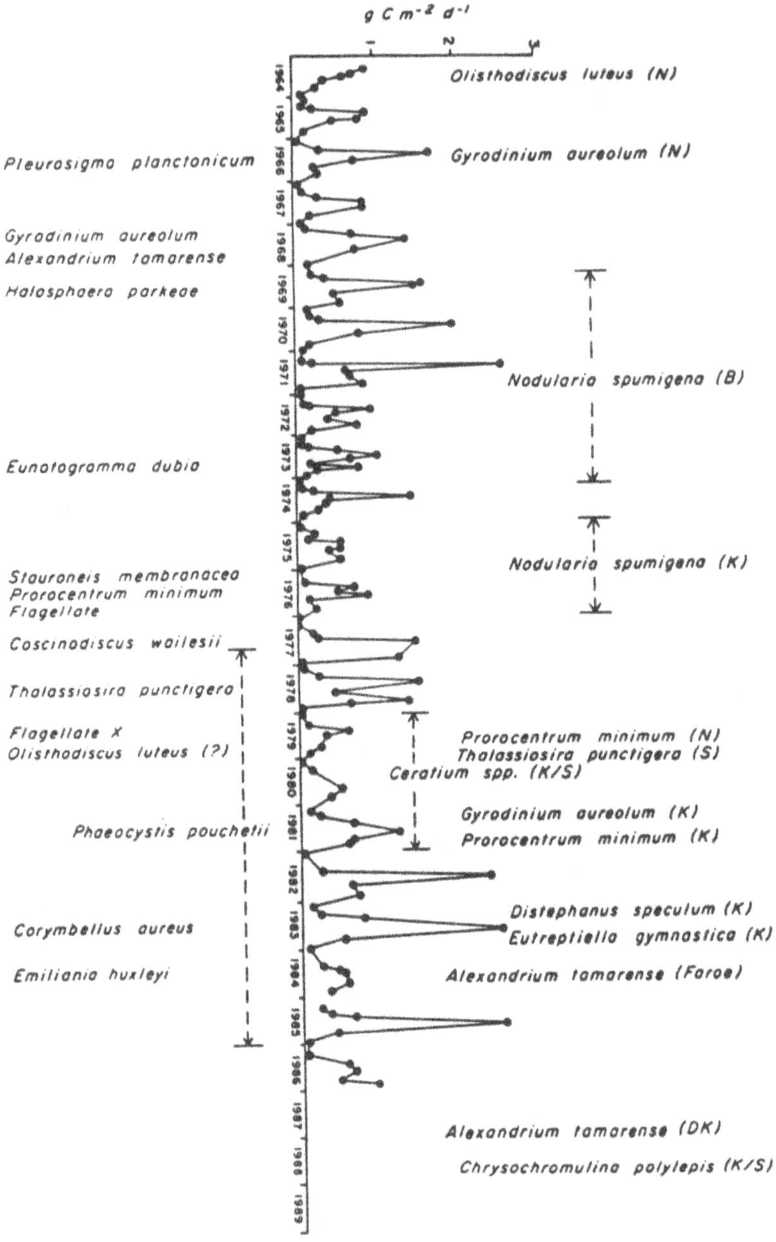

Figure 13.3. Right Panel: First occurrences and/or persistence of some novel, exceptional phytoplankton species' blooms in Norwegian (N) coastal waters, Danish fjords (DK), Kattegat (K), Skagerrak (S), and in the Baltic Sea (B) during 1964-1988. Left Panel: As for right panel, except annual calendar records species' occurrences in the Dutch Wadden Sea and southern North Sea. Middle Panel: Monthly survey of primary production rates in western English Channel (From Smayda, 1989a).

between 1964 and 1989 which were previously unimportant or have spread into the North Sea and Scandinavian coastal waters. One interpretation of this remarkable phenomenon, given the progressive, long term nutrient enrichment and increased productivity detectable in these waters (Smayda, 1989a,b), is that such plankton communities may be in temporary disequilibrium adjusting to altered nutrient conditions. This is accompanied by disruptions in upper trophic level dynamics (Smayda, 1990).

These various characteristics of red-tide blooms suggest that rather than being relatively benign aspects of phytoplankton growth in the sea such blooms have trophic aspects requiring elucidation. The following will describe the diverse modes of action and trophic pathways through which red-tide blooms can negatively impact, or alter various trophic components and processes. The issue of recruitment will be focused upon.

Trophic Coupling to Red-Tide Blooms and Species

Nutritional Use of Dinoflagellates by Larval Stages

My assessment of nuisance phytoplankton blooms and food-web dynamics will focus on fisheries' impacts, notably recruitment, given the thrust of this symposium. The basic paradigm that primary producers regulate fishery stocks is unassailable. The linkage between clupeoid stocks and upwelling (Cushing, 1971); annual production rates and fisheries' yields (Nixon, 1988); and the reductions in Nile River runoff, nutrient inputs, primary production, and sardine stocks following Aswan Dam construction (Dowidar, 1984) illustrate this trophic linkage. A pronounced nutritional dependence on dinoflagellates, a major component of nuisance phytoplankton blooms, appears to have evolved, particularly during first feeding and late larval stages. Last (1980), for example, reports that the larvae of 15 of 20 fish species sampled from natural populations began to feed on dinoflagellates while still in the yolk sac stage. Herring, sprat, and sand lance continued to ingest dinoflagellates until 8 mm in length. In northern anchovy, *Engraulis mordax*, first feeding stages ingested the four test dinoflagellates provided: diatoms and naked flagellates were rejected (Scura and Jerde, 1977). In Last's surveys, first feeding larvae ingested diatoms primarily during the spring bloom, whereas dinoflagellates were ingested year-round. Not all dinoflagellates are nutritionally adequate, however. *Engraulis mordax* larvae were reared successfully on *Gymnodinium splendens*, but not on the non-toxic *Gonyaulax polyedra* and *Prorocentrum micans* (Scura and Jerde, 1977; Lasker and Zweifel, 1978). Nor is the food quality fixed; although *G. splendens* is a good food source for *E. mordax* and *E.*

ringens larvae, adult *E. ringens* exhibits reduced growth rate, fat content, and spawning when prey-switching to this dinoflagellate during its red-tide blooms in Peruvian upwelling systems (de Mendiola, 1979). This situation provides the interesting situation where *G. splendens* blooms may enhance recruitment and year class strength when coincident with *E. ringens* larval swarms, but reduce fecundity when coincident with adult stages.

Given this nutritional dependence, the continuously changing mix of toxic and benign species of dinoflagellates and other flagellates in the sea supports Hjort's (1914) view that the availability of suitable food is variable and may affect the differential mortality of first feeding larvae and subsequent year class strength. His postulate that larval feeding is also dependent on suitable concentrations (patches) of food organisms, both spatially and in time, and the demonstration of a nutritional dependence on dinoflagellates have been eloquently strengthened by Lasker and Zweifel (1978). They showed that northern anchovy, *Engraulis mordax*, is attracted to dinoflagellate patches and stimulated into visually based feeding on cells >20 μm, if above a threshold density of 30 cells ml^{-1}. Wind induced breakup of the patch concentration below threshold feeding levels leads to starvation within 1.5 days in post yolk sac larvae (Lasker et al., 1970), if the patch is not reestablished within the critical period.

The apparent dependence of first feeding stages on dinoflagellates is supplemented through another mechanism: grazing upon tintinnids. Tintinnids comprised from 17% to 87% of the larval stage stomach contents in the 20 species examined by Last (1980). Cod larvae <6 mm, for example, primarily ingested tintinnids; larger specimens supplemented their diet with nauplii and copepodites. Mackerel, *Scomber scombrus*, also fed on cladocerans. The significance of such larval dietary preferences is that both tintinnids (Prakash, 1963) and cladocerans are major consumers of dinoflagellates, and represent trophic linkages between the phytoplankton and nekton via the microbial food web and herbivorous zooplankton, respectively. It also indicates that toxic dinoflagellates can impact fish larvae directly through ingestion or vectored through intermediate grazers.

Mortality of Larval Stages Ingesting Toxic Dinoflagellates

Experimental and field evidence indicates that in at least seven species of commercially important fish larval mortality has occurred following ingestion of toxic dinoflagellates (Table 13.2). This supplements White's (1977) observations that toxic dinoflagellate blooms can cause mortality in adult fish stocks, such as the 1976 and 1977 herring dieoffs in the Bay of Fundy. Experimental studies have also demonstrated the remarkable sensitivity of fish to dinoflagellate

Table 13.2. Mortality of fish species attributable to toxic phytoplankton exposure, exclusive of anoxic events (A = adult; L = larvae).

Species	Stage	Common name	Source
Ammodytes sp.	A	sand lance	(9)
Brevoortia tyrannus	A	menhaden	(9)
Clupea harengus harengus	L,A	herring	(2,6,7,8)
Engraulis japonicus	L	Japanese anchovy	(11)
Gadus morhua	A	cod	(1,8)
Hypomesus pretiosus japonicus	A	surf smelt	(4)
Leiognathus nuchalis	L	?	(12)
Lophius americanus	A	monkfish	(10)
Mallotus villosus	L	capelin	(2)
Merlangius merlangus	L	whiting	(1)
Onchorhyncus mykiss	A	rainbow trout	(1)
Pollachius virens	A	pollock	(1,8)
Pagrus major	L	red sea bream	(11)
Pomatomus saltatrix	A	bluefish	(9)
Pseudopleuronectes americanus	L,A	winter flounder	(3,8)
Raja sp.	A	skate	(10)
Salmo salar	A	salmon	(1,8)
Seriola quinqueradiata	A	yellowtail	(5)
Squalus acanthias	A	spiny dogfish	(10)

NOTE: (1) Dundas et al., 1989; (2) Gosselin et al., 1989; (3) Mills and Klein-MacPhee, 1979; (4) Ogata and Kodama, 1986; (5) Okaichi, 1989; (6) White, 1977; (7) White, 1980; (8) White, 1981b; (9) White, 1982; (10) White, 1984; (11) White et al., 1989; (12) Yuki and Yoshimatsu, 1989.

toxins similar to other vertebrates. This contrasts with the marked insensitivity of filter feeding bivalves to toxins which allows their accumulation without apparent physiological impairment (White, 1982). For example, the LD_{50} for herring, cod, flounder, pollock, and salmon is 4 to 12 μg toxin per kg body weight when introduced intraperitoneally. Shellfish commonly acquire 6000 μg toxin per 100 μg body weight without physiological impairment; i.e., a body burden about 5,000 times greater than that tolerated by fish and other vertebrates (White, 1984). One consequence is that the accumulation of large quantities of paralytic shellfish poison (= saxitoxin) by post-larval shellfish poses a human health hazard, whereas intoxicated fish may not because they die at low doses. An exception to this is ciguatera poisoning caused by consuming fish which have ingested toxic dinoflagellates symbiotic in their macroalgal diet. Ciguatera poisoning has been reported to affect up to 20,000 people annually, particularly in the South Pacific region (Legrand et al., 1989). PSP toxins from *Pyrodinium bahamense* var. *compressum* accumulated in the guts of planktivorous fish, and when eaten whole, have caused illness or death in Southeast Asia (White, pers. commun.). Fish flesh apparently does not accumulate such toxins.

Representative survival patterns of larval fish exemplified by capelin, *Mallotus villolus*, when fed both toxic and non-toxic strains of *Protogonyaulax* (= *Alexandrium*) *tamarensis* are shown in Figure 13.4. Capelin are unable to distinguish between toxic and non-toxic strains; both clonal types were ingested. Daily mortality rate was influenced by abundance of the toxic cells, larval age and the percentage of larvae feeding, with older larvae becoming progressively less susceptible at prey densities of <500 cells ml^{-1}. At 1,500 cells ml^{-1}, which is comparable to local bloom levels, mortality exceeded 90% d^{-1} and was independent of larval age. Herring larvae exhibited similar responses, differing from capelin by a greater resistance to toxic ingestion; six-day-old larvae were relatively invulnerable (Gosselin et al., 1989). Mortality rates of capelin and herring reached 92% and 77% d^{-1}, respectively, when fed the toxic strain in contrast to 30% d^{-1} when fed the non-toxic strain.

Neurotoxic effects and behavioral modification accompany ingestion of toxic cells. Capelin and herring larvae exhibit symptoms of paralysis; swim erratically, sink, and remain immobile prior to death (Gosselin et al., 1989). Similar behavioral modifications were reported for intoxicated larval winter flounder, red sea bream, and Japanese anchovy (Mills and Klein-MacPhee, 1979; White et al., 1989). The latter two species lost equilibrium, swam on their sides, upside down and/or in circles prior to paralysis. Such behavioral dysfunction poses a fundamental constraint on the searching behavior of fish larvae for food. Hunter and Thomas (1974) have described the swimming behavior of *Engraulis mordax* within patches of *Gymnodinium splendens*, to which yolk-sac and post-yolk-sac larvae are attracted by chemical, acoustical and/or tactile stimuli.

In summary, there is provocative evidence based on toxic prey experiments that the quality, density and patchiness of prey, and larval behavioral feeding mechanisms are indeed significant to successful larval feeding process and growth, as postulated by Hjort (1914).

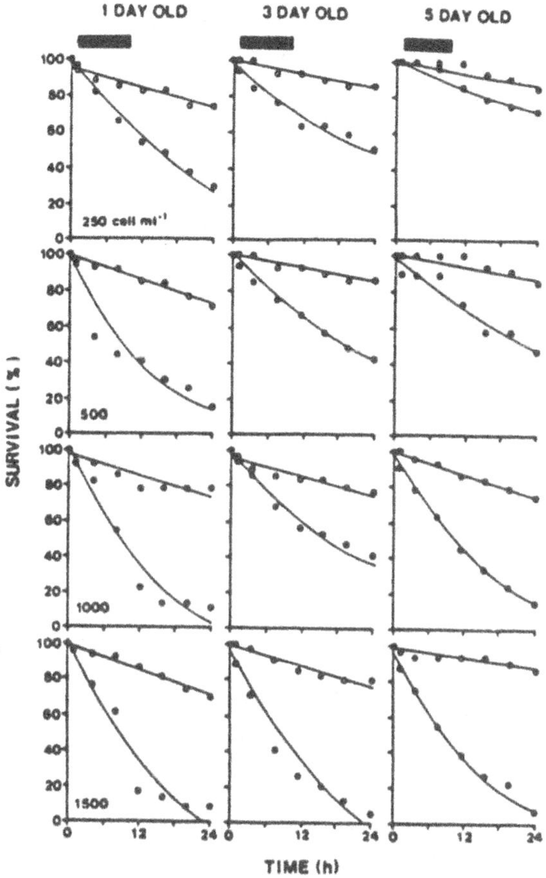

Figure 13.4. Time course of survival of capelin larvae, *Mallotus villosus*, of different ages when exposed to toxic (open circles) and non-toxic (closed circles) clones of *Alexandrium tamarense* at different cell concentrations. Darkened bars represent dark period of the 24-h experimental duration (From Gosselin et al., 1989).

Toxin Vectoring

Direct dependence on dinoflagellates as food items lasts for only about the first week of larval feeding, as shown for *Engraulis mordax* (Hunter and Thomas, 1974), or slightly longer. A gradual metamorphic progression towards larger prey capture occurs; the amount and size of the prey taken increasing with larval mouth size (Last, 1980). For 9-10 mm larvae, the mean diameter of prey was about

2.5-fold greater than that captured by <2 mm larvae; for 12-13 mm larvae, it was about 3-fold greater. This shift in prey capture away from the phytoplankton size classes towards tintinnids, nauplii, copepodites, copepods, cladocerans, etc. (i.e., from an early life stage herbivory to the carnivorous or omnivorous diet characteristic of many species, does not reduce susceptibility to intoxification). Vectorial intoxification through food-web transfer can occur, as shown by White (1977, 1980, 1981a, 1982, 1984) to be the cause of herring dieoffs in the Bay of Fundy. In these instances, saxitoxin produced by dinoflagellates was vectored to the herring larvae through their ingestion of tintinnids, cladocerans, copepods, and pteropods which had grazed upon toxic dinoflagellates. Experimental studies have confirmed that vectored toxins can be lethal. A daily mortality of about 30% d^{-1} occurred in larval herring (16 mm) and other species following their ingestion of zooplankton which fed upon toxic dinoflagellates (White, 1982; White et al.,1989; Gosselin et al., 1989). Thus, there is a broad spectrum of food-web routes available for vectoring toxins to larval and juvenile fish through the zooplankton comunity network. White's (1982) observation that a herring dieoff did not accompany the large 1980 toxic bloom in the Bay of Fundy when zooplankton levels were very low, unlike during 1976 and 1979, is consistent with this toxin routing mechanism.

Rhaphidophycean Ichthyotoxins

Table 13.2 lists some of the adult fish species which have exhibited dieoffs during toxic phytoplankton blooms; planktivorous, omnivorous, and piscivorous species are included. Although these dieoffs accompanied red-tide outbreaks, carcass toxin concentrations usually were not determined, and, thus, the evidence is circumstantial that endotoxins vectored through the food web were the cause of some of these dieoffs. More convincing are White's (See Table 13.2) field and laboratory studies linking fish dieoffs to vectored dinoflagellate toxins. Planktivorous herring, sand lance, and menhaden, exhibited high levels of paralytic shellfish toxins.

Table 13.2 also includes dieoffs resulting from a third type of toxic effect: ichthyotoxins, which are directly toxic without vectoring through the zooplankton community. The production of brevetoxin during *Gymnodinium breve* blooms, a frequent cause of massive fishkills since the 1800s in the Gulf of Mexico (not listed in Table 13.2), is the most renowned ichthyotoxic event (see Roberts, 1979). In recent years, blooms of naked flagellates collectively known as rhaphidophyceans increasingly have caused serious ichthyotoxic dieoffs. *Chattonella antiqua*, *Ch. marina*, and *Ch. subsala*, which appear to have spread from the Mediterranean Sea via Indian coastal waters to the Seto Inland Sea, and *Heterosigma akashiwo* have devastated cultured fisheries stocks in the latter region (Okaichi, 1989. *Chrysochromulina polylepis* exhibited an unprecedented bloom during 1988 in southern Norwegian and Swedish coastal waters, killing off much of the O-year group of cod, whiting, pollock, and saithe (Dundas et al., 1989). The toxic *Prymnesium parvum*, which also exhibited an unprecedented lethal bloom in Norwegian waters during the 1989 summer (Johnsen and Lien, 1989), is a well known ichthyotoxic species throughout its distributional range.

The ichthyotoxins produced by these flagellates are a complex of chemical structures having different physical and toxicological properties. They are water or lipid soluble, potent as excreted exotoxins or upon contact, and have cytolytic, haemolytic, or neurotoxic activity (Shimizu, 1989; Yasumoto, 1989). The gill filaments rather than prey ingestion function as the primary uptake sites. In the case of dieoffs of yellowtail, *Seriola quinqueradiata*, during *Chattonella* blooms, death is sometimes preceded by copious mucous production and histological changes thought to interfere with osmoregulated processes and gaseous exchange leading to asphyxiation (Endo et al., 1985; Onoue and Nozawa, 1989; Toyoshima et al., 1989). The very common, globally distributed toxic red-tide species, *Noctiluca miliaris*, a holozoic dinoflagellate which grazes upon copepod eggs and fish eggs and larvae, produces a remarkably simple ichthyotoxin: NH_4 (Okaichi and Nishio, 1976). The positive buoyancy of this species is due to its NH_4 content (Smayda, 1970) stored in its acidic vacuole following deamination of ingested prey. Okaichi and Nishio report a 10-fold range in NH_4 concentrations and associated ichthyotoxicity. This suggests the interesting situation that *Noctiluca* becomes progressively more toxic with greater food intake! In summary, a broad spectrum of ichthyotoxin modalities has evolved in a variety of taxa allowing direct impact on the toxin-sensitive nekton, in addition to their vulnerability to vectored toxin ingestion.

Marine Mammal Dieoff

Geraci et al. (1989) have recently reported a most remarkable dieoff: the first apparent mortality of marine mammals accompanying ingestion of dinoflagellate toxin (= saxitoxin). Fourteen humpback whale, *Megaptera novaeangliae*, died in Cape Cod waters after eating mackerel, *Scomber scombrus*, which contained saxitoxin. The intermediary vector(s) between the toxic dinoflagellates and mackerel were not established. Wyatt (1980) previously suggested that a massive dieoff of Cape fur seal, *Arctocephalus pusillus*, in the early 1800s resulted from a toxic bloom. The occurrence of other lethal factors complicates assessment of marine mammal dieoffs. For example, the massive 1988 dieoff (>18,000) of harbor seal, *Phoca vitulina*, in European waters has been attributed to viral attack (Dietz et al., 1989). Nonetheless, there is now reason to believe that marine mammals may also be vulnerable to toxic red-tide blooms. The apparent global increase in, and spreading of such blooms, accordingly, take on added significance.

Clearly, upper trophic levels have been demonstrated to suffer mortality and/or recruitment failure because of the direct effects of both dinoflagellate and other phytotoxins, or through the vectored food-web transmission of dinoflagellate toxins. A linkage between toxic phytoplankton species → zooplankton → planktivorous fish → piscivorous fish + marine mammals has now been unequivocably established. This is a major revision of the classical paradigm that toxic phytoplankton routing was restricted to: toxic species → shellfish → humans.

Concentrations, Tempo, Duration of Toxic Effects

The entrainment of fish and shellfish in cage and raft cultures, in contrast to the vagility of natural stocks, facilitates their chronic exposure to toxic bloom events. Hence, their dieoffs might not be unexpected, whereas the avoidance mechanisms of unrestrained, natural stocks may reduce exposure to lethal toxin levels. It might be concluded, therefore, that experimental studies of such effects have limited relevance to *in situ* events. The evidence to date strongly indicates, however, the remarkably small prey doses having toxic effects and the speed with which they are manifested. Experimentalists have generally been careful to use toxic prey densities similar to those occurring during blooms. White et al. (1989) estimated that only 6 to 11 cells of the toxic *Alexandrium tamarense* ingested by first feeding sea bream larvae would deliver a lethal dose. Gosselin et al. (1989), who worked with a more toxic strain, estimated that only one cell would be lethal to capelin (~0.13 mg) and herring (~0.17 mg) larvae. Toxic red-tide blooms commonly exceed 1,000 cells ml^{-1}. Clearly, excessive cell levels are not needed to achieve mortality. In fact, adventitious feeding on normally rejected prey may be a serious problem under certain conditions, given the extreme sensitivity of fish and the low doses required for intoxification.

The experimental data uniformily indicate that the toxic impact on larvae usually occurs within hours of exposure, rather than days. In adult herring, pollock, winter flounder, Atlantic salmon, and cod irregular, jerky swimming movements occurred within 5 mins following toxic injection; loss of equilibrium within 10-15 min; death within 20-60 min (White, 1984). Juvenile sea bream died within 4-10 min of exposure to the neurotoxic fraction from three red-tide species, and within 20-50 min after exposure to the hemolysin and hemoagglutinin fractions (Onoue and Nozawa, 1989). Sea bream larvae lost equilibrium within 1-2 h of toxin exposure (White et al., 1989), with mortality occurring within 70-130 mins of exposure to another toxic species (Matsusata and Kobayashi, 1974). Larval *Leiognathus nuchalis* died within 3-8 h of exposure to toxic *Cochlodinium* spp. (Yuki and Yoshimatsu, 1989). The longest lag period reported between toxic cell ingestion and dysfunction is three days in winter flounder (Mills and Klein-MacPhee, 1979). The high sensitivity of finfish to toxins is manifested in the low prey biomass of toxic shellfish and zooplankton (~1 g) that would yield a potentially toxic dose to finfish (White, 1984). The similar sensitivity of marine mammals, low dose tolerance and rapidity with which the toxins operate are further evident in the quick death, 90 min, following intoxification of humpback whale (Geraci et al., 1989). The latter calculated that a whale eating 4% of its weight as mackerel daily would consume saxitoxin at a dosage of 3.2 μg per kg body weight, a level suspected to cause illness.

The insensitivity to, and accumulation of, toxins by shellfish and certain zooplankton can provide a source of toxins for a considerable period following the disappearance of a bloom. Zooplankton stocks in the Bay of Fundy, for example, remained toxic for up to three weeks post-bloom (White, 1981a).

Toxic and Noxious Phytoplankton Avoidance

These dose/speed-of-response characteristics, coupled with the sensitivity and vulnerability of the nekton to toxins, and adventitious feeding, reveal the need for selective feeding mechanisms to have evolved to minimize lethal intoxification, particularly during the larval stages. Allelochemic responses or other mechanisms, such as visual avoidance cues based on cellular pigmentation (see Lasker and Zweifel, 1978), or patch avoidance reactions seem necessary. Potts and Edwards (1987) report that during noxious *Gyrodinium aureolum* blooms fish descend to bottom waters aggregating into 50-cm bands above the sea floor. Larval stages exhibit exit migrations during the bloom, reappearing following its demise. Lenanton et al. (1985) have described the emigration of adult stocks during dense blooms of the noxious blue-green species, *Nodularia spumigena*. The emigration cues are unknown, but may be triggered by reduced oxygen levels, toxin production, and/or inhibitory effects of the dense *Nodularia* clumps and patches on respiratory and swimming activities.

Despite such behavioral responses, the need to detect and reject toxic cells would appear to be necessary. The toxic population levels triggering avoidance migrations or dispersion would appear to be greater than lethal population concentrations, given the high sensitivity of the nekton. Yamamori et al. (1988) reported that rainbow trout, *Salmo gairdneri*, and Arctic char, *Salvelinus alpinus*, exhibited gustatory responses when stimulated by tetrodotoxin (TTX) and saxitoxin (STX). They suggested that fish may have gustatory receptors which detect toxins allowing their avoidance or rejection of toxic prey. Such a mechanism, clearly, would have adaptive significance if confirmed, particularly given the rate constants of the speed of toxic effects and the low toxin doses discussed above. Such a gustatory mechanism would provide millisecond discrimination given the sodium channel effects of STX, a time course suitable to the need for extremely rapid detection of toxic cells. It is unclear how this mechanism would allow detection of toxins vectored in prey, and whether it requires secretion of toxins (see below) for gustatory detection.

As Hunter and Thomas (1974) have pointed out, the laboratory food requirements of larval fish, which appear to exceed natural food densities, can indeed be found in-situ in patches which larvae seek out, find and remain in during the critical first feeding stages. Thus, there is the dual need for attraction to, and retention within, dinoflagellate patches, and for avoidance of individual cells and patches of toxic prey species. How this is achieved and the rate of successful avoidance are largely unknown aspects of the recruitment problem.

Zooplankton Trophic Component

The effect of red-tide blooms on the zooplankton trophic component is complex, a function of community structure, bloom species, and whether it is toxic. Certain zooplankton components appear to have a low sensitivity to toxin accumulation, serving as toxin vectors to higher trophic levels, as discussed above. Copepods, pteropods, tintinnids, and cladocerans are demonstrated vectors (White, 1982). Toxins vectored by the cladoceran, *Evadne nordmanni*, for example, have led to dieoffs of herring and adult sand lance. However, failure of the cladoceran community to develop during the 1985 brown tide bloom of *Aureococcus anophagefferens* in Narragansett Bay was a conspicuous feature (Figure 13.1). Zooplankton grazing during red-tide events is frequently extensive. Watras et al. (1985), for example, found intense grazing occurred during a bloom of the toxic *Alexandrium tamarense*, including by polychaete larvae, *Polydora ligni*, and tintinnids, *Favella* sp. In contrast, there are also numerous reports of diminished zooplankton grazing during bloom events, such as Fiedler's (1982) observations. Migratory cladoceran, copepod, and larvacean species avoided a dense, sub-surface layer patch of *Gymnodinium splendens*; feeding inhibition also occurred. Filtration rates of the five dominant copepod species were two- to six-fold lower in "patch" water. Grazing of *Calanus pacificus* was suppressed in a *Gymnodinium flavum* bloom; its filtration rate was 30-fold lower than when fed a small diatom (Huntley, 1982).

It has been shown experimentally that although certain copepods actively graze, grow, and reproduce on dinoflagellate prey, profound negative effects can also occur detrimental to zooplankton recruitment. Of 21 different dinoflagellate species/clones provided as prey to four copepod species, nine species were rejected--seven of which were toxin producers (Huntley et al., 1986; Uye and Takamatsu, 1990). Daily egg production rates also varied considerably with prey species, indicative of differing interspecific nutritional value. Uye and Takamatsu ranked the dinoflagellates in their experiments as "good," "intermediate," or "bad" food sources. For rhaphidophycean flagellates, whose blooms are increasingly a cause of fish kills in the sea, their effects on copepods were equally devastating (Uye and Takamatsu, 1990). All five rhaphidophyceans tested were either rejected as food, or supported very low egg production rates. Thus, blooms of rhaphidophyceans have the potential of negatively impacting both zooplankton and upper trophic components simultaneously.

The relevant aspect of these experiments is their demonstration that growth and recruitment of zooplankton, and hence upper trophic level food supply, can also be impaired by toxic dinoflagellate and rhaphidophycean blooms. In fact, the zooplankton community may be the pivotal trophic link determining the food-web consequences of red-tide blooms. Depending on the bloom species and zooplankton community structure, the latter may benefit nutritionally from such blooms, with or without vectoring of toxins to higher trophic levels, or suffer significant recruitment failure during such blooms. Moreover, to the extent that potential bloom species are acceptable prey items, zooplankton community structure may regulate, through grazing, whether a red-tide bloom will develop, its magnitude and duration. It has been suggested that nuisance blooms and other mass occurrences of phytoplankton in the sea fundamentally reflect the failure of normal grazing processes (Smayda and Villareal, 1989).

In the copepod experiments considered above not all toxin producing species were rejected; in fact, even non-toxic species (*Scrippsiella trochoidea*) were rejected. Moreover, ingestion of toxic cells led to significant behavioral and physiological responses (Huntley et al., 1986; Sykes and Huntley, 1987). Ingestion of *Gymnodinium breve*, the cause (= brevetoxin) of massive fish kills in the Gulf of Mexico, by *Calanus pacificus* led to increased heartbeat (>400 bpm), erratic mouth part movements, twitching, and loss of motor control, followed by 100% mortality after three days. Regurgitation also occurred following ingestion of *G. breve* and *Protoceratium reticulatum*. These abnormal physiological and behavioral responses were reversible; normal feeding patterns resumed when a nutritionally adequate prey, *Gyrodinium resplendens*, was provided. *Calanus* purportedly tests the suitability of prey through sample ingestions (Huntley et al., 1986).

Exudates as Toxins and the Microbial Food Web

Some dinoflagellates release exudates inhibitory to copepod grazing (Huntley et al., 1986; Uye and Takamatsu, 1990). However, these allelopathic, antipredation exudates may not be saxitoxin, the active principle in the PSP toxins suite, since some PSP toxin-producing species are competent food while others which do not produce toxins are rejected. The classical view that saxitoxin is restricted intracellularly in dinoflagellates is now doubtful, however. White (1981a) reported the extracellular occurrence of saxitoxin in grazing experiments conducted with *Alexandrium tamarense*. However, fish appear to be insensitive to dissolved PSP toxins; only the ingestion of intact, toxic cells delivers lethal doses (Gosselin et al., 1989; White et al., 1989). Ogata and Kodama (1986), unable to confirm earlier reports that dissolved PSP toxins were active in situ, suggested that metabolites produced by bacteria were the ichthyotoxic component.

Soluble PSP toxins may have a different effect, however, at the micro-zooplankton community (= microbial food web) and benthic trophic levels. The tintinnid, *Favella ehrenbergii*, ingests and grows on toxic *Alexandrium tamarense* at low prey concentrations (Hansen, 1989). However, with increasing abundance of this dinoflagellate or during its senescent stages, increasing concentrations of an exudate thought to include PSP substances toxic to *Favella* are produced. Thus, both the growth and grazing of an important microzooplankton component within the microbial food web can be inhibited by a red-tide species. This trophic level may be impacted in other ways. Regurgitation of undesirable dinoflagellate prey by copepods has been described (Huntley et al., 1986). Sykes and Huntley (1987) have suggested that this detrital production may stimulate the microbial food web and thereby modify normal energy flow pathways. Dinoflagellate production in this instance is deflected away from upper trophic levels via copepods and from benthic communities through reduced fecal pellet delivery.

Red-tide blooms of *Noctiluca*, a holozoic dinoflagellate component of the microzooplankton community, illustrate the subtle trophic responses that may accompany such blooms. Sekiguchi and Kato (1976) established that *Noctiluca scintillans* outcompeted larval and juvenile sand lance, *Ammodytes personatus*, in grazing the egg production of the copepod *Acartia hudsonica*. *Noctiluca* blooms consumed 74% of the egg production during May in Ise Bay. Thus, not only did a large increase in a microzooplankton red-tide component regulate the cohort size of an important copepod prey, but reduced its availability as an important food source for local fish stocks. That is, these *Noctiluca* blooms had a dual effect, impacting both herbivorous and carnivorous trophic levels. Clearly, microbial food-web dynamics are complex, but the salient feature of these results is that this trophic component is also influenced by red-tide events.

Benthic Community Responses

The high tolerance to, and accumulation of PSP toxins ingested by filter feeding bivalves have been mentioned. Andrasi's (1985) report that edible mussel, *Mytilus edulis*, can become toxic when exposed to the filtrate of toxic dinoflagellates, however, is a novel uptake mechanism. If confirmed, it provides additional evidence that PSP toxins may be excreted during red-tide blooms and become available for trophic routing in a dissolved state. White's (1981a) query as to the extent to which liberation of soluble PSP toxins is the result of passive leakage, from cells damaged during grazing, or secreted by grazers, remains to be established.

More relevant to the present analysis is the question whether intoxification or phytotoxic impacts on the benthic community can impair recruitment, either directly or, at upper trophic levels, through vectoring. There is substantial evidence that benthic recruitment can be impaired by red-tide blooms. The 1988 *Chrysochromulina polylepis* bloom in southern Scandinavian waters caused widespread mortality of many invertebrate species (Underdal et al., 1989). Granmo et al. (1988) have shown experimentally that egg fertilization and successful embryo development in the ascidian, *Ciona intestinalis*, and mussel, *Mytilus edulis*, were completely inhibited when exposed to *C. polylepis*; the first demonstration of toxicity for this flagellate. The authors point out that reduced larval settlement and recruitment of these and other benthic components probably resulted. Mortality of larval oyster, *Crassostrea gigas*, and a reduction in abundance of larval scallop, *Pecten maximus*, occurred during *Gyrodinium aureolum* blooms (Helm et al., 1974). Reduced Ca^{++} uptake and mortality of larval *Crassostrea virginica* occurred during *Cochlodinium heterolobatum* blooms (Ho and Zubkoff, 1979). The latter effects were attributed to physical injury accompanying collisions between larval oyster and dense *Cochlodinium* populations. These observations suggest that both allelopathic and physical antagonisms during red-tide blooms can cause benthic larval mortality.

Barnacle larvae remained toxic for several days following exposure to *Alexandrium tamarense*, their intoxication accompanied by diminished grazing (White, 1981a). This reveals the potential for toxin vectoring by benthic larvae. Significant dieoffs of post-juvenile populations of benthic species have also accompanied bloom events. In Narragansett Bay, *Mytilus edulis* exhibited up to 100% mortality and spawning failure during the 1985 brown tide outbreak of *Aureococcus anophagefferens*, which reached population levels of 10^9 cells L^{-1} (Tracey, 1988; Smayda and Villareal, 1989). Starvation may have been the cause of death, since the mussels ceased filtering at a concentration of about 250×10^6 cells L^{-1}. Extensive histological damage and mortality of *Mytilus edulis planulatus*, *Pecten alba* and *Ostrea angasi* accompanied blooms of the diatom *Rhizosolenia chunii* in New Zealand coastal waters (Parry *et al.*, 1989). This is the first report of a diatom bloom causing shellfish mortality; the specific, lethal mechanism (it was not anoxia) was not resolved.

Given the high tolerance and accumulation levels of PSP toxins in benthic bivalves, vectoring of lethal doses to groundfish might be expected, since consumption of only 1 g shellfish by a 100-g fish would be sufficient to supply a potentially lethal dose (White, 1984). While such an effect remains to be established, toxin vectoring between benthic infaunal and epifaunal components has been shown. The accumulation of PSP toxins in the crab, *Cancer irroratus*, following ingestion of the clam, *Mya arenaria*, and in gastropods following toxic bivalve consumption has been reported (Foxall et al., 1979; Tufts, 1979). Adverse effects are not mentioned, although *Cancer* upon parenteral injection became lethargic, lost mobility, and chelae grasping reflexes; gill ventilation ceased and the heartbeat slowed. Lobster, *Homarus americanus*, in contrast, did not accumulate PSP following consumption of toxic *Mya arenaria*. Drills and whelks are reported to accumulate PSP from shellfish prey, as was the filter feeding crab, *Emerita analoga*, following grazing on toxic *Alexandrium* cells (Foxall et al., 1979). The crab, *Zosemus aeneus*, exemplifies the remarkable resistance to, and accumulation of high PSP toxin levels by benthic species (Noguchi et al., 1985). This species secretes copious amounts of PSP toxins when touched, by predators for example, suggesting its use as an anti-predatory substance.

The butter clam, *Saxidomus giganteus*, can retain large quantities of PSP toxins for up to two years (Tufts, 1979), demonstrating the considerable potential for toxin vectoring by benthic components, including routing to birds. While bird and duck mortality has been reported to accompany PSP outbreaks, some interesting avian counter-measures to the ingestion of toxic shellfish have evolved. Gulls will eat *Spisula solidissima* beached during storms, but avoid toxic specimens (Tufts, 1979). Resident eider ducks shun toxic mussels ingested by and lethal to migratory black ducks. Eider forced fed toxic mussels promptly regurgitate.

The significance of these diverse results is that although shellfish are quite resistant to, and accumulate toxins available for food-web vectoring, benthic components also exhibit fecundity and recruitment failure similar to other trophic levels in response to red-tide events.

A Food-web Toxin Routing and Effects Paradigm

White (1981b, 1984) formulated a model of some potential routes through which *Alexandrium excavatum* toxins may cause fish kills, based on his studies of in situ vectoring of PSP toxins and fish sensitivity. In Figure 13.5, I present a general trophic routing and impact model for toxins produced during nuisance (= toxic, noxious) phytoplankton blooms independent of causative species. The selective criteria used were whether toxin transfer and/or deleterious effects on viability, growth, fecundity, and recruitment have occurred within some component of that trophic level, either through the direct or vectored transmission of PSP toxins or other phytotoxins. The model indicates that in all trophic components of the food-web, individual population and communal intoxification and/or impairment, dysfuncion and/or disruption may occur in reponse to red-tide events. That this model looks remarkably similar to an energy flow diagram should, in retrospect, not be surprising. It shows that noxious or toxic phytoplankton bloom events impact ecosystem trophodynamics following the same first principles governing classical energy flow and food-web processes. The difference is one of a nutritionally adequate energy flow versus nutritionally inimical or toxic energy flow. The model also reveals that the classical paradigm that bloom toxins are primarily routed to benign storage in the highly resistant post-juvenile, benthic community (excluding episodes of human toxicity) is invalid. Rather, the impact of toxin producing blooms, through grazing or secreted metabolites, is potentially networked throughout the food web. The extent to which such blooms are deleterious will reflect the degree of synchronicity between bloom events and the reproductive and migration cycles, predator-prey relationships, etc. Red-tide events, in fact, may also be a consequence of upper trophic level activities, given the central role of grazing in marine food-web dynamics (Smayda and Villareal, 1989).

The model also indicates that two major types of negative events can accompany nuisance blooms. The primary one is due to toxic effects as a result of direct or vectored impacts; these tend to be water column events. The other type, anoxic events, accompany under-utilized, decomposing blooms of species which are generally large-sized, non-toxic, and ungrazed. Anoxic outbreaks are generally benthic events, resulting in massive dieoffs of invertebrates, demersal fish, and macroalgae. Blooms of *Ceratium* spp. are often the cause of such anoxic events, such as the 1976 bloom of *Ceratium tripos* in the New York Bight which resulted in a dieoff of commercially important fish and shellfish valued at least $250 million (Figley et al., 1979).

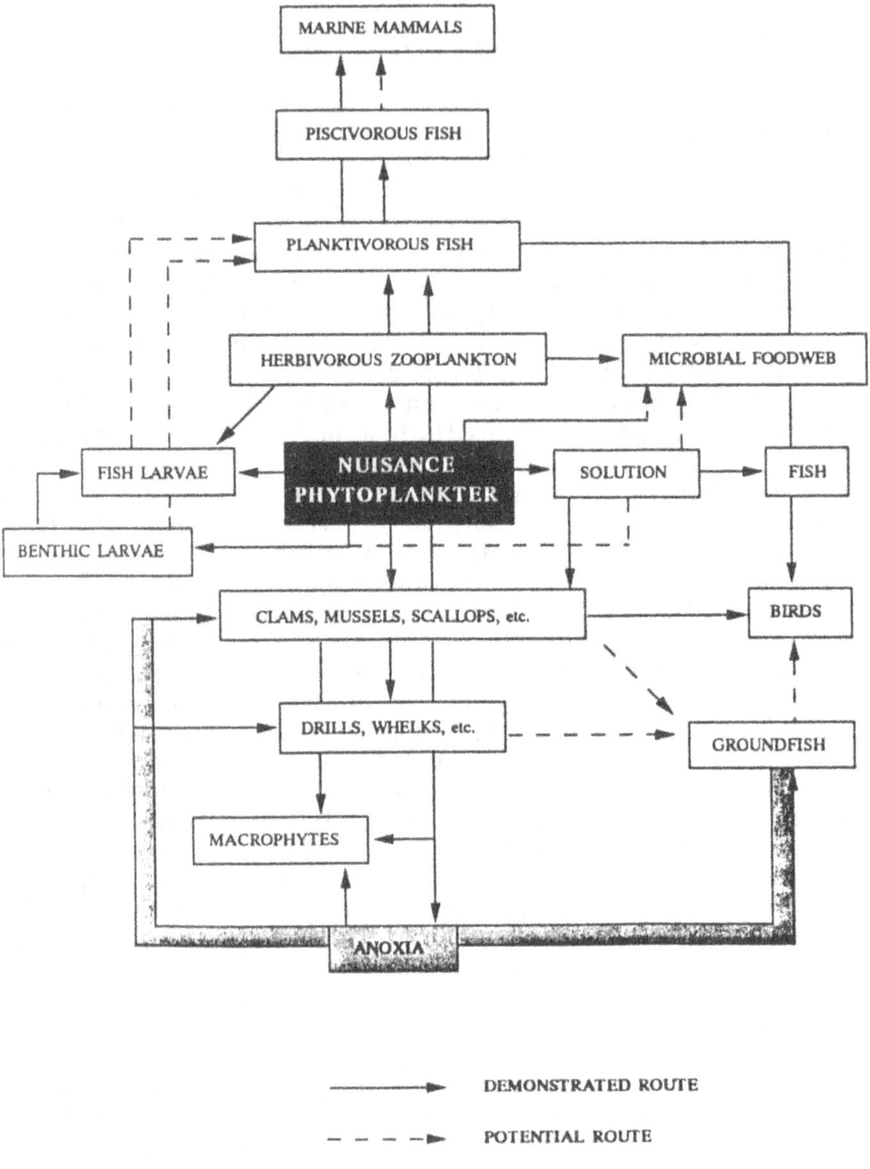

Figure 13.5. Trophic linkages between nuisance phytoplankton blooms and species illustrating some direct and vectored routes through which toxins, inimical species, and anoxia impact the depicted trophic compartments and may cause recruitment failure and/or mortality.

The inclusion of macrophytes (= macroalgae) in the model requires special note. Their inclusion is exclusive of ciguatera toxicity leading to human illness vectored through consumption of phytophagous fish which ingest toxic dinoflagellates symbiotic in their algal diet. The massive increase in macroalgal biomass in reponse to nutrification of the Baltic Sea, however, is relevant to the model, since Aneer (1987) has described what appears to be a new phenomenon. Unsually high mortality of herring eggs, >75% versus 10% normally, occurs where the usual spawning substratum has become modified through aggregation of the brown alga, *Pilayella littoralis*, which releases toxic exudates lethal to the eggs. In Narragansett Bay, a macroalgal dieoff occurred when the mussels to which laminarians were attached lost byssal contact and sank into the aphotic zone following their dieoff during the 1985 brown tide.

Bloom Events and Recruitment Patterns

The quantitative significance of noxious bloom events on recruitment and year-class and cohort strength cannot be assessed for any trophic level. The data are not available. There can be no doubt, however, that diminished recruitment, survival rates, and altered predator-prey interactions can accompany such blooms (Figure 13.5). Although catastrophic anoxic dieoffs in aquacultural and natural communities are exaggerated effects, they illustrate the disruption of normal food-web pathways, and probably mirror similar events that occur in more subtle fashion during lesser blooms. Upper trophic level responses to red-tide blooms, therefore, need to be established given their potential for serious recruitment failure.

This issue takes on greater interest, given the apparent global increase in the frequency, intensity, duration, and spreading of noxious blooms (Smayda, 1989a,b). It must be recognized that the major trophic level components and their feeding strategies evolved prior to the evolutionary appearance of the diatoms about 100 million years ago. That is, their grazing patterns evolved during the period when flagellates were the major primary producers. The remarkable nutritional dependency of first feeding larvae on dinoflagellates (Last, 1980) and evolved behavioral patterns used to detect and remain within prey patches, as evident from the grazing strategy of *Engraulis mordax* on *Gymnodinium splendens* (Hunter and Thomas, 1974; Lasker and Zweifel, 1978), may reflect this evolutionary background. Despite the demonstrated virulence of certain dinoflagellate and rhaphidophycean species, it must be recognized that only relatively few flagellates (perhaps <50) have been demonstrated to be noxious to fish. Most flagellates appear to be nutritionally adequate; benign if not utilized or, if toxic, rarely reach bloom densities to become troublesome. The problem lies in the apparent increased tendency of this small clutch of toxic species to exhibit bloom events in global coastal waters. This phenomenon, the changing character of red-tide blooms generally, and the increased occurrence of novel species blooms (Figure 13.3) may represent a major change in the phytoplankton dynamics of coastal waters. The classical (and stable) condition of an annual cycle in which the spring bloom is followed by a sequence of secondary blooms during the summer appears to be transforming to one in which the spring bloom is now increasingly followed by a multiple series of major, often toxic flagellate blooms. This altered pattern is reminiscent of catastrophy theory invoked to explain the sudden transition from one stable state to another, and specifically referable to cusp theory (Deniseger et al., 1986). This theory posits that when two major factors control a system (assume nutrients and grazing), an increase in one factor to stress conditions (eutrophication or, as during aquaculture, increased grazing) followed by a significant increase in the other factor can lead to selection of a doubly tolerant species (to elevated nutrients or grazing avoidance in our example), resulting in a catastrophic bloom. The multiple factors regulating phytoplankton dynamics, species selection, their blooms, and factor interactions present a bewildering array of factor combinations determining which, and when, a given species will bloom. The high stochastic aspect of red-tide bloom events is a manifestation of this complicated environmental control.

The transition from a stable, relatively predictable bloom cycle (i.e., the spring bloom event), to a highly unpredictable one (i.e., summer red-tide blooms), poses a serious problem to upper trophic level components whose survival strategy is linked to, and dependent upon more or less predictable bloom events of nutritionally adequate prey. Observations on the viviparous, demersal reef fish *Heteroclinus* sp., illustrate the importance of such temporal dependence on suitable bloom development (Thresher et al., 1989). Fully competent grazing larvae released asynchronously every 17 d become demersal after feeding pelagically for three weeks on zooplankton. Neonates starve within 12 h without adequate food. Successful development and demersal settlement, however, are linked to variations in phytoplankton pulses occurring seven to nine weeks earlier. These pulses stimulate production of the zooplankton preyed upon. Over a three year period no demersal "settlement pulses" of *Heteroclinus* occurred without being preceded by the lagged chlorophyll pulse. The occurrence and variability of these blooms were linked to wind induced mixing events causing nutrient enrichment triggering the blooms. Thus, temporal variability in bloom events two trophic levels away affected recruitment of *Heteroclinus* sp.

Another example of the importance of such timing is available from the Baltic Sea (Aneer, 1987). Blooms of the brown alga, *Pilayella littoralis*, have increased its biomass 10-fold in the last decade in response to increased nutrient levels. During the period when herring spawn, this alga begins to detach, collect into loose aggregates and drifts onto the spawning grounds, becoming the major available substratum onto which herring deposit their eggs. Toxic exudates from this alga result in 75% mortality in contrast to a normal egg mortality of 10%. Clearly, a slight temporal displacement in the coincidence of these two events would favor herring recruitment.

Gosselin et al. (1989) have pointed out that regional fish spawning strategies probably evolved partly in response to whether or not toxic phytoplankton blooms have been a regular component over an evolutionary time scale. Recurrent larval dieoffs during toxic blooms would provide the selective pressure to evolve bloom avoidance strategies. The extent to which this has occurred is very relevant when viewed in the context of the apparent global increase and regional spreading of noxious blooms presently occurring (Smayda, 1989a,b). In regions where toxic blooms have become new or more frequent, recruitment may now become increasingly diminished because the evolved, locally effective spawning and feeding strategies no longer, or less often, protect against spawning synchronously with toxic bloom events. In regions where blooms now occur with increasing frequency or duration, a similar problem may occur. Thus, in agreement with Gosselin et al., there is reason for concern that the present expansion of toxic phytoplankton blooms in the sea may become a threat to larval survival and recruitment in coastal fisheries by "narrowing the spatio-temporal window" within which sucessful spawning and recruitment occur. Sherman (1988) has pointed out that in nine large, coastal marine ecosystems (>200,000 km^2) globally "large scale biomass flips in the fish components . . . have occurred in which a dominant species within a relatively short span [<10 yr] has been reduced to a subordinate level . . . ". While this phenomenon and the toxic bloom expansion are probably not related, it is notable that major changes may be occurring at two or more trophic levels globally, in coastal waters and inland seas.

Sharp (1987) has summarized the potential causes of variability at different life history stages of fish (Figure 13.6). Among the potential causes, "local food availability" is identified as impacting the success of first feeding stages only. The egg, hatch and pre-feeding stages, and growth maturation and spawning processes are listed as independent of food availability. The present analysis has shown, however, that these stages and processes can, in fact, be negatively impacted during toxic blooms. If Sharp restricted food availability to mean prey density, then this potential cause of variability should either be expanded to include "food suitability", or the latter listed as a separate category. In either case, these additional stages and processes should be included in Sharp's matrix, given the

Potential Cause	Egg	Hatch	Prefeeding development	First feeding	Mobility w.r.t. turbulence	Transformation	Schooling	Filter feeding or?	Growth maturation	Spawning	Interspawning period
	Stage										
Predation	+	+	+	+	+	+	+	+	+	+	+
Transport	+	+	+	+	-	-	-	-	-	-	-
Local food availability	-	-	-	+	-	-	-	-	-	-	-
Feeding success rate	-	-	-	+	+	-	-	-	-	-	-
Temperate-related metabolism	+	+	+	+	+	+	+	+	+	+	+
Density-dependence (peers)	-	-	-	+	-	+	+	-	+	+	+
Food patch size	-	-	-	+	+	+	+	+	+	+	+
Interpatch distances	-	-	-	+	+	+	+	+	+	-	+
Cannibalism	+	+	+	+	+-	-	-	-	-	-	+
Developmental progression	+	+	+	+	+	+	+	+	+	-	+
Distribution and behaviour	+	+	+	+	+	+	+	+	+	+	+
	Plankton Environmentally dominated				Nekton-recruitment Environment and fishery effects combined						

Figure 13.6. Critical potential causes and conditions of variations at different life history stages of nektonic species (From Sharp, 1987).

impact of direct and vectored dinoflagellate toxins and other phytotoxins on larval and adult fish demonstrated experimentally and during nuisance phytoplankton blooms.

Acknowledgments

This work was completed in part during my tenure as a Visiting Fellow at the Ocean Research Institute, University of Tokyo, supported by the Japanese Ministry of Education Science and Culture. I wish to thank the late Prof. Takahisa Nemoto for making this affiliation possible. I also acknowledge the kindness and help of Dr. Makoto Terazaki of the Ocean Research Institute in making my visit rewarding. Ms. Cindy Heil helped with the graphic preparation of text Figure 13.5. Ms. Blanche Coyne drafted Figures 13.1-13.3 and provided secretarial services.

References

Anderson, D. M. 1989. Toxic algal blooms and red tides. A global perspective. Pp. 11-16. *In* T. Okaichi, D. M. Anderson, and T. Nemoto (Eds.) Red tides: Biology, environmental science and toxicology. Elsevier, New York.

Andrasi, A. 1985. Uptake of dissolved *Gonyaulax catenella* toxins from seawater by *Mytilus edulis* Linne. Pp. 401-406. *In* D. M. Anderson, A. W. White, and D. G. Baden (Eds.) Toxic dinoflagellates. Elsevier, New York.

Aneer, G. 1987. High natural mortality of Baltic herring (*Clupea harengus*) eggs caused by algal exudates? Mar. Biol. 94:163-169.

Bricelj, V. M., and S. H. Kuenstner. 1989. The feeding physiology and growth of bay scallops and mussels. Pp. 491-509. *In* E. M. Cosper, V. M. Bricelj, and E. J. Carpenter (Eds.) Novel phytoplankton blooms. Springer-Verlag, Berlin.

Cho, C. H. 1979. Mass mortalitys of oyster due to red tide in Jinhae Bay in 1978. Bull. Korean Fish. Soc. 12:27-33.

Cushing, D. M. 1971. Upwelling and the production of fish. Adv. mar. Biol. 9:255-334.

Deniseger, J., A. Austin, M. Roch, and M. J. R. Clark. 1986. A persistent bloom of the diatom *Rhizosolenia eriensis* (Smith) and other changes associated with decreases in heavy metal contamination in an oligotrophic lake, Vancouver Island. Environ. Expl. Bot. 26:217-226.

Dennison, W. C., G. J. Marshall, and G. Wigand. 1989. Effect of "brown tide" shading on eelgrass (*Zostera marina* L.) distribu-ions. Pp. 675-692. *In* E. M. Cosper, V. M. Bricelj, and E. J.

Carpenter (Eds.) Novel phytoplankton blooms. Springer-Verlag, Berlin.

Dietz, R., M. P. H. Jørgensen, and T. Härkönen. 1989. Mass deaths of harbor seals (*Phoca vitulina*) in Europe. Ambio 18:258-264.

Dowidar, N. M. 1984. Phytoplankton biomass and primary productivity of the southeastern Mediterranean. Deep-Sea Res. 31:983-1000.

Dundas, I., O. M. Johannessen, G. Berge, and B. Heimdal. 1989. Toxic algal bloom in Scandinavian waters. Oceanography 2:9-14.

Endo, M., T. Sakai, and A. Kuroki. 1985. Histological and histochemical changes in the gills of the yellowtail *Seriola quinqueradiata* exposed to Rhaphidophycean flagellate *Chattonella marina*. Mar. Biol. 87:193-197.

Fiedler, P. C. 1982. Zooplankton avoidance and reduced grazing responses to *Gymnodinium splendens* (Dinophyceae). Limnol. Oceanogr. 27:961-965.

Figley, W., B. Pyle, and B. Halgren. 1979. Socioeconomic impacts. Chapter 14. *In* Oxygen depletion and associated benthic mortalities in New York Bight, 1976. NOAA Professional Paper No. 11:315-322.

Foxall, T. L., N. H. Shoptaugh, M. Ikawa, and J. S. Sasner, Jr. 1979. Secondary intoxification with PSP in *Cancer irroratus*. Pp. 413-418. *In* D. L. Taylor and H. H. Seliger (Eds.) Toxic dinoflagellate blooms. Elsevier/North Holland, New York.

Geraci, J. R., D. M. Anderson, R. J. Timperi, D. J. St. Aubin, G. A. Early, J. H. Prescott, and C. A. Mayo. 1989. Humpback whales (*Megaptera novaeangliae*) fatally poisoned by dinoflagellate toxin. Can. J. Fish. Aquat. Sci. 46:1895-1898.

Gosselin, S., L. Fortier, and J. A. Gagné. 1989. Vulnerability of marine fish larvae to the toxic dinoflagellate *Protogonyaulax tamarensis*. Mar. Ecol. Prog. Ser. 57:1-10.

Granmo, Å., J. Havenhand, K. Magnusson, and I. Svane. 1988. Effects of the planktonic flagellate *Chrysochromulina polylepis* Manton et Park on fertilization and early development of the ascidian *Ciona intestinalis* (L.) and the blue mussel *Mytilus edulis* L. J. Exp. Mar. Biol. Ecol. 124:65-71.

Hallegraeff, G. M., D. A. Steffensen, and R. Wetherbee. 1988. Three estuarine Australian dinoflagellates that can produce paralytic shellfish toxins. J. Plankton Res. 10:533-541.

Hansen, P. J. 1989. The red tide dinoflagellate *Alexandrium tamarense* : Effects on behaviour and growth of a tintinnid ciliate. Mar. Ecol. Prog. Ser. 53:105-116.

Helm, M. M., B. T. Hepper, B. E. Spencer, and P. R. Walne. 1974. Lugworm mortalities and a bloom of *Gyrodinium aureolum*

Hulburt in the Eastern Irish Sea, autumn 1971. J. mar. biol. Ass. U.K. 54:857-869.

Hjort, J. 1914. Fluctuations in the great fisheries of northern Europe. Rapp. Proc. P.-v. Cons. int. Explor. Mer 20:1-228.

Ho, M. S., and P. L. Zubkoff. 1979. The effects of a *Cochlodinium heterolobatum* bloom on the survival and calcium uptake by larvae of the American oyster, *Crassostrea virginica*. Pp. 409-412. *In* D. L. Taylor and H. H. Seliger (Eds.) Toxic dinoflagellate blooms. Elsevier/North Holland, New York.

Hunter, J. R., and G. L. Thomas. 1974. Effect of prey distribution and density on the searching and feeding behaviour of larval anchovy *Engraulis mordax* Girard. Pp. 559-574. *In* J. H. S. Blaxter (Ed.) The early life history of fish. Springer-Verlag, New York.

Huntley, M. E. 1982. Yellow water in La Jolla Bay, California, July 1980. II. Suppression of zooplankton grazing. J. expl. mar. Biol. Ecol. 63:81-91.

Huntley, M., P. Sykes, S. Rohan, and V. Marin. 1986. Chemically-mediated rejection of prey by the copepods *Calanus pacificus* and *Paracalanus parvus*: Mechanism, occurrence and significance. Mar. Ecol. Prog. Ser. 28:105-120.

Johnsen, T. M., and T. E. Lien. 1989. *Prymnesium parvum* Carter (Prymnesiophyceae) in association with macroalgae in Ryfylke, Southwestern Norway. Sarsia 74:277-281.

Lam, C. W. Y., and K. C. Ho. 1989. Red tides in Tolo Harbour, Hong Kong. Pp. 49-52. *In* T. Okaichi, D. M. Anderson, and T. Nemoto (Eds.) Red tides: Biology, environmental science and toxicology. Elsevier, New York.

Lasker, R., and J. R. Zweifel. 1978. Growth and survival of first-feeding northern anchovy larvae (*Engraulis mordax*) in patches containing different proportions of large and small prey. Pp. 329-354. *In* J. H. Steele (Ed.) Spatial patterns in plankton communities. Plenum Press, New York.

Lasker, R., H. M. Feder, G. H. Theilacker, and R. C. May. 1970. Feeding, growth and survival of *Engraulis mordax* larvae reared in the laboratory. Mar. Biol. 5:345-353.

Last, J. M. 1980. The food of twenty species of fish larvae in the west-central North Sea. Fish. Res. Tech. Rep., MAFF Direct. Fish. Res., Lowestoft, 60:1-44.

Legrand, A. M., M. Litaudon, J. N. Genthon, R. Bagnis, and T. Yasumoto. 1989. Isolation and some properties of ciguatoxin. J. Appl. Phycol. 1:183-188.

Lenanton, R. C. J., N. R. Loneragan, and I. C. Potter. 1985. Blue-green algal blooms and the commercial fishery of a large Australian estuary. Mar. Pollut. Bull. 16:477-482.

Matsusata, H., and H. Kobayashi. 1974. Studies on death of fish caused by red tide. Bull. Nansei Reg. Fish. Lab. 7:43-67.

McLean, J. L. 1984. Indo-Pacific toxic red-tide occurrences, 1972-1984. Pp. 92-97. *In* A. W. White, M. Anraku, and K. K. Hooi (Eds.) Toxic red tides and shellfish toxicity in Southeast Asia. Southeast Asian Fisheries Development Center, Singapore.

de Mendiola, B. R. 1979. Red tide along the Peruvian Coast. Pp. 183-90. *In* D. L. Taylor and H. H. Seliger (Eds.) Toxic dinoflagellate blooms. Elsevier/North Holland, New York.

Mills, L. J., and G. Klein-MacPhee. 1979. Toxicity of the New England red tide dinoflagellate to winter flounder larvae. Pp. 389-394. *In* D. L. Taylor and H. H. Seliger (Eds.) Toxic dinoflagellate blooms. Elsevier/North Holland, New York.

Nixon, S. 1988. Physical energy inputs and the comparative ecology of lake and marine ecosystems. Limnol. Oceanogr. 33:1005-1025.

Noguchi, T., K. Daigo, O. Arakawa, and K. Hashimoto. 1985. Release of paralytic shellfish poison from the exoskeleton of a xanthid crab. Pp. 293-298. *In* D. M. Anderson, A. W. White, and D. Baden (Eds.) Toxic dinoflagellates. Elsevier, New York.

Ogata, T., and M. Kodama. 1986. Ichthyotoxicity found in cultured media of *Protogonyaulax* spp. Mar. Biol. 92:35-43.

Okaichi, T. 1989. Red tide problems in the Seto Inland Sea, Japan. Pp. 137-142. *In* T. Okaichi, D. M. Anderson, and T. Nemoto (Eds.) Red tides: Biology, environmental science and toxicology. Elsevier, New York.

Okaichi, T., and S. Nishio. 1976. Identification of ammonia as the toxic principle of red tide of *Noctiluca miliaris*. Bull. Plankton Soc. Jpn. 23:25-30.

Onoue, Y., and K. Nozawa. 1989. Separation of toxins from harmful red tides occurring along the coast of Kagoshima Prefecture. Pp. 371-374. *In* T. Okaichi, D. M. Anderson, and T. Nemoto (Eds.) Red tides: Biology, environmental science and toxicology. Elsevier, New York.

Parry, G. D., J. S. Langdon, and J. M. Huisman. 1989. Toxic effects of a bloom of the diatom *Rhizosolenia chunii* on shellfish in Port Phillip Bay, Southeastern Australia. Mar. Biol. 102:25-41.

Partensky, F., and A. Sournia. 1986. Le dinoflagellé *Gyrodinium* cf. *aureolum* dans le plancton de l'Atlantique nord: Identification écologie, toxicité. Cryptogamie Algologie 7:251-271.

Potts, G. W., and J. M. Edwards. 1987. The impact of a *Gyrodinium aureolum* bloom on inshore young fish populations. J. mar. biol. Ass. U.K. 67:293-297.

Prakash, A. 1963. Source of paralytic shellfish toxin in the Bay of Fundy. J. Fish. Res. Board Can. 20:983-996.

Roberts, B. S. 1979. Occurrence of *Gymnodinium breve* red tides along the west and east coasts of Florida during 1976 and 1977.

Pp. 199-202. *In* D. L. Taylor and H. H. Seliger (Eds.) Toxic dinoflagellate blooms. Elsevier/North Holland, New York.

Scura, E. D., and C. W. Jerde. 1977. Various species of phytoplankton as food for larval northern anchovy, *Engraulis mordax*, and relative nutritional value of the dinoflagellates *Gymnodinium splendens* and *Gonyaulax polyedra*. Fish. Bull., U.S. 75:577-583.

Sekiguchi, H., and T. Kato. 1976. Influence of *Noctiluca*'s predation on the *Acartia* population in Ise Bay, Central Japan. J. Oceanogr. Soc. Jpn 32:195-198.

Sharp, G. D. 1987. Climate and fisheries: Cause and effect or managing the long and short of it all. S. Afr. J. mar. Sci. 5:811-838.

Sherman, K. 1988. Large marine ecosystems as global units for recruitment experiments. Pp. 459-476. *In* B. J. Rothschild (Ed.) Toward a theory on biological-physical interactions in the World Ocean. NATO ASI Series C, Vol. 239. Kluwer Academic Publ., Dordrecht, The Netherlands.

Shimizu, Y. 1989. Toxicology and pharmacology of red tides: An overview. Pp. 17-22. *In* T. Okaichi, D. M. Anderson, and T. Nemoto (Eds.) Red tides: Biology, environmental science and toxicology. Elsevier, New York.

Shirota, A. 1989. Red tide problem and countermeasures (1). Int. J. Aq. Fish. Technol. 1:25-38.

Smayda, T. J. 1970. The suspension and sinking of phytoplankton in the sea. Oceanogr. Ann. Rev. Mar. Biol. 8:353-414.

Smayda, T. J. 1980. Phytoplankton species succession. Pp. 493-570. *In* I. Morris (Ed.) The physiological ecology of phytoplankton. Blackwell Scientific Publications, Oxford.

Smayda, T. J. 1989a. Novel and nuisance phytoplankton blooms in the sea: Evidence for a global epidemic. Pp. 29-40. *In* E. Granéli, B. Sundström, L. Edler, and D. M. Anderson (Eds.) Toxic marine phytoplankton, Elsevier, New York.

Smayda, T. J. 1989b. Primary production and the global epidemic of phytoplankton blooms in the sea: A linkage? Pp. 449-483. *In* E. M. Cosper, V. M. Bricelj, and E. J. Carpenter (Eds.) Novel phytoplankton blooms. Springer Verlag, New York.

Smayda, T. J. 1990. *Chrysochromulina polylepis* and exceptional phytoplankton blooms in the sea: Some patterns. *In* O. Skulberg (Ed.) Nordic Symposium on Toxin Producing Algae. (in press).

Smayda, T.J., and P. Fofonoff. 1989. An extraordinary noxious brown-tide in Narragansett Bay. II. Inimical effects. Pp. 133-136. *In* T. Okaichi, D. M. Anderson, and T. Nemoto (Eds.) Red tides: Biology, environmental science and toxicology. Elsevier, New York.

Smayda, T. J., and T. A. Villareal. 1989. The 1985 "brown-tide" and the open phytoplankton niche in Narragansett Bay during the summer. Pp. 159-187. *In* E. M. Cosper, V. M. Bricelj, and E. J. Carpenter (Eds.) Novel phytoplankton blooms. Springer Verlag, New York.

Sykes, P. F., and M. E. Huntley. 1987. Acute physiological reactions of *Calanus pacificus* to selected dinoflagellates: Direct observations. Mar. Biol. 94:19-24.

Thresher, R. E., G. P. Harris, J. S. Gunn, and L. A. Clementson. 1989. Phytoplankton production pulses and episodic settlement of a temperate marine fish. Nature 341:641-643.

Toyoshima, T., M. Shimada, H. S. Ozaki, T. Okaichi, and T. H. Murakami. 1989. Histological alteration to gills of the yellowtail *Seriola quinqueradiata* following exposure to the red tide species *Chattonella antiqua*. Pp. 439-442. *In* T. Okaichi, D. M. Anderson, and T. Nemoto (Eds.) Red tides: Biology, environmental science and toxicology. Elsevier, New York.

Tracey, G. A. 1988. Feeding reduction, reproductive failure, and mortality in the mussel, *Mytilus edulis*, during the 1985 "brown tide" in Narragansett Bay, Rhode Island. Mar. Ecol. Prog. Ser. 50:73-81.

Tufts, N. 1979. Molluscan transvectors of paralytical shellfish poisoning. Pp. 403-408. *In* D. L. Taylor and H. H. Seliger (Eds.) Toxic dinoflagellate blooms. Elsevier/North Holland, New York.

Underdal, B., O. M. Skulberg, E. Dahl, and T. Aune. 1989. Disastrous bloom of *Chrysochromulina polylepis* (Prymnesiophyceae) in Norwegian coastal waters 1988- mortality in marine biota. Ambio 18:265-270.

Uye, S. I., and K. Takamatsu. 1990. Feeding interactions between planktonic copepods and red tide flagellates from Japanese coastal waters. Mar. Ecol. Prog. Ser. 49:97-107.

Wardle, W. J., S. M. Ray, and A. S. Aldrich. 1975. Mortality of marine organisms associated with offshore summer blooms of the toxic dinoflagellate *Gonyaulax monilata* Howell at Galveston, Texas. Pp. 257-263. *In* V. R. LoCicero (Ed.) Proceedings of The First International Conference on Toxic Dinoflagellate Blooms. Massachusetts Science and Technology Foundation, Wakefield, MA, U.S.A.

Watras, C. J., V. C. Garcon, R. J. Olson, S. W. Chisholm, and D. M. Anderson. 1985. The effect of zooplankton grazing on estuarine blooms of the toxic dinoflagellate *Goniaulax tamarensis*. J. Plankton Res. 7:891-908.

White, A. W. 1977. Dinoflagellate toxins as probable cause of an Atlantic herring (*Clupea harengus harengus*) kill, and pteropods as apparent vector. J. Fish. Res. Board Can. 34:2421-2424.

White, A. W. 1980. Recurrence of kills of Atlantic herring (*Clupea harengus harengus*) caused by dinoflagellate toxins transferred through herbivorous zooplankton. Can. J. Fish. Aquat. Sci. 37:2262-2265.

White, A. W. 1981a. Marine zooplankton can accumulate and retain dinoflagellate toxins and cause fish kills. Limnol. Oceanogr. 26:103-109.

White, A. W. 1981b. Sensitivity of marine fishes to toxins from red-tide dinoflagellate *Gonyaulax excavata* and implications for fish kills. Mar. Biol. 65:255-260.

White, A. W. 1982. The scope of impact of toxic dinoflagellate blooms on finfish in Canada. Can. tech. Rep Fish. aquat. Sci. No. 1063:1-5.

White, A. W. 1984. Paralytic shellfish toxins and finfish. Pp. 171-180. *In* E. P. Ragelis (Ed.) Seafood toxins, ACS Symposium Series No. 262, Washington, D.C. American Chemical Society.

White, A. W., O. Fukuhara, and M. Anraku. 1989. Mortality of fish larvae from eating toxic dinoflagellates or zooplankton containing dinoflagellate toxins. Pp. 395-398. *In* T. Okaichi, D. M. Anderson, and T. Nemoto (Eds.) Red tides: Biology, environmental science and toxicology. Elsevier, New York.

Wong, P. S., and R. S. S. Wu. 1987. Red tides in Hong Kong: Problems and management strategy with special reference to the mariculture industry. J. Shoreline Management 3:1-21.

Wyatt, T. 1980. Morrell's seals. J. Cons. int. Explor. Mer 39:1-6.

Yamamori, K., M. Nakamura, T. Matsui, and T. J. Hara. 1988. Gustatory responses to tetrodotoxin and saxitoxin in fish: A possible mechanism for avoiding marine toxins. Can. J. Fish. Aquat. Sci. 45:2181-2186.

Yasumoto, T. 1989. Marine microorganisms toxins--an overview. Pp. 3-8. *In* E. Granéli, B. Sundström, L. Edler and D. M. Anderson (Eds.) Toxic marine phytoplankton. Elsevier, New York.

Yuki, K., and S. Yoshimatsu. 1989. Two fish-killing species of *Cochlodinium* from Harimada Nada, Seto Inland Sea, Japan. Pp. 451-454. *In* T. Okaichi, D. M. Anderson, and T. Nemoto (Eds.) Red tides: Biology, environmental science and toxicology. Elsevier, New York.

Index